己酉嘉平之月寫此爲振鐸兄大著補白

Dyadic Green's Functions in Electromagnetic Theory

Dyadic Green's Functions in Electromagnetic Theory

CHEN-TO TAI

Professor of Electrical Engineering
The University of Michigan

INTEXT EDUCATIONAL PUBLISHERS

College Division of Intext

Scranton - San Francisco - Toronto - London

The Intext *Monograph Series in*

ELECTRICAL ENGINEERING

Consulting Editor

DAVID K. CHENG

Syracuse University

ISBN 0-7002-2345-2

 Printed in the United States of America by The Haddon Craftsmen, Inc., Scranton, Pennsylvania.

Library of Congress Catalog Card Number: 70-151648

To

DR. CHIH KUNG JEN

An inspiring teacher of science and humanity

Foreword

With the publication of this book, a monograph series in electrical engineering is launched. It is intended that the books in this series will be authoritative expositions of special or advanced topics in electrical engineering which will be useful to practicing engineers and graduate students. They are not to be sketchy outlines or expanded versions of research reports. In this respect, it is indeed hard to imagine a better beginning for this series than Professor Tai's book *Dyadic Green's Functions in Electromagnetic Theory*.

The dyadic Green's function technique is an elegant and very powerful method for solving boundary-value problems. Professor Tai, who has done much original work in this area, is eminently qualified to write a comprehensive book on this topic. The book is self-contained and classroom-tested. It starts with a review of the relevant mathematics and the basic structure of electromagnetic theory, introduces the concepts and properties of Green's functions, and gradually delves into the outstanding features and important applications of dyadic Green's functions. Along the way the reader will encounter many sophisticated problems, all of which are properly formulated, logically solved, and lucidly explained. Many of the topics are appearing in book form for the first time.

Professor Tai has been a close personal friend since the days when we were fellow graduate students at Harvard almost thirty years ago. I recall the words by Roscommon, "*Choose an author as you choose a friend*," and doubt very much that I could do better in choosing either.

David K. Cheng
Syracuse University

Preface

The dyadic Green's function technique of treating electromagnetic boundary-value problems was first formulated by Schwinger in the early 1940's. The application of this technique to waveguide problems was known to many of his associates working then at the Radiation Laboratory of the Massachusetts Institute of Technology. Unfortunately, not much of this valuable work has been published. The theory of electromagnetic wave diffraction by an aperture in an infinite plane conducting screen was formulated by Levine and Schwinger (1950) using this technique. By means of an operational method, they derived the expression for the free-space dyadic Green's function in a very elegant manner. The subject matter was subsequently described by Morse and Feshbach (1953) in their well-known treatise on the methods of theoretical physics. These two authors introduced the dyadic Green's function by way of the vector Helmholtz equation so that the result can be used for both electromagnetic problems and elasticity problems. In their presentation of the eigenfunction expansion of various dyadic Green's functions three sets of vector wave functions, originally introduced by Hansen (1935), are needed. Their approach has been followed by several authors of books on electromagnetic theory. For electromagnetic problems alone it is considerably more simple to deal directly with the dyadic Green's functions pertaining to the vector wave equation satisfied by the electric or the magnetic field. By so doing only two sets of vector wave functions are needed in the eigenfunction expansion. The treatment found in this book is based on this approach.

The principal material contained in this book was compiled by the author when he was associated with Stanford Research Institute from 1949 to 1954. The problem, which he first resolved using the then relatively new dyadic Green's function technique, was prompted by a puzzling remark made by Sommerfeld in his book on Optics (1949) concerning the electromagnetic radiation by a spherical source in the presence of a half-sheet. This remark will be referred to later in Chapter 10. A technical report was written on this research (Tai, 1954a) giving the detailed treatment of that problem. Before his departure from Stanford Research Institute another report (Tai, 1954b) was prepared listing a glossary of dyadic Green's functions but without any derivation. The list includes the functions pertaining to perfectly conducting circular cylinders, elliptical cylinders, wedges, spheres, cones, and the flat earth. These two reports now form the nucleus of this book. Over the years the author has compiled more materials of a similar nature on this general subject, including the dyadic Green's functions for rectangular waveguide, circular waveguide, the dielectric cylinder, and the dielectric sphere. The functions associated with some radially stratified media and the free-space dyadic Green's function for moving isotropic media have also been found. All these are now contained in this book.

Most of the material has been used in a graduate course on electromagnetic theory offered in the Department of Electrical Engineering at The University of Michigan. One main objective of the course is to emphasize the foundation and the techniques of formulating electromagnetic boundary-value problems. A part of the material has also been used in a course on antenna theory. The students taking these courses are supposed to have had an intermediate course in electromagnetic theory using a text at the level of, for example, Reitz-Milford (1960) or Ramo-Whinnery-Van Duzer (1965). It is hoped that the book may be used for similar purposes at other institutions offering these kinds of courses. A number of exercises have been chosen to supplement the instruction. A few difficult ones can be assigned to students majoring in electromagnetic theory, hence requiring a more thorough training. As a reference book it may be of aid to those who seek to formulate certain problems by just making use of some of the general formulas provided in this book.

In regard to the composition of this book, Chapter 1 starts with a brief review of the mathematics which are needed in subsequent chapters. It is followed by a condensed version of the basic structure of electromagnetic theory (Chapter 2), including a quite unconventional presentation of the integral form of Maxwell's equations in order to facilitate the derivation of the boundary conditions. In Chapter 3, the concept of Green's functions is introduced by way of transmission-line theory, since most of the readers using this book are assumed to have had adequate knowledge of this subject. The classification of various kinds of Green's functions and the articulation of the Ohm-Rayleigh method will find their counterparts later in dealing with the vector wave equation.

The proper material concerning the dyadic Green's function technique starts with Chapter 4. The fundamentals of the dyadic algebra and the dyadic calculus are treated from the very beginning. No previous knowledge of this subject is assumed. The free-space dyadic Green's function is then derived from the known results obtained from potential theory, and again from an operational method due to Levine and Schwinger. The classification of various kinds of dyadic Green's functions is taken up next. The symmetrical properties of these functions are shown in detail. The function of the first kind, or the second kind, is illustrated by one associated with a half-space bounded by a plane conducting screen.

Chapter 5 begins with a general discussion of the construction of the vector wave functions. The eigenfunction expansions of the dyadic Green's functions pertaining to rectangular waveguides are treated next. Chapter 6 treats cylindrical waveguides. Dyadic Green's functions for exterior boundary-value problems are introduced in Chapter 7, using the circular cylinder as a diffracting body. Chapter 8 gives a reformulation of Sommerfeld's famous work using the dyadic Green's function technique. It should give a quite clear picture regarding the different approaches between the classical method and the method based on the dyadic Green's functions. Conducting elliptical cylinders as diffracting bodies are covered in Chapter 9.

Chapter 10 is probably the most comprehensive chapter of the book. Using the half-sheet as an example, we have applied the dyadic Green's function technique to illustrate various radiation problems. The chapter also presents many numerical calculations which are available to us. Plane-wave diffraction is treated as an asymptotic solution of a spherical-wave diffraction problem. Spherical structures and perfectly conducting cones are the subjects of Chapter 11. In addition to the standard method of deriving a dyadic Green's function for a sphere, we also include therein an algebraic method. It is a tedious method but it reveals some interesting properties of several types of spherical vector wave functions.

The last chapter is devoted to stratified media and moving media. The electromagnetic theory of spherical Luneberg lenses is treated in detail mainly because some of the wave functions involved are not commonly known. Using the rectangular waveguide as an example, we show that the method of eigenfunction expansion is still applicable to the electromagnetic wave equation in the presence of a moving isotropic medium. We applied very effectively Levine and Schwinger's operational method in the derivation of the free-space dyadic Green's function for such a medium.

The references quoted in the book are listed at the end of the book in Appendix E. When a specific item is quoted from a text, we also list the page number, such as (Stratton, p. 401). An equation number referred to in a different section is identified by the section number as well, such as (3-26). It represents Equation 26 of Section 3. The section number is omitted if the equation referred to is in the same section.

CHEN-TO TAI

Ann Arbor, Michigan
June, 1971

Acknowledgments

It is indeed a pleasure for me to acknowledge my indebtedness to the many friends who have given me constant encouragement at the time when most of the material contained in this book was being compiled. They are my former teacher, Professor Ronald W. P. King of Harvard University; Dr. John V. Granger and the late Dr. John T. Bolljahn formerly of Stanford Research Institute: Professor John D. Kraus of The Ohio State University; Professors William G. Dow, Ralph E. Hiatt and C. Bruce Sharpe of The University of Michigan.

Mr. Burton A. Harrison of The University of Michigan-Dearborn has kindly helped me in the editing of the manuscript. Professor David H. S. Cheng of the University of Missouri corrected many errors in the original manuscript. Many former and current graduate students have contributed significantly in the preparation of this book. In particular, the notes on my lectures taken by Dr. Rudolph M. Kalafus and Mr. Carl F. Stubenrauch have been very helpful. I also wish to thank Mr. Stubenrauch for assisting in the proofreading. Without the tender care of Mrs. Claire F. White, the manuscript of the book would have not been in such an orderly fashion for the publisher.

I wish to thank Mr. Larry Hager, the Engineering Editor of Intext Educational Publishers for the careful attention given to the production of the book. My fine arts tutor, Mr. S. C. Chiang of the Palace Museum, Taiwan, China, has kindly provided a painting to decorate the book. The painting has been an inspiration for my lessons with him. Finally, I wish to express my sincere gratitude to my old friend Professor David K. Cheng of Syracuse University, the editor of this series, for writing the foreword and for his brotherly advice on all phases of this work. He had, years ago, urged me to complete my thesis and his urging was similarly instrumental in the completion of this book.

The consideration and the support for this endeavor which I received from my wife, Ming, and from our children have been most helpful.

C-T. T.

Contents

chapter **1**

General Theorems and Formulas

In this chapter we review some of the important theorems and formulas needed in the subsequent chapters. It is assumed that the reader has had an adequate course in advanced calculus, including vector analysis, Fourier series and integrals, and the theory of complex variables. Our review will contain sufficient material so that references to other books will be kept to a minimum. We sacrifice to some extent the mathematical rigor that may be required in a more thorough treatment. For example, we use quite freely the integral representation of the delta function, assuming that an exponential function with imaginary argument is Fourier-transformable. Whenever necessary, adequate references* will be given to strengthen any plausible statement, or to remove possible ambiguity.

1. VECTOR NOTATIONS AND THE COORDINATE SYSTEMS

A vector quantity or a vector function will be denoted by $\bar{F}$. A letter with a cap "^", such as $\hat{x}$, is used to denote a unit vector in the direction of the covered letter. In most cases, these letter correspond to the variables in a coordinate system. The scalar product of two vectors is denoted by $\bar{A} \cdot \bar{B}$ and the vector product by $\bar{A} \times \bar{B}$. The three commonly used systems in this book are:

(a) rectangular, or Cartesian, x, y, z

(b) circular cylindrical or simply cylindrical, r, ϕ, z

(c) spherical, R, θ, ϕ.

* References are given in Appendix E. The author's name is used as the identification. If a book, either the section number or the pages will be cited.

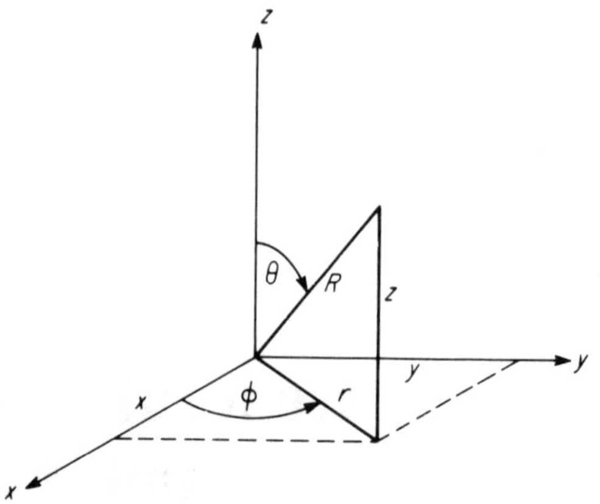

FIG. 1-1. Three commonly used coordinate systems.

The spatial variables associated with these systems are shown in Fig. 1-1. It should be pointed out that the same ϕ-variable is used for both the cylindrical and the spherical systems. The unit vectors pertaining to these

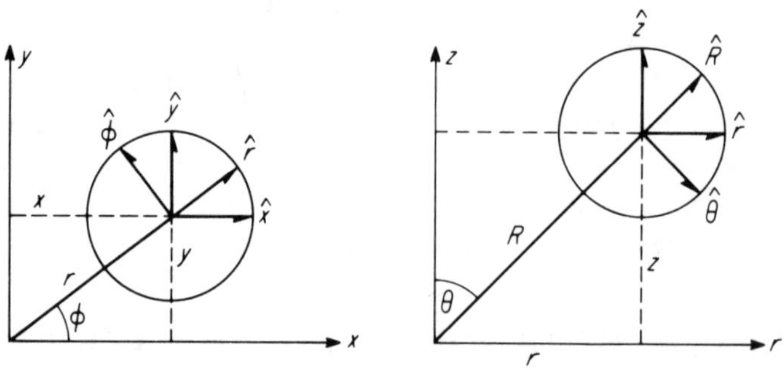

FIG. 1-2. The unit vectors in three commonly used coordinate systems.

systems are displayed in Fig. 1-2 in two cross-sectional views. The relation between these unit vectors are summarized in Tables 1-1 and 1-2.

TABLE 1-1

Relations Between the Unit Vectors in the Rectangular and the Cylindrical Coordinate Systems

	$\hat{x}$	$\hat{y}$	$\hat{z}$
$\hat{r}$	$\cos\phi$	$\sin\phi$	0
$\hat{\phi}$	$-\sin\phi$	$\cos\phi$	0
$\hat{z}$	0	0	1

TABLE 1-2

Relations between the Unit Vectors in the Rectangular and the Spherical Coordinate Systems

	$\hat{x}$	$\hat{y}$	$\hat{z}$
$\hat{R}$	$\sin\theta\cos\phi$	$\sin\theta\sin\phi$	$\cos\theta$
$\hat{\theta}$	$\cos\theta\cos\phi$	$\cos\theta\sin\phi$	$-\sin\theta$
$\hat{\phi}$	$-\sin\phi$	$\cos\phi$	0

To express unit vector $\hat{x}$ in terms of the unit vectors in the spherical system, one uses the coefficients in the first column of Table 1-2 which gives

$$\hat{x} = \sin\theta\cos\phi\,\hat{R} + \cos\theta\cos\phi\,\hat{\theta} - \sin\phi\,\hat{\phi}.$$

Likewise, the second row gives

$$\hat{\theta} = \cos\theta\cos\phi\,\hat{x} + \cos\theta\sin\phi\,\hat{y} - \sin\theta\,\hat{z}.$$

The reader can verify for himself that the above tables apply equally well to the transform of the components of a vector, for example:

$$A_\theta = \cos\theta\cos\phi\,A_x + \cos\theta\sin\phi\,A_y - \sin\theta\,A_z.$$

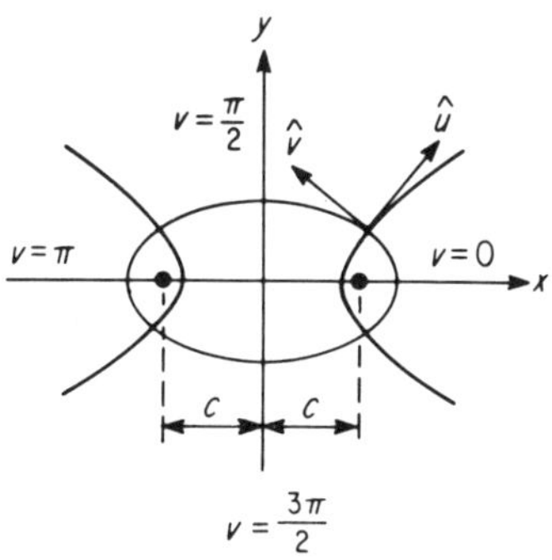

FIG. 1-3. A cross-sectional view of the elliptical coordinate system.

Another coordinate system will be used in this book—one associated with an elliptical cylinder. One set of variables that can be used in this system is designated by (u,v,z). A cross-sectional view of a plane perpendicular to the z-axis is shown in Fig. 1-3. The constant u contours and the constant v contours correspond respectively to a family of confocal ellipses and a family of confocal hyperbolas. The relations between (x,y) and (u,v) are

$$x = c\cosh u\cos v$$
$$y = c\sinh u\sin v$$

where $\infty > u \geqslant 0$, $2\pi \geqslant v \geqslant 0$. Two alternate variables which are used sometimes in place of (u,v) are defined by

$$\xi = \cosh u$$
$$\eta = \cos v$$

then $\infty > \xi \geqslant 0$, $1 \geqslant \eta \geqslant -1$. Table 1-3 contains the transformation coefficients between the unit vectors of the rectangular system and the elliptical cylinder system.

TABLE 1-3

Relations Between the Unit Vectors in the Rectangular System and the Elliptical Cylinder System

$\Delta = (\cosh^2 u - \cos^2 v)^{1/2}$

	$\hat{x}$	$\hat{y}$	$\hat{z}$
$\hat{u}$	$\frac{1}{\Delta}\sinh u \cos v$	$\frac{1}{\Delta}\cosh u \sin v$	0
$\hat{v}$	$-\frac{1}{\Delta}\cosh u \sin v$	$\frac{1}{\Delta}\sinh u \cos v$	0
$\hat{z}$	0	0	1

2. VECTOR ANALYSIS

The entire subject of vector analysis consists of three definitions, namely, the gradient, the divergence and the curl; a number of identities; and two theorems named after Gauss and Stokes. For convenient reference some of the identities and formulas are listed in Appendix A. We will not review here the elementary aspects of vector analysis, but rather will outline the two theorems and a number of useful lemmas that can be derived from these theorems.

Gauss' theorem states that for any vector function of position $\bar{F}$ with continuous first derivatives throughout a volume V and over the enclosing surface S,

(1) $$\iiint_V \nabla \cdot \bar{F}\, dV = \oiint_S \bar{F} \cdot d\bar{S}. \qquad \text{(Gauss' theorem)}$$

The ring around a surface integral is to emphasize the fact that the surface is a closed one. The same notation will be applied to a closed line integral. There are two versions of Stokes' theorem. The first one states that for any vector function of position with continuous first derivatives on an open surface S bounded by a contour C,

(2) $$\iint_S (\nabla \times \bar{F}) \cdot d\bar{S} = \oint_C \bar{F} \cdot d\bar{l}. \qquad \text{(Stokes' theorem)}$$

It is understood that the direction of the line integral and the direction of $d\bar{S}$ follows the right-hand screw rule. Another version of Stokes' theorem is

(3) $$\iiint_V (\nabla \times \bar{F})\, d\bar{V} = \oiint_S (\hat{n} \times \bar{F})\, dS \qquad \text{(vector Stokes' theorem)}$$

where $\hat{n}$ denotes the outward normal to the bounding surface S. It can be shown that (2) and (3) are equivalent and both versions are equally useful in our work. For identification purposes, (3) will be designated as the vector Stokes' theorem. It may be of interest to point out that the vector Stokes' theorem is closely related to Gauss' theorem, or can be derived from it. Thus if we let $\bar{A} = \bar{a} \times \bar{b}$, where $\bar{a}$ denotes a constant but arbitrary vector and $\bar{b}$ a vector function of position then $\nabla \cdot \bar{A} = -\bar{a} \cdot \nabla \times \bar{b}$ as a result of identity (7) of Appendix A. Upon substituting it into Gauss' theorem, one obtains

$$\bar{a} \cdot \iiint_V \nabla \times \bar{b}\, dV = \bar{a} \cdot \oiint_S (\hat{n} \times \bar{b})\, dS.$$

Since $\bar{a}$ is arbitrary, the following relation must hold true:

$$\iiint_V \nabla \times \bar{b}\, dV = \oiint_S (\hat{n} \times \bar{b})\, dS$$

which is the same as (3).

In addition to the two fundamental theorems quoted above, there is another theorem named after Green. Historically, Green's theorem is an independent and original theorem. However, because of its close relation to Gauss' theorem, it may be treated as a lemma. Thus if we define

$$\bar{A} = \Phi \nabla \psi - \psi \nabla \Phi$$

where Φ and ψ are two scalar functions of position, then in view of identity (5) of Appendix A,

$$\nabla \cdot \bar{A} = \Phi \nabla^2 \psi - \psi \nabla^2 \Phi.$$

It follows from Gauss' theorem that

(4) $$\iiint_V (\Phi \nabla^2 \psi - \psi \nabla^2 \Phi)\, dV = \oiint_S (\Phi \nabla \psi - \psi \nabla \Phi) \cdot d\bar{S}. \quad \text{(Green's theorem)}$$

Another more useful variant of the scalar Green's theorem for vector field problems is the vector Green's theorem. This is obtained by letting

$$\bar{A} = \bar{Q} \times \nabla \times \bar{P} - \bar{P} \times \nabla \times \bar{Q}$$

where $\bar{Q}$ and $\bar{P}$ denote two vector functions of position. By applying identity (7) of Appendix A, one funds

$$\nabla \cdot \bar{A} = \bar{P} \cdot \nabla \times \nabla \times \bar{Q} - \bar{Q} \cdot \nabla \times \nabla \times \bar{P}.$$

Upon substituting it into Gauss' theorem, we have

(5) $$\iiint_V (\bar{P} \cdot \nabla \times \nabla \times \bar{Q} - \bar{Q} \cdot \nabla \times \nabla \times \bar{P})\, dV$$

$$= \oiint_S (\bar{Q} \times \nabla \times \bar{P} - \bar{P} \times \nabla \times \bar{Q}) \cdot d\bar{S}. \quad \text{(vector Green's theorem)}$$

Another relation that finds occasional use is given below:

(6) $$\iiint_V \nabla \psi \, dV = \oiint_S \psi \hat{n}\, dS. \quad \text{(gradient identity)}$$

This identity can be derived from Gauss' theorem if we let $\bar{A} = \bar{a}\psi$, where $\bar{a}$ denotes a constant but arbitrary vector and ψ a scalar function of position. Then $\nabla \cdot \bar{A} = \bar{a} \cdot \nabla \psi$. Equation (6) follows immediately from Gauss' theorem.

3. FOURIER TRANSFORM AND HANKEL TRANSFORM

In this section we will review the basic formulas in the theory of the Fourier transform and in the theory of the Hankel transform or Fourier-Bessel transform. At the end of this section we will derive the integral representation of the delta functions weighted according to the dimension in which these functions are used. The meaning of a weighted delta function will be explained later.

The Fourier transform of a piecewise continuous function $f(\xi)$ is defined by

(1) $$g(h)=\int_{-\infty}^{\infty} f(\xi)e^{-ih\xi}\, d\xi .$$

The existence of $g(h)$ requires that $\int_{-\infty}^{\infty} |f(\xi)|\, d\xi$ be bounded. The inverse of (1) is given by

(2) $$f(x)=\frac{1}{2\pi}\int_{-\infty}^{\infty} g(h)\ e^{ihx}\, dh .$$

The Fourier transform can be extended to functions of many variables. In particular, for functions of two variables, we have the following two-dimensional Fourier transform pair

(3) $$g(h_1,h_2)=\int\int_{-\infty}^{\infty} f(\xi_1,\xi_2)e^{-i(h_1\xi_1+h_2\xi_2)}\, d\xi_1\, d\xi_2$$

(4) $$f(x_1,x_2)=\frac{1}{(2\pi)^2}\int\int_{-\infty}^{\infty} g(h_1,h_2)e^{i(h_1x_1+h_2x_2)}\, dh_1\, dh_2 .$$

The Hankel transform or Fourier-Bessel transform can be considered as a special case of the two-dimensional Fourier transform. It deals with a class of functions whereby $f(x_1, x_2)$ is a function of $(x_1^2+x_2^2)^{1/2}$ or $f(r)$ only when expressed in the radial cylindrical variable. To derive the Hankel transform pair from this point of view, we make the following changes of variables:

$$x_1=r\cos\phi \qquad x_2=r\sin\phi$$
$$\xi_1=\rho\cos\beta \qquad \xi_2=\rho\sin\beta$$
$$h_1=\lambda\cos\alpha \qquad h_2=\lambda\sin\alpha .$$

Equations (3) and (4) can then be written in the form

(5) $$g(\lambda,\alpha)=\int_0^{\infty}\rho\, d\rho\int_0^{2\pi} d\beta\, f(\rho,\beta)e^{-i\lambda\rho\cos(\beta-\alpha)}$$

(6) $$f(r,\phi)=\frac{1}{(2\pi)^2}\int_0^{\infty}\lambda d\lambda\int_0^{2\pi} d\alpha\, g(\lambda,\alpha)e^{i\lambda r\cos(\alpha-\phi)} .$$

Upon combining (5) and (6) we obtain

(7) $$f(r,\phi)=\frac{1}{(2\pi)^2}\int_0^{\infty}\rho\, d\rho\int_0^{\infty}\lambda d\lambda\int_0^{2\pi} d\beta\int_0^{2\pi} d\alpha\ .$$
$$\cdot f(\rho,\beta)e^{i[\lambda r\cos(\alpha-\phi)-\lambda\rho\cos(\beta-\alpha)]}$$

Now let $f(r,\phi)$ be a function in the form $F(r)\,e^{in\phi}$ where n is assumed to be a positive real constant not necessarily integer. Equation (7) then becomes

$$F(r)e^{in\phi}=\frac{1}{(2\pi)^2}\int_0^\infty \rho\,d\rho\int_0^\infty \lambda\,d\lambda\int_0^{2\pi} d\beta\int_0^{2\pi} d\alpha \cdot F(\rho)e^{in\beta}\,e^{i[\lambda r\cos(\alpha-\phi)-\lambda\rho\cos(\beta-\alpha)]}. \tag{8}$$

Dividing the entire equation by $e^{in\phi}$ and rearranging the terms in the exponential functions, with the anticipation that the integrals with respect to α and β have the appearance of the integral representations of Bessel functions, one can write the resultant equation in the form

$$F(r)=\frac{1}{(2\pi)^2}\int_0^\infty \rho\,d\rho\int_0^\infty \lambda\,d\lambda\int_0^{2\pi} d\beta\int_0^{2\pi} d\alpha\,. \cdot F(\rho)e^{i\left[\lambda r\cos(\alpha-\phi)+n\left(\alpha-\phi-\frac{\pi}{2}\right)-\lambda\rho\cos(\beta-\alpha)+n\left(\beta-\alpha+\frac{\pi}{2}\right)\right]} \tag{9}$$

Let us now consider the integral with respect to β first, or more specifically, the integral

$$\frac{1}{2\pi}\int_0^{2\pi} e^{i\left[-\lambda\rho\cos(\beta-\alpha)+n\left(\beta-\alpha+\frac{\pi}{2}\right)\right]}\,d\beta.$$

By changing the variable of integration to ω defined by $\omega=\beta-\alpha$, this becomes

$$\frac{1}{2\pi}\int_{-\alpha}^{2\pi-\alpha} e^{i\left[-\lambda\rho\cos\omega+n\left(\omega+\frac{\pi}{2}\right)\right]}\,d\omega.$$

If we judiciously choose the limit of integration such that the contour follows the path from $-\dfrac{3\pi}{2}+i\,\infty$ to $\dfrac{\pi}{2}+i\infty$, then the above integral becomes the integral representation of the Bessel function of order n which is assumed to be positive and real but not confined to integers—that is,

$$J_n(\lambda\rho)=\frac{1}{2\pi}\int_{-\frac{3\pi}{2}+i\infty}^{\frac{\pi}{2}+i\infty} e^{-i\left[\lambda\rho\cos\omega-n\left(\omega+\frac{\pi}{2}\right)\right]}\,d\omega. \tag{10}$$

The remaining integration with respect to α is now evaluated in a similar manner. We change the variable of integration to $\omega=\alpha-\phi$ and choose the path of integration from $-\dfrac{\pi}{2}+i\,\infty$ to $\dfrac{3\pi}{2}+i\infty$, which yields

$$J_n(\lambda r)=\frac{1}{2\pi}\int_{-\frac{\pi}{2}+i\infty}^{\frac{3\pi}{2}+i\infty} e^{i\left[\lambda r\cos\omega+n\left(\omega-\frac{\pi}{2}\right)\right]}\,d\omega. \tag{11}$$

The final form for (9) after this reduction of the angular integrals becomes

$$F(r)=\int_0^\infty \lambda\, d\lambda \int_0^\infty \rho\, d\rho\, F(\rho)J_n(\lambda\rho)J_n(\lambda r). \tag{12}$$

This can be separated in a pair of identities by letting

$$G(\lambda)=\int_0^\infty F(\rho)J_n(\lambda\rho)\rho\, d\rho \tag{13}$$

then

$$F(r)=\int_0^\infty G(\lambda)J_n(\lambda r)\lambda\, d\lambda. \tag{14}$$

Equations (13) and (14) constitute the pair of Hankel transforms which are valid for Bessel functions of any order. It should be remarked that the derivation which we have presented here follows very closely the one described by Sommerfeld (*Partial Diff.*, pp. 109–111). However, he derived these expressions under the condition that n is an integer. Later, he applied these formulas to nonintegral values of n without further elaboration (p. 211). In fact, when the Hankel transform is applied to spherical problems, the Bessel functions involved are of half-integer order, or more precisely, the spherical Bessel functions. For this reason it is more convenient to modify (13) and (14) so that they would contain these functions directly. To obtain these desired expreessions, we let $n=m+1/2$ and change the notation r to R in order to conform to spherical nomenclature. Now, the spherical Bessel function is defined by

$$j_m(x)=\left(\frac{\pi}{2x}\right)^{1/2} J_{m+1/2}(x). \tag{15}$$

Equations (13) and (14) can then be written in the form

$$G(\lambda)=\int_0^\infty \left(\frac{2\lambda\rho}{\pi}\right)^{1/2} F(\rho)j_m(\lambda\rho)\,\rho\, d\rho \tag{16}$$

$$F(R)=\int_0^\infty \left(\frac{2\lambda R}{\pi}\right)^{1/2} G(\lambda)j_m(\lambda R)\lambda\, d\lambda. \tag{17}$$

To recast (16) and (17) in a symmetrical form, we introduce two new functions $f(R)$ and $g(\lambda)$ defined by

$$G(\lambda)=\lambda^{1/2}g(\lambda)$$

and

$$F(R)=R^{1/2}f(R).$$

The pair of Hankel transforms for $f(R)$ and $g(\lambda)$ in terms of the spherical Bessel function then has the form

(18) $$g(\lambda)=\left(\frac{2}{\pi}\right)^{1/2}\int_0^\infty f(\rho)j_m(\lambda\rho)\rho^2\,d\rho$$

(19) $$f(R)=\left(\frac{2}{\pi}\right)^{1/2}\int_0^\infty g(\lambda)j_m(\lambda R)\lambda^2\,d\lambda.$$

These two expressions have previously been derived by Stratton (1941, pp. 411–412) using a different technique. The present derivation appears to be less formal, but perhaps simpler.

We will now apply the Fourier transform pair (1) and (2), and the Hankel transform pairs (13) and (14) and (18) and (19) to derive the integral representation of the delta function weighted according to the dimension in which the function is used. These weighted delta functions are defined as follows:

(20) one-dimensional: $\delta(x-x')$

(21) two-dimensional: $\delta(r-r')/r$

(22) three-dimensional: $\delta(R-R')/R^2$

The integral property of a delta function implies that

(23) $$\int_{-\infty}^\infty f(x)\,\delta(x-x')\,dx=f(x')$$

hence, for the weighted delta function in the two-dimensional case

$$\int_0^\infty f(r)\left[\frac{\delta(r-r')}{r}\right]r\,dr=f(r');$$

likewise, for the three-dimensional case,

$$\int_0^\infty f(R)\left[\frac{\delta(R-R')}{R^2}\right]R^2\,dR=f(R').$$

Upon substituting $f(x)=\delta(x-x')$ or $f(\xi)=\delta(\xi-x')$ into (1) and (2), one finds

(24) $$\delta(x-x')=\frac{1}{2\pi}\int_{-\infty}^\infty e^{ih(x-x')}\,dh.$$

Similarly, by letting $F(r)=\delta(r-r')/r$ or $F(\rho)=\delta(\rho-r')/\rho$ in (13) and (14), we obtain

$$\frac{\delta(r-r')}{r}=\int_0^\infty J_n(\lambda r)J_n(\lambda r')\lambda\, d\lambda. \tag{25}$$

Finally, by applying (18) and (19) to the weighted delta function in the three-dimensional case, we obtain

$$\frac{\delta(R-R')}{R^2}=\frac{2}{\pi}\int_0^\infty j_n(\lambda R)j_n(\lambda R')\lambda^2\, d\lambda. \tag{26}$$

Expressions (24). (25) and (26) will be used later very often in solving the vector wave equation by the method of continuous eigenfunction expansion.

4. SADDLE-POINT METHOD OF INTEGRATION AND THE EVALUATION OF SOME SEMI-INFINITE INTEGRALS INVOLVING THE PRODUCT OF TWO BESSEL FUNCTIONS

Complex integrals of the type

$$F(\rho)=\int_{-\infty}^{\infty} f(h)e^{i\rho\phi(h)}\, dh \tag{1}$$

will occur frequently in our work. When certain conditions are met, the integral can be evaluated approximately by the method of saddle-point integration. The key conditions are that ρ be a large number compared to unity and $\phi(h)$, whose magnitude is of the order of unity, have an extreme value at a certain point h_0, so that $\phi'(h_0)=0$. The function $f(h)$ is assumed to be a slowly varying function in the neighborhood of h_0. We consider $\phi(h)$ to be an analytic function of the complex variable $h=\xi+i\eta$ so that

$$\phi(h)=u(\xi,\eta)+iv(\xi,\eta) \tag{2}$$

then u and v satisfy the Cauchy-Riemann relations

$$\frac{\partial u}{\partial \xi}=\frac{\partial v}{\partial \eta}, \qquad \frac{\partial u}{\partial \eta}=-\frac{\partial v}{\partial \xi}. \tag{3}$$

A three-dimensional plot of the surface $z=v(\xi,\eta)$ shows that in the neighborhood of the point $h=h_0$ or $\xi=\xi_0$ and $\eta=\eta_0$ the surface has the shape of a saddle because

$$\left[\frac{\partial^2 v}{\partial \xi^2}\cdot\frac{\partial^2 v}{\partial \eta^2}-\left(\frac{\partial^2 v}{\partial \xi\partial \eta}\right)^2\right]_{\xi=\xi_0,\ \eta=\eta_0}<0$$

as a result of the Cauchy-Reimann relations (Courant, Vol. II, p. 205). The family of curves described by

$$v(\xi,\eta)=c$$

for different values of c has the appearance shown in Fig. 4-1 in the neighborhood of the saddle point, where $v_0=v(\xi_0,\eta_0)$.

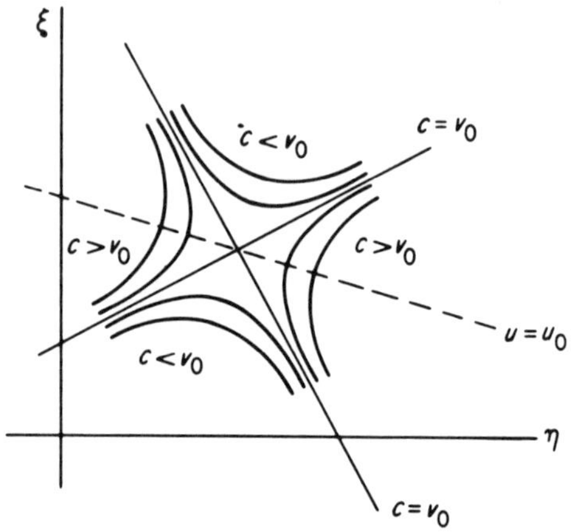

FIG. 4-1. The family of curves described by $v(\xi,\eta)=c$.

The above description also applies to the function $u(\xi,\eta)$ and the family of curves corresponding to $u(\xi,\eta)=$const. are orthogonal to those shown in Fig. 4-1. We now change the path of integration such that the contour would pass through h_0 and follow a path corresponding to

$$u(\xi,\eta)=u_0$$

in the neighborhood of the saddle point. A section of this path is shown in Fig. 4-1 by the dotted line. Under that condition the original integral can be written in the form

$$F(\rho)=\int_c f(h)e^{i\rho(u_0+iv)}\,dh=e^{i\rho u_0}\int_c f(h)e^{-\rho v}\,dh.$$

The function $e^{-\rho v}$ has a significant value only along a small segment of the path near the saddle point. Thus we may approximate the function $\phi(h)$ by a series expansion in the neighborhood of h_0 by retaining only the first three terms, that is,

(4)
$$\phi(h)=\phi(h_0)+\phi'(h_0)(h-h_0)+\frac{1}{2}\phi''(h_0)(h-h_0)^2=\phi(h_0)+\frac{1}{2}\phi''(h_0)(h-h_0)^2.$$

For the slowly varying function $f(h)$ it can be replaced by $f(h_0)$. The original integral, then, can be approximated by

$$F(\rho)=e^{i\rho\phi(h_0)}f(h_0)\int_{h_0-\Delta h}^{h_0+\Delta h} e^{i\frac{1}{2}\rho\phi''(h_0)(h-h_0)^2}\,dh. \tag{5}$$

This can be simplified if we let

$$h-h_0=se^{i\alpha}$$

$$\phi''(h_0)=|\phi''(h_0)|\,e^{i\beta}$$

then

$$i\frac{1}{2}\rho\phi''(h_0)(h-h_0)^2=\frac{1}{2}\rho|\phi''(h_0)|s^2e^{i\left(2\alpha+\beta+\frac{\pi}{2}\right)}.$$

In order to confine the path of integration along the contour $u=u_0$, the angle α must be so chosen that

$$\alpha=\frac{1}{2}\left(\frac{\pi}{2}-\beta\right).$$

Under this condition, (5) reduces to

$$F(\rho)=e^{i\left[\rho\phi(h_0)-\frac{\beta}{2}+\frac{\pi}{4}\right]}f(h_0)\int_{-\tau}^{\tau} e^{-\frac{1}{2}\rho\phi''(h_0)s^2}\,ds \tag{6}$$

where $\tau=|\Delta h|$. The last integral can be evaluated by a change of variable to

$$t=\left[\frac{\rho|\phi''(h_0)|}{2}\right]^{\frac{1}{2}}s$$

and in the limit as

$$\left[\frac{\rho|\phi''(h_0)|}{2}\right]^{\frac{1}{2}}\tau$$

becomes very large we obtain

$$F(\rho)=\left[\frac{2\pi}{\rho|\phi''(h_0)|}\right]^{\frac{1}{2}}f(h_0)e^{i\left[\rho\phi(h_0)-\frac{\beta}{2}+\frac{\pi}{4}\right]}. \tag{7}$$

This is the asymptotic formula for (1) under the conditions we have stated.

As an example, let us consider an integral of the form

$$F(r,z)=\int_{-\infty}^{\gamma}\frac{G(h)}{\sqrt{2\pi\lambda r}}\,e^{i(hz+\lambda r)}\,dh \tag{8}$$

where

$$\lambda = \sqrt{k^2 - h^2}$$

and r, z denote two of the variables in the cylindrical coordinate system, k being a constant. Thus we have

$$\rho\phi(h) = hz + \sqrt{k^2 - h^2}\, r$$

$$h_0 = \frac{kz}{\sqrt{r^2 + z^2}} = k \cos\theta$$

$$\rho\phi(h_0) = k\sqrt{r^2 + z^2} = kR$$

$$\rho\phi''(h_0) = \frac{-k^2 r}{(k^2 - h_0^2)^{3/2}} = -\frac{R}{k \sin^2\theta}$$

where R, θ, denote two of the variables defined in the corresponding spherical coordinate system. When kR is very large, the original integral is well approximated by

$$F(R,\theta) = \frac{G(k\cos\theta)}{R}\, e^{i\left(kR - \frac{\pi}{4}\right)}. \tag{9}$$

This approximation will be used later in finding the asymptotic expressions for various dyadic Green's functions.

Another integral which will be encountered often in our work is of the form

$$f(r,r') = \int_0^\infty g(\lambda)\, \frac{J_\nu(\lambda r) J_\nu(\lambda r')}{\lambda^2 - k^2}\, d\lambda \tag{10}$$

where $J_\nu(\lambda r)$ and $J_\nu(\lambda r')$ denote two Bessel functions of order ν, not necessarily integers, and $g(\lambda)$ is an odd function of λ with no poles in the complex λ-plane. The evaluation of this type of integral has been described by Sommerfeld (*Partial Diff.*, p. 197) for the case where the product is made of two spherical Bessel functions. We shall adopt Sommerfeld's method, but put no restriction on the values of ν. Using the well-known relation between a Bessel function and the two Hankel functions, we can write (10) in the form

$$f(r,r') = \frac{1}{2}\int_0^\infty \frac{g(\lambda) J_\nu(\lambda r)}{\lambda^2 - k^2}\, [H_\nu^{(1)}(\lambda r') + H_\nu^{(2)}(\lambda r')]\, d\lambda = f_1 + f_2\,.$$

By changing the variable of integration from λ to $\lambda e^{-i\pi}$ we have

$$f_2 = \frac{1}{2}\int_0^{-\infty} \frac{g(\lambda e^{-i\pi})}{\lambda^2 - k^2} J_\nu(\lambda r e^{-i\pi}) H_\nu^{(2)}(\lambda r' e^{-i\pi})(-d\lambda).$$

According to the circulation relations of the Bessel functions (Sommerfeld, p. 314), we have

$$J_\nu(\lambda r e^{-i\pi}) = e^{-i\nu\pi} J_\nu(\lambda r)$$

$$H_\nu^{(2)}(\lambda r e^{-i\pi}) = -e^{i\nu\pi} H_\nu^{(1)}(\lambda r).$$

Since $g(\lambda) = -g(-\lambda)$, we obtain

$$f_2 = \frac{1}{2}\int_{-\infty}^{0} \frac{g(\lambda)}{\lambda^2 - k^2} J_\nu(\lambda r) H_\nu^{(1)}(\lambda r')\, d\lambda.$$

The original integral, therefore, can be written in the form

$$f(r,r') = \frac{1}{2}\int_{-\infty}^{\infty} \frac{g(\lambda)}{\lambda^2 - k^2} J_\nu(\lambda r) H_\nu^{(1)}(\lambda r')\, d\lambda. \tag{11}$$

By a similar reasoning, (10) can alternatively be expressed in the form

$$f(r,r') = \frac{1}{2}\int_{-\infty}^{\infty} \frac{g(\lambda)}{\lambda^2 - k^2} H_\nu^{(1)}(\lambda r) J_\nu(\lambda r')\, d\lambda. \tag{12}$$

These two integrals can be evaluated in a closed form by completing the contour of integration along a semi-infinite circular path in the upper λ plane. As a result of the residue theorem, we find

$$f(r,r') = \frac{i\pi g(k)}{2k} \begin{cases} J_\nu(kr) H_\nu^{(1)}(kr'), & r < r' \\ H_\nu^{(1)}(kr) J_\nu(kr'), & r > r' \end{cases} \tag{13}$$

In the case of spherical and conical problems, we will encounter integrals of the type

$$F(R,R') = \int_0^{\infty} \frac{G(\lambda) j_\nu(\lambda R) j_\nu(\lambda R')}{\lambda^2 - k^2}\, d\lambda \tag{14}$$

where $j_\nu(\lambda R)$ and $j_\nu(\lambda R')$ denote two spherical Bessel functions and $G(\lambda)$ is an even function of λ. Using the relation that

$$j_\nu(x) = \left(\frac{\pi}{2x}\right)^{1/2} J_{\nu+1/2}(x)$$

(14) can be written in the form

$$F(R,R') = \frac{\pi}{2(RR')^{1/2}} \int_0^{\infty} \frac{G(\lambda)}{\lambda(\lambda^2 - k^2)} J_{\nu+1/2}(\lambda R) J_{\nu+1/2}(\lambda R')\, d\lambda. \tag{15}$$

This is of the same form as (10) provided $G(\lambda)/\lambda$ has no poles in the λ plane. Applying the general result described by (13), we obtain

$$F(R,R') = \frac{i\pi^2 G(k)}{4k^2(RR')^{1/2}} \begin{cases} J_{\nu+1/2}(kR)H^{(1)}_{\nu+1/2}(kR'), & R<R' \\ H^{(1)}_{\nu+1/2}(kR)J_{\nu+1/2}(kR'), & R>R' \end{cases} \tag{16}$$

$$= \frac{i\pi G(k)}{2k} \begin{cases} j_\nu(kR)h^{(1)}_\nu(kR'), & R<R' \\ h^{(1)}_\nu(kR)j_\nu(kR'), & R>R' \end{cases}$$

where $h^{(1)}_\nu(kR)$ denotes the spherical Hankel function of the first kind, that is

$$h^{(1)}_\nu(x) = \left(\frac{\pi}{2x}\right)^{1/2} H^{(1)}_{\nu+1/2}(x).$$

Equation (16) holds true for functions of any order, not necessarily for ν equal to integers. In fact for conical problems ν is in general fractional.

chapter **2**

Maxwell's Equations

Historically, Maxwell's theory of electromagnetism was founded on the basic laws available at his time. His main contribution was to supplement Ampere's law by a now famous term called the *displacement current* to make it compatible with the equation of continuity and with Gauss' law. Nowadays it is more practical to present the theory in its entire form without following the historical course. However, it is important to distinguish the dependent equations from the independent ones of the entire set of equations and also to understand the significance of the definite form in contrast to the indefinite form. Our review will emphasize these two aspects of the theory. In addition, the boundary conditions for both the electric and the magnetic fields will be derived with the aid of the vector Stokes' theorem that appears to be more appealing than the conventional approach. The remaining parts of the chapter are conventional, but we will need them in the subsequent chapters.

5. THE INDEPENDENT AND DEPENDENT EQUATIONS, AND THE INDEFINITE AND DEFINITE FORMS OF MAXWELL'S SYSTEM OF EQUATIONS

There are three *independent* equations in Maxwell's theory of electromagnetism:

(1) $$\nabla \times \bar{E} = -\frac{\partial \bar{B}}{\partial t} \qquad \text{(Faraday's law)}$$

(2) $$\nabla \times \bar{H} = \bar{J} + \frac{\partial \bar{D}}{\partial t} \qquad \text{(Maxwell-Ampere law)}$$

(3) $$\nabla \cdot \bar{J} = -\frac{\partial \rho}{\partial t} \qquad \text{(equation of continuity)}$$

where $\bar{E}$ = electric field (volt/meter)
$\bar{D}$ = electric flux density (coulomb/meter^2)
$\bar{H}$ = magnetic field (ampere/meter)
$\bar{B}$ = magnetic flux density (weber/meter^2)
$\bar{J}$ = electric current density (ampere/meter^2)
ρ = electric charge density (coulomb/meter^3)

It is understood that all the field quantities, including the current density and the charge density, are functions of position and time. By taking the divergence of (1) and setting the constant of integration with respect to time equal to zero we obtain

(4) $$\nabla \cdot \bar{B} = 0 \qquad \text{(Gauss' law-magnetic).}$$

In a similar manner, upon taking the divergence of (2) and eliminating $\bar{J}$ between the resultant equation and (3), we have

(5) $$\nabla \cdot \bar{D} = \rho \qquad \text{(Gauss' law).}$$

Since (4) and (5) are considered to be derivable from (1) through (3), these two equations should be treated as auxiliary or *dependent* equations in the entire system of equations 1 through 5. An alternative view is to take (1), (2), and (5) as independent equations while treating (3) and (4) as dependent equations. Such an alternative choice does not change the basic point of view.

The three independent equations described by (1) through (3) actually consist of seven scalar differential equations inasmuch as one vector equation is equivalent to three scalar equations. Now each vector function has three components. We have, therefore, sixteen unknown scalar functions altogether. It is obvious that the three independent equations are not sufficient to form a complete system of equations to solve for the unknown functions. For clarity, we shall designate (1) through (3) as Maxwell's equations in the *indefinite form* as long as the constitutive relations between the field quantities are unknown or unspecified. Under such a condition, many alternative forms can be used to describe Maxwell's theory. One common form is to introduce two material field vectors, $\bar{P}$ and $\bar{M}$ which are defined by

(6) $$\bar{D} = \varepsilon_0 \bar{E} + \bar{P}$$

(7) $$\bar{B} = \mu_0 (\bar{H} + \bar{M})$$

where $\bar{P}$ = polarization (coulomb/meter^2)
$\bar{M}$ = magnetization (ampere/meter)

$$\begin{aligned}
\varepsilon_0 &= \text{the electric constant} \\
&= 8.854 \times 10^{-12} \text{ (farad/meter)} \\
&\simeq \frac{1}{36\pi} \times 10^{-9} \text{ (farad/meter)} \\
\mu_0 &= \text{the magnetic constant} \\
&= 4\pi \times 10^{-7} \text{ (henry/meter)}.
\end{aligned}$$

When $\bar{E}$, $\bar{H}$, $\bar{P}$, and $\bar{M}$ are used, the number of unknowns and the number of equations remain the same, and the essential characteristics of Maxwell's equations are not altered. This is the invariant property of Maxwell's equations. We shall not elaborate here this aspect of the electromagnetic theory since it is irrelevant to the present work. In any event, to make the Maxwell's equation definite we need more information. This additional information is provided by the constitutive relations between the field quantities. For example, in a simple isotropic medium, the field quantities are related as follows.

$$\bar{D} = \varepsilon \bar{H} \tag{8}$$

$$\bar{B} = \mu \bar{H} \tag{9}$$

$$\bar{J} = \sigma \bar{E} \tag{10}$$

where ε, μ, and σ denote, respectively, the permittivity, permeability, and conductivity of the medium. Equations 8 through 10 provide nine more scalar relations that make the number of unknowns and the number of equations compatible. When the constitutive relations between the field quantities are known, Maxwell's equations become *definite.* In many boundary-value problems, the constitutive relations between $\bar{D}$, $\bar{B}$, $\bar{E}$, and $\bar{H}$ are usually known while the current density function $\bar{J}$ is treated as a source term. In that case, we are interested in finding the solutions for $\bar{E}$ and $\bar{H}$ in terms of $\bar{J}$ that satisfy certain boundary conditions. Much of the work discussed in this book falls into this case. Thus if the medium under consideration is air, which is practically a vacuum, the definite form of Maxwell's equations becomes

$$\nabla \times \bar{E} = -\mu_0 \frac{\partial \bar{H}}{\partial t} \tag{11}$$

$$\nabla \times \bar{H} = \bar{J} + \varepsilon_0 \frac{\partial \bar{E}}{\partial t} \tag{12}$$

$$\nabla \cdot \bar{J} = -\frac{\partial \rho}{\partial t} \tag{13}$$

The two dependent equations are

(14) $$\nabla \cdot (\mu_0 \bar{H}) = 0$$

(15) $$\nabla \cdot (\varepsilon_0 \bar{E}) = \rho .$$

Similar equations for more complex media will be introduced later.

6. INTEGRAL FORMS OF MAXWELL'S EQUATIONS AND THE BOUNDARY CONDITIONS

Although the integral forms of Maxwell's equations are not needed in the present work, we will give a brief review here mainly because certain features of these forms are not discussed in standard books on electromagnetic theory, and they are useful in deriving the boundary conditions for the field quantities.

In order to give a complete description of the boundary conditions, we start with the indefinite form of Maxwell's equations. Upon integrating (5-1) through (5-5) through a volume V with an enclosing surface S we obtain

(1) $$\iiint \nabla \times \bar{E}\, dV = -\iiint \frac{\partial \bar{B}}{\partial t}\, dV$$

(2) $$\iiint \nabla \times \bar{H}\, dV = \iiint \left(\bar{J} + \frac{\partial \bar{D}}{\partial t}\right) dV$$

(3) $$\iiint \nabla \cdot \bar{J}\, dV = -\iiint \frac{\partial \rho}{\partial t}\, dV$$

(4) $$\iiint \nabla \cdot \bar{B}\, dV = 0$$

(5) $$\iiint \nabla \cdot \bar{D}\, dV = \iiint \rho\, dV.$$

Applying the vector Stokes' theorem, (2-3) to (1) and (2) and Gauss' theorem (2-1) to (3) through (5), these equations can be written in the form

(6) $$\oiint (\hat{n} \times \bar{E})\, dS = -\iiint \frac{\partial \bar{B}}{\partial t}\, dV$$

(7) $$\oiint (\hat{n} \times \bar{H})\, dS = \iiint \left(\bar{J} + \frac{\partial \bar{D}}{\partial t}\right) dV$$

(8) $$\oiint (\hat{n} \cdot \bar{J})\, dS = -\iiint \frac{\partial \rho}{\partial t}\, dV$$

(9) $$\oint\!\!\oint (\hat{n} \cdot \bar{B})\, dS = 0$$

(10) $$\oint\!\!\oint (\hat{n} \cdot \bar{D})\, dS = \iiint \rho\, dV.$$

It may be recalled that the integral forms of Faraday's law and the Maxwell-Ampere law are usually written as

(11) $$\oint \bar{E} \cdot d\bar{l} = \iint \frac{\partial \bar{B}}{\partial t} \cdot d\bar{S}$$

(12) $$\oint \bar{H} \cdot d\bar{l} = \iint \left(\bar{J} + \frac{\partial \bar{D}}{\partial t} \right) \cdot d\bar{S}.$$

However, (6) and (7) are simpler to use as far as the derivation of boundary conditions is concerned. As we are going to show, the method and the model are the same for all these equations 6 through 10. The same approach was used very effectively by Costen and Adamson (1965) to formulate the electromagnetic jump conditions and momentum-energy laws at a moving boundary.

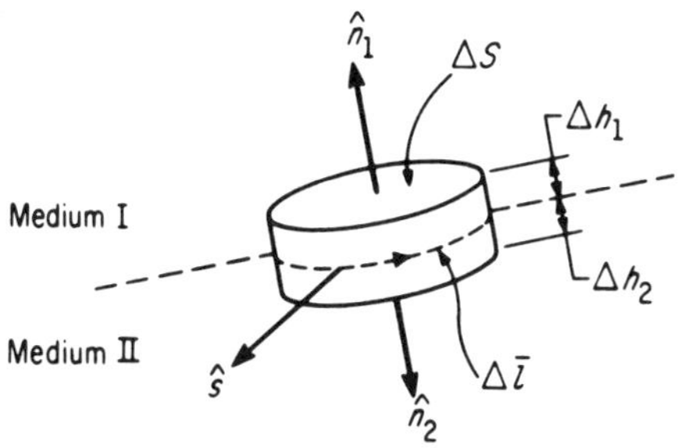

FIG. 6-1. Two media in contact and a small pillbox across the boundary.

We consider two media in contact as shown in Fig. 6-1. The integral relations are now evaluated through a small pillbox shaped volume across the boundary. We assume that the field vectors $\bar{B}$ and $\bar{D}$ are finite at the boundary surface. They may, of course, be discontinuous. The boundary conditions resulting from (6) as Δh_1 and $\Delta h_2 \to 0$ yields

(13) $$\hat{n}_1 \times \bar{E}_1 + \hat{n}_2 \times \bar{E}_2 = 0,$$

since

$$\lim_{\Delta h \to 0} (\Delta h_1 \bar{B}_1 + \Delta h_2 \bar{B}_2) = 0.$$

In evaluating (7) and (8) we assume that the boundary may support a surface current and a surface charge whose densities are defined, respectively, by

(14) $$\bar{K} = \lim_{\Delta h \to 0} (\bar{J}_1 \Delta h_1 + \bar{J}_2 \Delta h_2) \text{ amp/meter},$$

and

(15) $$\rho_s = \lim_{\Delta h \to 0} (\rho_1 \Delta h_1 + \rho_2 \Delta h_2) \text{ coulomb/meter}^2.$$

The limiting value of (7) then gives

(16) $$(\hat{n}_1 \times \bar{H} + \hat{n}_2 \times \bar{H}_2) = \bar{K}.$$

In deriving the boundary condition from (8) we observe that the flow of current through the top and bottom surfaces as well as through the side surface contributes to the surface integral. We obtain first

(17)
$$\oint \hat{s} \cdot (\bar{J}_1 \Delta h_1 + \bar{J}_2 \Delta h_2)\, dl + (\hat{n}_1 \cdot \bar{J}_1 + \hat{n}_2 \cdot \bar{J}_2) \Delta S = -\left(\frac{\partial \rho_1}{\partial t} \Delta h_1 + \frac{\partial \rho_2}{\partial t} \Delta h_2 \right) \Delta S.$$

To find the limiting value of (17), we have to introduce the two-dimensional divergence or surface divergence which is defined by

(18) $$\nabla_2 \cdot \bar{F} = \lim_{\Delta S \to 0} \frac{\oint (\hat{s} \cdot \bar{F})\, dl}{\Delta S}$$

Thus, as Δh_1, Δh_2 and $\Delta S \to 0$, (17), after division by ΔS, reduces to

(19) $$\nabla_2 \cdot \bar{K} + (\hat{n}_1 \cdot \bar{J}_1 + \hat{n}_2 \cdot \bar{J}_2) = -\frac{\partial \rho_s}{\partial t}.$$

Applying the same technique to (9) and (10), we find

(20) $$(\hat{n}_1 \cdot \bar{B}_1 + \hat{n}_2 \cdot \bar{B}_2) = 0$$

and

(21) $$(\hat{n}_1 \cdot \bar{D} + \hat{n}_2 \cdot \bar{D}_2) = \rho_s.$$

This completes our derivation of the boundary conditions. For convenience, we tabulate these results in Table 6-1. It also includes two special cases which are frequently encountered. The characteristic of an ideal perfect conductor is that it cannot sustain a field inside.

TABLE 6-1

Boundary Conditions

Differential Equation	The General Boundary Condition	Neither of the Media Being a Perfect Conductor	Medium II being a Perfect Conductor
$\nabla \times \bar{E} = -\frac{\partial \bar{B}}{\partial t}$	$\hat{n}_1 \times \bar{E}_1 + \hat{n}_2 \times \bar{E}_2 = 0$	$\hat{n}_1 \times \bar{E}_1 + \hat{n}_2 \times \bar{E}_2 = 0$	$\hat{n}_1 \times \bar{E}_1 = 0$
$\nabla \times \bar{H} = \bar{J} + \frac{\partial \bar{D}}{\partial t}$	$\hat{n}_1 \times \bar{H}_1 + \hat{n}_2 \times \bar{H}_2 = \bar{K}$	$\hat{n}_1 \times \bar{H}_1 + \hat{n}_2 \times \bar{H}_2 = 0$	$\hat{n}_1 \times \bar{H}_1 = \bar{K}$
$\nabla \cdot \bar{J} = -\frac{\partial \rho}{\partial t}$	$\nabla_s \cdot \bar{K} + n_1 \cdot \bar{J}_1 + \hat{n}_2 \cdot \bar{J}_2 = -\frac{\partial \rho_s}{\partial t}$	$\hat{n}_1 \cdot \bar{J}_1 + \hat{n}_2 \cdot \bar{J}_2 = -\frac{\partial \rho_s}{\partial t}$	$\nabla_s \cdot \bar{K} + \hat{n}_1 \cdot \bar{J}_1 = -\frac{\partial \rho_s}{\partial t}$
$\nabla \cdot \bar{D} = \rho$	$\hat{n}_1 \cdot \bar{D}_1 + \hat{n}_2 \cdot \bar{D}_2 = \rho_s$	$\hat{n}_1 \cdot \bar{D}_1 + \hat{n}_2 \cdot \bar{D}_2 = \rho_s$	$\hat{n}_1 \cdot \bar{D}_1 = \rho_s$
$\nabla \cdot \bar{B} = 0$	$\hat{n}_1 \cdot \bar{B}_1 + \hat{n}_2 \cdot \bar{B}_2 = 0$	$\hat{n}_1 \cdot \bar{B}_1 + \hat{n}_2 \cdot \bar{B}_2 = 0$	$\hat{n}_1 \cdot \bar{B}_1 = 0$

7. HARMONICALLY OSCILLATING FIELDS IN FREE SPACE

When the field quantities in Maxwell's equations are harmonically oscillating functions with a single frequency of oscillation, f, the system of equations can be simplified considerably. In order to avoid a possible confusion of notation due to the two possible choices of the complex time function, we will give an example. In the first place, we shall use the cosine function to describe the time-varying part. Thus the electric field is represented by

(1) $$\begin{aligned}\bar{E}(x,y,z;t) &= E_{x0} \cos(\omega t - \Phi_x)\,\hat{x} \\ &\quad + E_{y0} \cos(\omega t - \Phi_y)\,\hat{y} \\ &\quad + E_{z0} \cos(\omega t - \Phi_z)\,\hat{z}\end{aligned}$$

where

$$\omega = 2\pi f = \text{angular frequency}.$$

The amplitude functions and the phase functions are, in general, functions of position. We introduce the complex vector function $\bar{E}(x,y,z)$ defined by

(2) $$\bar{E}(x,y,z) = E_{x0}e^{i\Phi_x}\hat{x} + E_{y0}e^{i\Phi_y}\hat{y} + E_{z0}e^{i\Phi_z}\hat{z}$$

then

(3) $$\bar{E}(x,y,z;t) = Re[\bar{E}(x,y,z)e^{-i\omega t}].$$

We use, therefore, the time function $e^{-i\omega t}$ in our work. In terms of the complex functions, Maxwell's equations in free space with a source function $\bar{J}$ can be described by

$$\nabla \times \bar{E} = i\omega\mu_0 \bar{H} \tag{4}$$
$$\nabla \times \bar{H} = \bar{J} - i\omega\varepsilon_0 \bar{E} \tag{5}$$
$$\nabla \cdot \bar{J} = i\omega\rho \tag{6}$$
$$\nabla \cdot (\varepsilon_0 \bar{E}) = \rho \tag{7}$$
$$\nabla \cdot (\mu_0 \bar{H}) = 0 \tag{8}$$

where the spatial functional dependence (x,y,z) has been omitted for simplicity. By eliminating $\bar{H}$ or $\bar{E}$ between (4) and (5), we obtain

$$\nabla \times \nabla \times \bar{E} - k^2 \bar{E} = i\omega\mu_0 \bar{J} \tag{9}$$

and

$$\nabla \times \nabla \times \bar{H} - k^2 \bar{H} = \nabla \times \bar{J} \tag{10}$$

where $k = \omega\sqrt{\mu_0\varepsilon_0} = 2\pi/\lambda$, and λ denotes the free-space wavelength. For identification purposes, equations of the type described by (9) or (10) will be designated as *inhomogeneous vector wave equations.* The entire subject of dyadic Green's technique is developed mainly to find the solutions for this type of equation under the constraint of various boundary conditions. When the domain under consideration is infinite, there are several distinct methods of finding the solution for (9) and (10). The classical method of potentials is reviewed in the following section. In Chapter 4, we will commence our discussion on dyadic Green's function technique based on this classical solution.

Before closing this section we would like to mention an equivalent formulation which is useful to find the field due to small current loops. In principle, once the appropriate dyadic Green's function is known, the electromagnetic field due to any arbitrary current distribution can be found by evaluating some definite integrals involving the current distribution function. In practice, it may be a very tedious task to evaluate these integrals. When the current distribution is in the form of current loops, it is more convenient to introduce an equivalent magnetization vector $\bar{M}$ to replace the current function $\bar{J}$. Thus, if we write Maxwell's equations in the form

$$\nabla \times \bar{E} = i\omega\mu_0 (\bar{H} + \bar{M}) \tag{11}$$
$$\nabla \times \bar{H} = -i\omega\varepsilon_0 \bar{E}, \tag{12}$$

an elimination of $\bar{H}$ and $\bar{E}$ between these two equations yields

$$\nabla \times \nabla \times \bar{E} - k^2 \bar{E} = i\omega\mu_0 \nabla \times \bar{M} \tag{13}$$

$$\nabla \times \nabla \times \bar{H} - k^2 \bar{H} = k^2 \bar{M}. \tag{14}$$

Comparing (9) with (13), we see that the current density function can be replaced by an equivalent magnetization density function $\bar{M}$ such that

$$\bar{J} = \nabla \times \bar{M} .$$

As shown by Stratton (p. 235) for a loop carrying a uniform current, a convenient measure of the strength of the current loop is in terms of the magnetic dipole moment $\bar{m}$. The relation between the magnetic dipole moment and the loop current and its area is given by

$$\iiint \bar{M}\, dV = \bar{m} = I\bar{A} . \tag{15}$$

It will be shown in Chapter 4 and illustrated in Chapter 10 that for a short current element (electric dipole), it is relatively simple to find the solution for (9) with the appropriate dyadic Green's function. Likewise, for a small current loop it is equally simple to find the solutions for (10) with the corresponding dyadic Green's function. In essence, this alternative formulation is just to bypass an otherwise tedious task of evaluating some integrals that contain the current distribution function of a loop as a part the integrand.

8. INTEGRATION OF MAXWELL'S EQUATIONS BY MEANS OF THE METHOD OF POTENTIALS IN AN INFINITE DOMAIN

In view of (7-8) and the vector identity, $\nabla \cdot \nabla \times \bar{A} = 0$, we can define a vector potential function $\bar{A}$ such that

$$\mu_0 \bar{H} = \nabla \times \bar{A} . \tag{1}$$

Upon substituting it into (7-4), and recognizing the identity, that $\nabla \times \nabla \psi = 0$, we can introduce a scalar potential function ψ such that

$$\bar{E} = i\omega \bar{A} - \nabla \psi . \tag{2}$$

Substituting both (1) and (2) into (7-5), one finds that

$$\nabla \times \nabla \times \bar{A} = \mu_0 \bar{J} + k^2 \bar{A} + i\omega\mu_0\varepsilon_0 \nabla \psi . \tag{3}$$

Using identity (10) of Appendix A, we can change (3) into

$$-\nabla^2 \bar{A} + \nabla\nabla \cdot \bar{A} = \mu_0 \bar{J} + k^2 \bar{A} + i\omega\mu_0\varepsilon_0 \nabla \psi . \tag{4}$$

Now we impose the condition that

$$\nabla \cdot \bar{A} = i\omega\mu_0\varepsilon_0 \psi . \tag{5}$$

Equation 4 then is reduced to a vector differential equation containing the vector potential function $\bar{A}$ only—that is,

(6) $\nabla^2 \bar{A} + k^2 \bar{A} = -\mu_0 \bar{J}$ (inhomogeneous vector Helmholtz equation)

In contrast to the vector wave equation defined by (7-9), we will designate (6) as the inhomogeneous vector Helmholtz equation. The differential equation for ψ can be obtained by taking the divergence of (3) and making use of (7-6) and (5), one obtains

(7) $$\nabla^2 \psi + k^2 \psi = -\frac{\rho}{\varepsilon_0}$$ (inhomogeneous scalar wave equation)

Solutions for (6) and (7) that correspond to outgoing waves from the source are given by

(8) $$\bar{A}(\bar{R}) = \mu_0 \iiint \bar{J}(\bar{R}') G_0(\bar{R}|\bar{R}')\, dV'$$

and

(9) $$\psi(\bar{R}) = \frac{1}{\varepsilon_0} \iiint \rho(\bar{R}') G_0(\bar{R}|\bar{R}')\, dV'$$

where

(10) $$G_0(\bar{R}|\bar{R}') = \frac{e^{ik|R-R'|}}{4\pi|\bar{R}-R'|}, \quad |\bar{R}-\bar{R}'| = [(x-x')^2 + (y-y')^2 + (z-z')^2]^{1/2}.$$

The function $G_0(\bar{R}|\bar{R}')$ is called the free-space Green's function pertaining to a three-dimensional scalar wave equation, where $\bar{R}'$ denotes the position vector of a source point and $\bar{R}$ denotes the position vector of a field point or that of an observer. A more detailed description of this function is given in Chapter 3, where the origin of this function is shown. Once we know the solution for $\bar{A}(\bar{R})$, the electromagnetic field vectors $\bar{E}$ and $\bar{H}$ can be found. As a result of (2) and (5) we have

(11) $$\bar{E} = i\omega\left(\bar{A} + \frac{1}{k^2} \nabla\nabla \cdot \bar{A}\right)$$

where (1) gives

(12) $$\bar{H} = \frac{1}{\mu_0} \nabla \times \bar{A}.$$

Before we conclude this section a brief review will be given to the characteristics of the far-zone field. Two conditions are imposed on $\bar{R}$ to derive the expressions for the far-zone field; they are

$$|\bar{R}| \gg |\bar{R}'|$$

and

$$kR \gg 1, \text{ or } R \gg \lambda .$$

Under these conditions the function $G_0(\bar{R}|\bar{R}')$ is given approximately by

$$G_0(\bar{R}|\bar{R}') = \frac{e^{ik(R - \bar{R}' \cdot \hat{R})}}{4\pi R} \tag{13}$$

hence

$$\bar{A}(\bar{R}) = \frac{\mu_0 e^{ikR}}{4\pi R} \iiint \bar{J}(\bar{R}') e^{-ik\bar{R}' \cdot \hat{R}}\, dV' = \frac{\mu_0 e^{ikR}}{4\pi R} \bar{N} . \tag{14}$$

The function $\bar{N}$, called the *radiation vector*, is a function of the spherical angular variables (θ,ϕ) only and is not a function of R. Upon substitution of (14) into (11) and (12) and neglecting higher-order terms, one finds that

$$\bar{E} = \frac{i\omega\mu_0 e^{ikR}}{4\pi R} \bar{N}_t = \frac{ik\eta_0 e^{ikR}}{4\pi R} \bar{N}_t \tag{15}$$

and

$$\bar{H} = \frac{ike^{ikR}}{4\pi R} \hat{R} \times \bar{N}_t = \frac{1}{\eta_0} \hat{R} \times \bar{E}, \tag{16}$$

where $\bar{N}_t$ denotes the transverse part, with respect to $\hat{R}$, of the radiation vector—that is,

$$\bar{N}_t = \bar{N} - N_R \hat{R} . \tag{17}$$

The constant η_0 denotes the free-space wave impedance, being equal to $(\mu_0/\varepsilon_0)^{\frac{1}{2}} \simeq 120\pi$ ohms. The terms which have been neglected are of the order of $1/kR^2$ or higher as compared to $1/R$. Thus, the far-zone electromagnetic field due to any current distribution satisfies the condition that

$$\lim_{R\to\infty} R \left[\nabla \times \begin{pmatrix} \bar{E} \\ \bar{H} \end{pmatrix} - ik\hat{R} \times \begin{pmatrix} \bar{E} \\ \bar{H} \end{pmatrix} \right] = 0 . \tag{18}$$

This is referred to as the *radiation condition* for an electromagnetic field in free space.

chapter **3**

Green's Functions Pertaining to the Scalar Wave Equations

As an introduction to the terminology, the concept, and the method of deriving various types of the dyadic Green's function, we shall first review the transmission line theory from the point of view of the Green's function technique since the subject matter, presumably, is already familiar to most of the readers. Much of the material covered here finds its analogy later when we deal with the vector wave function. The Green's functions pertaining to the two-dimensional and three-dimensional scalar wave equations will also be introduced but no detailed treatment will be given. Those who are interested in grasping more thoroughly this technique for the scalar-wave equation can fill some of the gaps by working the exercises assigned for this chapter.

9. TRANSMISSION-LINE WITH A DISTRIBUTED CURRENT SOURCE: CLASSIFICATION OF GREEN'S FUNCTIONS

We consider a transmission line excited by a distributed current source, $K(x)$, as sketched in Fig. 9-1. The line may be finite or infinite and it may be terminated at either end with an impedance or by another line. For a harmonically oscillating source, the voltage and the current on the line satisfy the following pair of equations:

$$\frac{dV(x)}{dx} = i\omega L I(x) \tag{1}$$

$$\frac{dI(x)}{dx} = i\omega C V(x) + K(x) \tag{2}$$

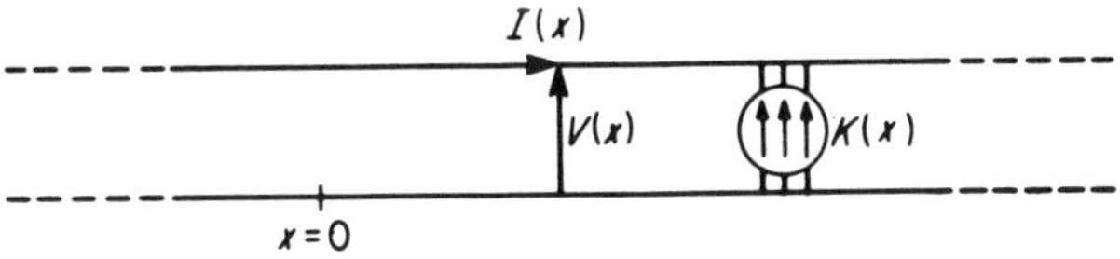

FIG. 9-1. Transmission-line excited by a distributed current source.

where L and C denote, respectively, the distributed inductance and capacitance of the line. Our problem is to find $V(x)$ and $I(x)$ for certain terminations of the line. Because of (1), it is sufficient for us to find $V(x)$ only. By eliminating $I(x)$ between (1) and (2), we obtain

$$\frac{d^2 V(x)}{dx^2} + k^2 V(x) = i\omega L K(x) \tag{3}$$

where $k = \omega\sqrt{LC}$ denotes the propagation constant of the line. Equation (3) has been designated as an inhomogeneous one-dimensional scalar wave equation. The Green's function method has been developed particularly to solve this type of equation in a rather elegant way. The method has an analogy in circuit theory whereby the response of a network to any input function can be determined by an integration based on the impulse response of the network. The Green's function for a spatial problem plays the same role as the impulse response function in a time-domain problem. For transient field problems, the Green's function may be constructed to include the impulsive time characteristics as well. By definition, the Green's function pertaining to a one-dimensional scalar wave equation, denoted by $g(x|x')$, is a solution of the following equation.

$$\frac{d^2 g(x|x')}{dx^2} + k^2 g(x|x') = -\delta(x - x') \tag{4}$$

where $\delta(x - x')$ represents a delta function already introduced in Chap. 1 (Sec. 3). The physical interpretation of (4) is that if we let

$$K(x) = \frac{i}{\omega L}\,\delta(x - x') \tag{5}$$

then

$$\int_{x'-\varepsilon}^{x'+\varepsilon} K(x)\,dx = \frac{i}{\omega L} \tag{6}$$

and (3) reduces to (4). Equations (5) and (6) imply that the line is excited by a localized current source of amplitude $i/\omega L$ placed at $x = x'$. It is known from the theory of differential equations that the solution for $g(x|x')$ satisfying (4) is not completely determined unless we specify the two boundary conditions which the function must satisfy at the extremities of the spatial domain in which the function is defined. The boundary conditions which must be satisfied by $g(x|x')$ are the same as those dictated by the original function which we intend to determine, namely, $V(x)$ in the

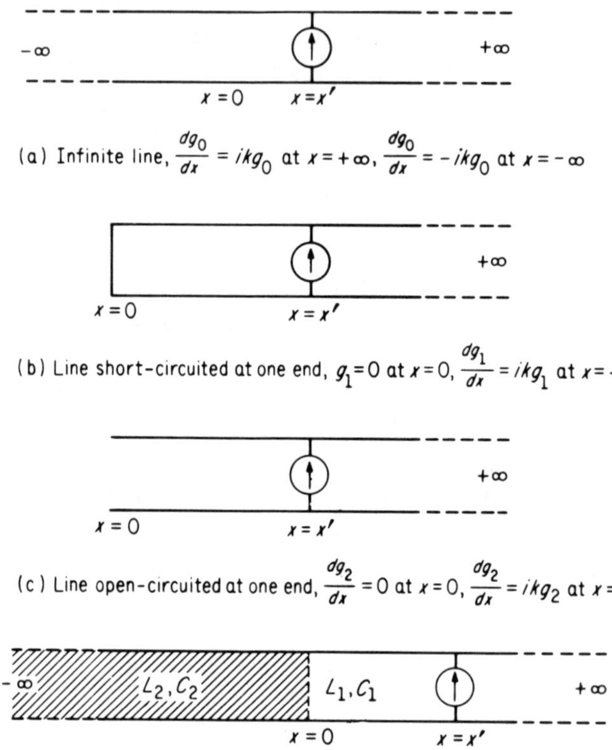

(a) Infinite line, $\dfrac{dg_0}{dx} = ikg_0$ at $x = +\infty$, $\dfrac{dg_0}{dx} = -ikg_0$ at $x = -\infty$

(b) Line short-circuited at one end, $g_1 = 0$ at $x = 0$, $\dfrac{dg_1}{dx} = ikg_1$ at $x = +\infty$

(c) Line open-circuited at one end, $\dfrac{dg_2}{dx} = 0$ at $x = 0$, $\dfrac{dg_2}{dx} = ikg_2$ at $x = +\infty$

(d) Two lines joining at $x = 0$, $\dfrac{1}{L_2}\dfrac{dg_3^{(2)}}{dx} = \dfrac{1}{L_1}\dfrac{dg_3^{(1)}}{dx}$, $g_3^{(2)} = g_3^{(1)}$ at $x = 0$;

$\dfrac{dg_3^{(2)}}{dx} = -ik_2 g_3^{(2)}$ at $x = -\infty$; $\dfrac{dg_3^{(1)}}{dx} = ik_1 g_3^{(1)}$ at $x = +\infty$.

FIG. 9-2. Classification of Green's functions according to the boundary conditions.

present case. For this reason, the Green's functions are classified according to the boundary conditions which they must obey. Some of the typical ones (for the transmission line) are illustrated in Fig. 9-2. To distinguish various types of functions satisfying different boundary conditions, we use a subscript to identify them. In general, the subscript 0 designates infinite domain so that the radiation condition prevails. Subscript 1 means that one of the boundary conditions satisfies the so-called Dirichlet condition while the other satisfies the radiation condition. When one of the boundary conditions satisfies the so-called Neumann condition we use subscript 2. Subscript 3 is reserved for the mixed type. The same nomenclature will be used later in our classification of the dyadic Green's functions. Actually, we should have used a double subscript for two distinct boundary conditions. For example, case (b) of Fig. 9-2 should be denoted by g_{01}, indicating that one radiation condition and one Dirichlet condition are involved. With such an understanding, the simplified notation should be acceptable. In case (d) a superscript becomes necessary because we have two sets of line voltage and current (V_1,I_1) and (V_2,I_2) in this problem, and the Green's function also has different algebraic forms in the two regions. Before we give the derivation of the explicit expressions for various types of the Green's functions, the main formula showing the application of the Green's function should be derived.

Let us consider a single line first and let the domain of x correspond to (x_1,x_2). The function $g(x|x')$ in (4) can represent any of the three types, g_0, g_1 and g_2, illustrated in Fig. 9-2. The treatment of case (d) is slightly different and it will be formulated later. By multiplying (3) by $g(x|x')$ and (4) by $V(x)$ and taking the difference of the two resultant equations, we obtain

$$(7) \qquad V(x)\,\frac{d^2g(x|x')}{dx^2} - g(x|x')\,\frac{d^2V(x)}{dx^2} = -\,V(x)\delta(x-x') - i\omega L\, g(x|x')K(x).$$

An integration of (7) through the entire domain of x yields

$$\int_{x_1}^{x_2} \left[V(x)\,\frac{d^2g(x|x')}{dx^2} - g(x|x')\,\frac{d^2V(x)}{dx^2}\right] dx$$

$$= -\int_{x_1}^{x_2} V(x)\delta(x-x')\,dx - i\omega L \int_{x_1}^{x_2} g(x|x')K(x)\,dx\,.$$

The first term at the right-hand side of the above equation is simply $V(x')$, and the term at the left side can be simplified by integration by parts which gives

$$V(x') = -\,i\omega L \int_{x_1}^{x_2} g(x|x')K(x)\,dx + \left[g(x|x')\,\frac{dV(x)}{dx} - V(x)\frac{dg(x|x')}{dx}\right]_{x_1}^{x_2}.$$

Since $g(x|x')$ and $V(x)$ satisfy the same boundary condition for g_0, g_1, and g_2 at x_1 and x_2, the last term vanishes. Thus we obtain the important formula

(8) $$V(x') = -i\omega L \int_{x_1}^{x_2} g(x|x')K(x)\,dx.$$

If we use the unprimed variable x to denote the position of a field point, as usually is the case, (8) can be changed to

(9) $$V(x) = -i\omega L \int_{x_1}^{x_2} g(x'|x)K(x')\,dx' = -i\omega L \int_{x_1}^{x_2} g(x|x')K(x')\,dx'.$$

The last identity is due to the symmetrical property of the Green's function that will be shown in the next section. The shifting of the primed and unprimed variables is often practiced in our work. For this reason, it is important to point out that $g(x'|x)$, by definition, satisfies the equation

(10) $$\frac{d^2 g(x'|x)}{dx'^2} + k^2\, g(x'|x) = -\delta(x' - x).$$

A comparison of (4) and (10) would help us to realize the importance of not arbitrarily altering the positions of x and x'.

10. DERIVATION OF $g(x|x')$ BY THE CONVENTIONAL METHOD AND THE OHM-RAYLEIGH METHOD

The expressions for various g's can be derived by the conventional method described in the theory of differential equations. An alternative approach is to apply the method of Ohm-Rayleigh,* or the method of eigenfunction expansion. This method is not needed for the problems involving a transmission line. However, it is a necessary tool in our forthcoming treatment of the dyadic Green's function. For this reason we introduce it here as a preparation for the future work. In the following, we will derive the various g's by the conventional method first and then apply the Ohm-Rayleigh method to rederive the free-space Green's function as a demonstration of the techniques involved in that method.

Case (a): Free-Space Green Function, $g_0(x|x')$

The general solution for (9-4) in the two regions (see Fig. 9-2 for the layout) are:

(1) $$g_0(x|x') = Ae^{ikx} \qquad x > x'$$

and

(2) $$g_0(x|x') = Be^{-ikx} \qquad x' > x.$$

* Sommerfeld (*Partial Diff.*, p. 179) introduced this terminology.

The choice of the exponential function with the proper sign assures us the satisfaction of boundary conditions at infinity. At $x=x'$, the function must be continuous, and its derivative is discontinuous. The physical interpretation of these two conditions is that the voltage at x' is continuous but the difference of the line currents at x' must be equal to the source current. Algebraically, if we integrate (9-4) in a small interval around x', we obtain, after identifying $g(x|x')$ as $g_0(x|x')$,

(3) $$\int_{x'-\varepsilon}^{x'+\varepsilon} \frac{d^2 g_0(x|x')}{dx^2}\, dx + k^2 \int_{x'-\varepsilon}^{x'+\varepsilon} g_0(x|x')\, dx = -1.$$

In the limit as $\varepsilon \to 0$, the second term approaches zero, assuming $g(x|x')$ is finite at $x=x'$, (3) then becomes

(4) $$\left[\frac{dg_0(x|x')}{dx}\right]_{x'-0}^{x'+0} = -1.$$

The continuity of $g(x|x')$ at x' means

(5) $$\left[g_0(x|x')\right]_{x'-0}^{x'+0} = 0.$$

Applying these two conditions to (1) and (2), one finds

(6) $$g_0(x|x') = \begin{cases} \dfrac{i}{2k}\, e^{ik(x-x')} & x \geqslant x' \\[2ex] \dfrac{i}{2k}\, e^{-ik(x-x')} & x' \geqslant x \end{cases}$$

A more compact expression for (6) is

(7) $$g_0(x|x') = \frac{i}{2k}\, e^{ik|x-x'|}.$$

Case (b): Green's Function of the First Kind, $g_1(x|x')$

For this case, we let

(8) $$g_1(x|x') = Ae^{ikx}, \qquad x \geqslant x'$$

and

(9) $$g_1(x|x') = B \sin kx, \qquad x' \geqslant x \geqslant 0.$$

The choice of the sine function assures us that the Dirichlet condition is satisfied at $x=0$. Applying (4) and (5), with the notation $g_0(x|x')$ changed to $g_1(x|x')$, we have

$$g_1(x|x') = \begin{cases} \frac{1}{k} \sin kx' \, e^{ikx} & x \geqslant x' \\ \frac{1}{k} \sin kx \, e^{ikx'}, & x' \geqslant x \geqslant 0. \end{cases} \tag{10}$$

Equation 10 may be written in the form

$$g_1(x|x') = \begin{cases} \frac{i}{2k} [e^{ik(x-x')} - {}^{ik(x+x')}], & x \geqslant x' \\ \frac{i}{2k} [e^{-ik(x-x')} - e^{ik(x+x')}], & x' \geqslant x \geqslant 0. \end{cases} \tag{11}$$

In view of (6), we may interpret (11) as consisting of an incident wave and a scattered wave, that is,

$$g_1(x|x') = g_0(x|x') + g_{1_s}(x|x')$$

where

$$g_{1s} = \frac{-i}{2k} e^{ik(x+x')}$$

Such a notion is not only physically useful, but mathematically offers a shortcut to finding a composite Green's function. Because once we know $g_0(x|x')$, it is just a matter of finding g_s so that the desired boundary condition is satisfied. It is observed that $g_{1_s}(x|x')$ or any $g_s(x|x')$ is a solution of the homogeneous differential equation

$$\frac{d^2 g_s}{dx^2} + k^2 g_s = 0,$$

and it does not have any discontinuous characteristics as does $g_0(x|x')$. An illustration is given below in determining $g_2(x|x')$ by this shortcut method, or the method of scattering superposition.

Case (c): Green's Function of the Second Kind, $g_2(x|x')$

The method of scattering superposition suggests that we can start with

$$g_2(x|x') = g_0(x|x') + Ae^{ikx}.$$

To satisfy the Neumann condition at $x = 0$, we require

$$\left[\frac{dg_0(x|x')}{dx} + ikAe^{ikx} \right]_{x=0} = 0,$$

or

$$\frac{1}{2}e^{ikx'} + ikA = 0,$$

hence

$$A = \frac{i}{2k}e^{ikx'}.$$

The complete expression for $g_2(x|x')$ is therefore given by

(12) $$g_2(x|x') = \frac{i}{2k}\begin{cases} e^{ik(x-x')} + e^{ik(x+x')}, & x \geqslant x' \\ e^{-ik(x-x')} + e^{ik(x+x')}, & x' \geqslant x \geqslant 0 \end{cases}$$

$$= \frac{i}{k}\begin{cases} \cos kx'\, e^{ikx}, & x \geqslant x' \\ \cos kx\, e^{ikx'}, & x' \geqslant x \geqslant 0. \end{cases}$$

Case (d): Green's Function of the Third Kind, $g_3(x|x')$

In this case, we have two differential equations to start with:

(13) $$\frac{d^2V_1(x)}{dx^2} + k_1^2V_1(x) = i\omega L_1K_1(x), \qquad x \geqslant 0$$

(14) $$\frac{d^2V_2(x)}{dx^2} + k_2^2V_2(x) = 0, \qquad 0 \geqslant x.$$

It is assumed that the source function is defined in region I (see Fig. 9-2). We introduce the Green's function $g_3(x|x')$ which has two parts. They are solutions of the following two equations:

(15) $$\frac{d^2g_3^{(1)}(x|x')}{dx^2} + k_1^2g_3^{(1)}(x|x') = -\delta(x-x'), \quad x \geqslant 0$$

and

(16) $$\frac{d^2g_3^{(2)}(x|x')}{dx^2} + k_2^2g_3^{(2)}(x|x') = 0, \qquad 0 \geqslant x.$$

At the junction corresponding to $x=0$, $g_3^{(1)}$ and $g_3^{(2)}$ satisfy the boundary condition that

(17) $$g_3^{(1)}(x|x')\big|_{x=0} = g_3^{(2)}(x|x')_{x=0}$$

(18) $$\frac{1}{L_1}\left.\frac{dg_3^{(1)}(x|x')}{dx}\right|_{x=0} = \frac{1}{L_2}\left.\frac{dg_3^{(2)}(x|x')}{dx}\right|_{x=0}$$

The last condition corresponds to the physical requirement that the current at the junction must be continuous. Again, by means of the method of scattering superposition, we write

(19) $$g_3^{(1)}(x|x') = g_0(x|x') + g_{3s}^{(1)}(x|x') = \frac{i}{2k_1}\begin{cases} e^{ik_1(x-x')} + Re^{ik_1(x+x')}, & x \geqslant x' \\ e^{-ik_1(x-x')} + Re^{ik_1(x+x')}, & x' \geqslant x \geqslant 0, \end{cases}$$

(20) $$g_3^{(2)}(x|x') = g_{3s}^{(2)}(x|x') = \frac{i}{2k_1} Te^{-i(k_2x - k_1x')}, \qquad 0 \geqslant x.$$

We have also put an extra term, $e^{ik_1x'}$, with the unknown coefficients R and T to make the final solutions in a more attractive form. Upon applying the boundary conditions (17) and (18) to these two functions, we find

(21) $$1 + R = T$$

(22) $$\frac{k_1}{L_1}(1 - R) = \frac{k_2}{L_2} T$$

or

(23) $$\frac{1}{Z_1}(1 - R) = \frac{1}{Z_2} T$$

where

$$Z_1 = \left(\frac{L_1}{C_1}\right)^{1/2}, \quad Z_2 = \left(\frac{L_2}{C_2}\right)^{1/2},$$

representing, respectively, the characteristic impedance of the lines. By solving for R and T from (21) and (23), we find

(24) $$R = \frac{Z_2 - Z_1}{Z_2 + Z_1}$$

(25) $$T = \frac{2Z_2}{Z_2 + Z_1}.$$

These are just the reflection and the transmission coefficient of a wave propagating from line I toward line II. Knowing $g_3^{(1)}(x|x')$ and $g_3^{(2)}(x|x')$, we can determine V_1 and V_2 by applying the one-dimensional Green's theorem to (13) through (16) as we did in deriving (9-8). Thus, from (13) and (14), we obtain

(26) $$V_1(x') = -i\omega L_1 \int_0^\infty g_3^{(1)}(x|x') K_1(x)\, dx + \left[g_3^{(1)}(x|x') \frac{dV_1(x)}{dx} - V_1(x) \frac{dg_3^{(1)}(x|x')}{dx} \right]_0^\infty.$$

From (14) and (16), we have

$$\left[g_3^{(2)}(x|x') \frac{dV_2(x)}{dx} - V_2(x) \frac{dg_3^{(2)}(x|x')}{dx} \right]_{-\infty}^{0} = 0 \,. \tag{27}$$

Because of the radiation condition at $x = -\infty$, and the fact that

$$\frac{dV_2(x)}{dx} = i\omega L_2 I_2(x),$$

$$\frac{dg_3^{(2)}(x|x')}{dx} = -ik_2 g_3^{(2)}(x|x'),$$

and

$$V_2(x) = Z_2 I_2(x)$$

identity (27) is indeed satisfied. Likewise the second term at the right-hand side of (26) vanishes for the similar reason. Thus we have

$$V_1(x') = -i\omega L_1 \int_0^\infty g_3^{(1)}(x|x') K_1(x)\, dx .$$

By shifting the variables, we obtain

$$V_1(x) = -i\omega L_1 \int_0^\infty g_3^{(1)}(x|x') K_1(x')\, dx', \qquad x \geqslant 0 \tag{28}$$

where we have already made use of the symmetrical property of the Green's function, that is, $g_3^{(1)}(x|x') = g_3^{(1)}(x'|x)$, the proof of which will be found in the next section. To determine $V_2(x)$, we write

$$V_2(x) = V_2(0)e^{-ik_2x} = V_1(0)e^{-ik_2x}$$

which is a solution of (14) that fulfills the radiation condition at $x = -\infty$. The value of $V_1(0)$ can be found from (28) by putting $x = 0$; thus the integral expression for $V_2(x)$ is given by

$$V_2(x) = \left(-i\omega L_1 \int_0^\infty g_3^{(1)}(0|x') K_1(x')\, dx' \right) e^{-ik_2x}. \tag{29}$$

In view of (17) through (20), we have

$$g_3^{(2)}(x|x') = g_3^{(2)}(0|x')e^{-ik_2x} = g_3^{(1)}(0|x')e^{-ik_2x}$$

thus (29) can be written in the form

$$V_2(x) = -i\omega L_1 \int_0^\infty g_3^{(2)}(x|x') K_1(x')\, dx', \qquad 0 \geqslant x \tag{30}$$

which has the same appearance as (28), except that $g_3^{(1)}(x|x')$ is replaced by $g_3^{(2)}(x|x')$ as the domains of the x variables in these two equations are different. This completes our discussion of the derivation and the usage of various kinds of Green's function based on the conventional method. As an introduction to the Ohm-Rayleigh method, we will now rederive the expression for $g_0(x|x')$ by this alternative approach. For convenience, the differential equation for $g_0(x|x')$ is repeated here.

$$\frac{d^2 g_0(x|x')}{dx^2} + k^2\, g_0(x|x') = -\,\delta(x-x'), \quad \infty > x > -\infty. \tag{31}$$

The key step in the Ohm-Rayleigh method is to expand $\delta(x-x')$ in terms of the eigenfunction of a homogeneous equation of the same type as (31). The eigenfunction in this case is e^{ihx} which is a solution of the equation

$$\frac{d^2 f}{dx^2} + h^2 f = 0 \tag{32}$$

where h is an arbitrary constant. Physically, the function e^{ihx} represents a wave that can exist on an infinite line with propagation constant equal to h. Since e^{ihx} is the spectral function used in the Fourier transform, the method of Ohm-Rayleigh in this case is equivalent to the method of Fourier transform in solving (31). Before we apply these methods to (31), a few words must be said about the question of rigor concerning the existence of the Fourier transform of a delta function. A more satisfactory approach to this subject would certainly be based on the theory of distributions which is outside the scope of this book. Our view is that as long as a correct result can be obtained by an operational method, a sacrifice of rigor can be tolerated. With this understanding in mind, we can solve (31) by the method of Fourier transform. Thus we define

$$f(h) = \int_{-\infty}^{\infty} g_0(x|x') e^{-ihx}\, dx \tag{33}$$

then

$$g_0(x|x') = \frac{1}{2\pi} \int_{-\infty}^{\infty} f(h)\, e^{ihx}\, dh. \tag{34}$$

Applying the Fourier transform to (31), and assuming that both g and dg/dx approach zero at $x = \pm\infty$, we obtain

$$\int_{-\infty}^{\infty} \left(\frac{d^2 g_0(x|x')}{dx^2} + k^2 g_0(x|x') \right) e^{-ihx}\, dx = (k^2 - h^2) f(h) = -\,e^{-ihx'},$$

or

$$f(h) = \frac{e^{-ihx'}}{h^2 - k^2} \tag{35}$$

hence

$$g_0(x|x') = \frac{1}{2\pi} \int_{-\infty}^{\infty} \frac{e^{ih(x-x')}}{h^2 - k^2} \, dh \, . \tag{36}$$

The assumption concerning the behavior of g and dg/dx at infinity can be justified if we let k be complex with $Im\,(k) > 0$. Such an artifice is commonly used, and it corresponds to a line with a loss. After the final expression for $g(x|x')$ is obtained we can restore the lossless condition by allowing $Im(k) \to 0$. The procedure also helps us to evaluate the integral representation of $g_0(x|x')$ as expressed by (36) in a relatively simple manner. With such an interpretation, the locations of the poles of the integrand in the h-plane are shown in Fig. 10-1 where the contour of integration is assumed to be along the real axis.

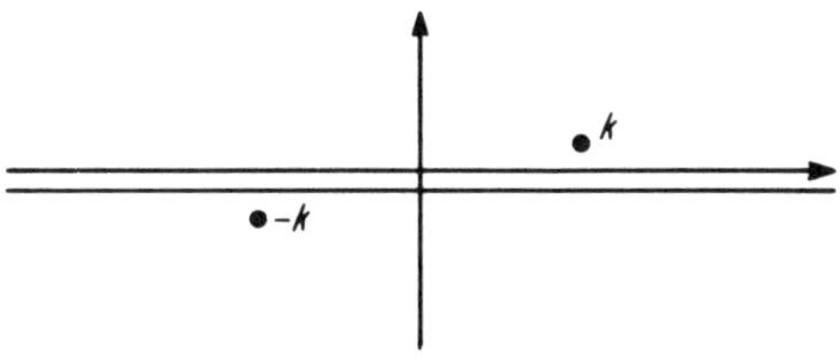

FIG. 10-1. Locations of the poles of the integral representation of $g_0(x|x')$.

For $x - x' \geqslant 0$, the contour can be closed by an infinite path in the upper half-plane without changing its value; for $x - x' \leqslant 0$, the contour can be closed at the lower half-plane. Applying Cauchy's residue theorem to the closed contour integrals we obtain

$$g_0(x|x') = \frac{i}{2k} \begin{cases} e^{ik(x-x')} & x \geqslant x' \\ e^{-ik(x-x')} & x' \geqslant x . \end{cases} \tag{37}$$

For real values of k, the same result is obtained by deforming the contour so that it is properly indented at the poles as shown in Fig. 10-2. Equation (37), of course, is identical to (6). If we had followed precisely the steps involved in the Ohm-Rayleigh method, the procedures are slightly different. First, we let

$$\delta(x - x') = \int_{-\infty}^{\infty} A(h) \, e^{ihx} \, dh \tag{38}$$

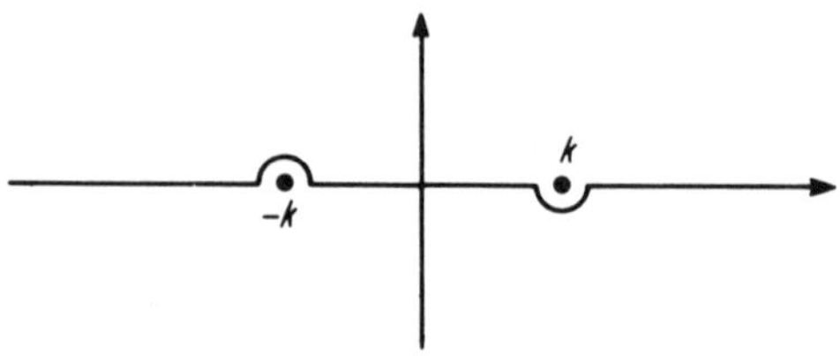

FIG. 10-2. Indentation of contour for real k.

whereby e^{ihx} is treated as an eigenfunction pertaining to the one-dimensional scalar wave equation for an infinite domain. Multiplying (38) by $e^{-ih'x}$ and integrating the resultant equation from $-\infty$ to ∞, we have

$$e^{-ih'x'} = \iint_{-\infty} A(h)\, e^{i(h-h')x}\, dh\, dx$$

where h' denotes an arbitrary constant. In view of (3-24), the integration with respect to x yields $2\pi\delta(h-h')$; thus,

$$e^{-ih'x'} = 2\pi \int_{-\infty}^{\infty} A(h)\delta(h-h')\, dh = 2\pi\, A(h')$$

hence

$$A(h) = \frac{1}{2\pi}\, e^{-ihx'}$$

or

(39) $$\delta(x-x') = \frac{1}{2\pi} \int_{-\infty}^{\infty} e^{ih(x-x')}\, dh.$$

Now we assume that $g_0(x|x')$ can be represented by a similar integral in terms of the eigenfunction e^{ihx}. More precisely, we let

(40) $$g_0(x|x') = \frac{1}{2\pi} \int_{-\infty}^{\infty} B(h) e^{ih(x-x')}\, dh.$$

Upon substituting (39) and (40) into (31), we find that

$$B(h)\,(k^2 - h^2) = -1$$

hence

$$g_0(x|x') = \frac{1}{2\pi} \int_{-\infty}^{\infty} \frac{e^{ih(x-x')}}{h^2 - k^2}\, dh.$$

Although the direct Fourier transform method and the Ohm-Rayleigh method are equivalent, the concepts behind these two methods are quite different. In the Ohm-Rayleigh method, we emphasize the notion that e^{ihx} is an eigenfunction of the differential equation under consideration so we treat (39) and (40) as the eigenfunction expansions of these two functions. The same procedure will later be followed in finding the eigenfunction expansion for the dyadic Green's function.

11. SYMMETRICAL PROPERTIES OF GREEN'S FUNCTIONS

The symmetrical properties of the Green's functions have already been used in deriving (9-9) and (10-28) without a proof. We will now supply the details. We consider two Green's functions corresponding to two different source positions, and the domain in which these two functions are defined is the same. Thus, denoting the source positions, respectively, by x_a and x_b and the domain of x by (x_1, x_2), we have

$$\frac{d^2 g(x|x_a)}{dx^2} + k^2 g(x|x_a) = -\delta(x - x_a) \tag{1}$$

$$\frac{d^2 g(x|x_b)}{dx^2} + k^2 g(x|x_b) = -\delta(x - x_b). \tag{2}$$

Upon multiplying (1) and (2), respectively, by $g(x|x_b)$ and $g(x|x_a)$ and taking the difference of two resultant equations and integrating it through the entire domain of x, we obtain

$$g(x_a|x_b) - g(x_b|x_a) = \left[g(x|x_b)\frac{dg(x|x_a)}{dx} - g(x|x_a)\frac{dg(x|x_b)}{dx} \right]_{x_1}^{x_2}. \tag{3}$$

If $g(x|x_a)$ and $g(x|x_b)$ represent any of the four kinds defined in Sec. 10, that is, g_0, g_1, g_2 and $g_3^{(1)}$, they would satisfy the same boundary conditions at the extremities, hence the term at the right-hand side of (3) vanishes. We therefore conclude that

$$g(x_a|x_b) = g(x_b|x_a) \tag{4}$$

which is the mathematical statement of the symmetrical property of the Green's functions g_0, g_1, g_2 and $g_3^{(1)}$. It is implied that x_a and x_b lie inside the domain where the function is defined. The symmetrical property of $g_3^{(2)}$ exhibits a slightly different form. In the first place, a double subscript must be used to distinguish the g's from the two locations of the source position. If we let x_a be located in region I $(x>0)$ and x_b be located in region II, the various g's satisfy the following equations:

$$\frac{d^2 g_3^{(11)}(x|x_a)}{dx^2} + k_1^2 g_3^{(11)}(x|x_a) = -\delta(x-x_a), \qquad x \geqslant 0 \tag{5}$$

$$\frac{d^2 g_3^{(21)}(x|x_a)}{dx^2} + k_2^2 g_3^{(21)}(x|x_a) = 0, \qquad 0 \geqslant x \tag{6}$$

$$\frac{d^2 g_3^{(22)}(x|x_b)}{dx^2} + k_2^2 g_3^{(22)}(x|x_b) = -\delta(x-x_b), \qquad 0 \geqslant x \tag{7}$$

$$\frac{d^2 g_3^{(12)}(x|x_b)}{dx^2} + k_1^2 g_3^{(12)}(x|x_b) = 0, \qquad x \geqslant 0. \tag{8}$$

The $g_3^{(11)}$ and $g_3^{(21)}$ here correspond to $g_3^{(1)}$ and $g_3^{(2)}$ designated before, as illustrated in Fig. 9-2. At $x=+\infty$, both $g_3^{(11)}$ and $g_3^{(12)}$ satisfy the same radiation conditions specified for region I, and at $x=-\infty$, both $g_3^{(22)}$ and $g_3^{(21)}$ satisfy the identical radiation condition as required for region II. At $x=0$, the boundary conditions are

$$\begin{cases} g_3^{(11)}(0|x_a) = g_3^{(21)}(0|x_a) \\ \dfrac{1}{L_1} \left.\dfrac{dg_3^{(11)}(x|x_a)}{dx}\right|_{x=0} = \dfrac{1}{L_2} \left.\dfrac{dg_3^{(21)}(x|x_a)}{dx}\right|_{x=0} \\ g_3^{(22)}(0|x_b) = g_3^{(12)}(0|x_b) \\ \dfrac{1}{L_2} \left.\dfrac{dg_3^{(22)}(x|x_b)}{dx}\right|_{x=0} = \dfrac{1}{L_1} \left.\dfrac{dg_3^{(12)}(x|x_b)}{dx}\right|_{x=0} \end{cases} \tag{9}$$

Upon multiplying (5) by $g_3^{(12)}$ and (8) by $g_3^{(11)}$ and integrating the difference of the two resultant equations from $x=0$ to $x=+\infty$, we obtain

$$g_3^{(12)}(x_a|x_b) = \left[g_3^{(11)}(x|x_a) \frac{dg_3^{(12)}(x|x_b)}{dx} - g_3^{(12)}(x|x_b) \frac{dg_3^{(11)}(x|x_a)}{dx} \right]_0^\infty$$

$$= \left[g_3^{(12)}(x|x_b) \frac{dg_3^{(11)}(x|x_a)}{dx} - g_3^{(11)}(x|x_a) \frac{dg_3^{(12)}(x|x_b)}{dx} \right]_{x=0}$$

The two terms within the brackets cancel each other at $x=+\infty$ because of the same radiation condition. Repeating a similar calculation for (6) and (7) in the negative x domain, we have

$$g_3^{(21)}(x_b|x_a) = \left[g_3^{(22)}(x|x_b) \frac{dg_3^{(21)}(x|x_a)}{dx} - g_3^{(21)}(x|x_a) \frac{dg_3^{(22)}(x|x_b)}{dx} \right]_{x=0}.$$

In view of (9), it can be shown readily that

(10) $$L_1 g_3^{(21)}(x_b|x_a) = L_2 g_3^{(12)}(x_a|x_b).$$

The same result can, of course be obtained from (10-20) by interchanging x with x', k_1 with k_2 and Z_1 with Z_2. However, the present proof does not require the explicit solutions for $g_3^{(12)}$ or $g_3^{(21)}$.

Another relationship that has a vector analogue later deals with g_1 and g_2. From (10-10) and (10-12) it is obvious that

(11) $$\frac{\partial g_1(x|x')}{\partial x} = -\frac{\partial g_2(x|x')}{\partial x'}.$$

This relationship can be derived directly from the differential equation for g_1 and g_2 without resorting to their explicit solutions.

12. FREE-SPACE GREEN'S FUNCTION PERTAINING TO THE THREE-DIMENSIONAL SCALAR WAVE EQUATION

The free-space Green's function $G_0(\bar{R}|\bar{R}')$, pertaining to a three-dimensional scalar wave equation has already been used in our previous discussion on the potential method. We will now show the origin of this function. The differential equation which governs $G_0(\bar{R}|\bar{R}')$ is

(1) $$\nabla^2 G_0(\bar{R}|\bar{R}') + k^2 G_0(\bar{R}|\bar{R}') = -\delta(\bar{R} - \bar{R}').$$

The function must satisfy the radiation condition that

(2) $$\lim_{R \to \infty} R \left[\frac{\partial G_0}{\partial R} - ikG_0\right] = 0.$$

If we make a change of the variable $\bar{R} - \bar{R}' = \bar{R}_1$ as shown in Fig. 12-1, the problem would have a spherical symmetry with respect to the new origin $0'$, hence the function G_0 would be a funtion of R_1 only. In terms of the new spherical coordinate system with origin at $0'$, the free space Green's function, henceforth denoted by $G_0(\bar{R}_1|0)$, satisfies the equation

(3) $$\frac{1}{R_1^2}\frac{d}{dR_1}\left[R_1^2 \frac{dG_0(\bar{R}_1|0)}{dR_1}\right] + k^2 G_0(\bar{R}_1|0) = -\frac{\delta(R_1 - 0)}{4\pi R_1^2}.$$

The weighting factor $1/4\pi R_1^2$ attached to $\delta(R_1 - 0)$ is to recognize the fact that

$$\iiint_V \delta(\bar{R} - \bar{R}')dV = 1$$

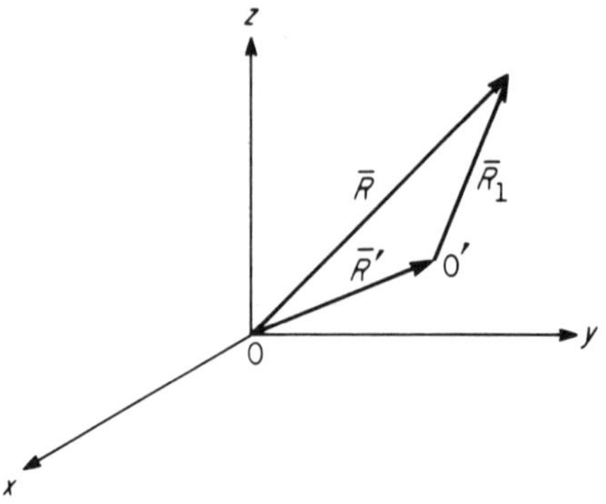

FIG. 12-1. The position vectors in $G_0(\bar{R}|\bar{R}')$.

where V encloses 0′. For $R_1 \neq 0$, the homogeneous equation is the same as the spherical Bessel equation of zeroth order. The appropriate solution for $G_0(\bar{R}_1|0)$, which satisfies the radiation condition at infinity, must therefore be proportional to the spherical Hankel function of the first kind of zeroth order (Stratton, p. 404) or

$$G_0(\bar{R}_1|0) = Ah_0^{(1)}(kR_1) = A\left[\frac{e^{ikR_1}}{ikR_1}\right]. \tag{4}$$

To determine the constant of proportionality A, we make use of the Gauss' theorem that

$$\iiint \nabla^2\psi\, dV = \oint\!\!\!\oint \nabla\psi \cdot \overline{dS}$$

thus a volume of integration of (3) through a small spherical region with center at 0′ gives

$$4\pi R_1^2 \left.\frac{dG_0(R_1|0)}{dR_1}\right|_{R_1=a} + k^2\int_0^a G_0(R_1|0)4\pi R_1^2\, dR_1 = -1\,. \tag{5}$$

Upon substituting (4) into (5) and letting $a \to 0$, we obtain $A = ik/4\pi$. The complete expression for $G_0(\bar{R}_1|0)$ is, therefore, given by

$$G_0(\bar{R}_1|0) = \frac{e^{ikR_1}}{4\pi R_1}\,.$$

Transforming back into the original coordinate system with center at 0, we obtain

$$G_0(\bar{R}|\bar{R}') = \frac{e^{ik|\bar{R}-\bar{R}'|}}{4\pi|\bar{R}-\bar{R}'|}\,. \tag{6}$$

The expression for $G_0(\bar{R}|\bar{R}')$ may be derived alternatively by applying a three-dimensional Fourier integral transform to (1) or a spherical Hankel transform to (3). These techniques are similar to the one described in Sec. 10 for $g_0(x|x')$. However, there is no need to go through these exercises for our future work.

It is worth mentioning that for a two-dimensional scalar wave equation, the corresponding Green's function is given by

$$G_0(\bar{r}|\bar{r}') = \frac{i}{4} H_0^{(1)}(k|\bar{r} - \bar{r}'|) \tag{7}$$

where $\bar{r}$ and $\bar{r}'$ denote the position vector in a cylindrical coordinate system, and the differential equation for G_0 is assumed to be independent of z. The function $G_0(\bar{r}|\bar{r}')$ will not be encountered in the main body of this book except in some exercises.

chapter **4**

Some General Properties of Dyadic Green's Functions

This chapter will introduce, for the first time, the notion of a dyadic function and the composition of a dyadic Green's function. The formation of a dyadic Green's function from the three vector Green's functions will be discussed in detail. An elegant derivation of the free-space dyadic Green's function due to Levine and Schwinger will be presented. Huygen's principle for electromagnetic fields is reformulated, using a combined field vector made of $\bar{E}$ and $\bar{H}$. Finally, the symmetrical properties of various kinds of dyadic Green's functions are shown.

13. DYADS FORMED BY VECTOR FUNCTIONS

The word "dyad" means a group of two or a pair of quantities. In its present usage, a dyad or a dyadic function, denoted by $\bar{\bar{D}}$, is formed by two vector functions. It is defined by

$$\bar{\bar{D}} = \bar{A}\bar{B}. \tag{1}$$

We call $\bar{A}$ the anterior element of $\bar{\bar{D}}$ and $\bar{B}$ the posterior element. A dyadic function by itself does not have any physical interpretation such as a vector function. However, when it acts upon another vector function the result may become more meaningful. A dyadic function, therefore, plays a role similar to that played by a matrix in algebra. There are two *scalar products* between a dyadic function and another vector function $\bar{C}$. The anterior scalar product is defined by

$$\bar{C} \cdot \bar{\bar{D}} = (\bar{C} \cdot \bar{A})\bar{B} = \bar{B}(\bar{C} \cdot \bar{A}) = \bar{B}(\bar{A} \cdot \bar{C}) \tag{2}$$

and the posterior scalar product is defined by

(3) $$\bar{\bar{D}} \cdot \bar{C} = \bar{A}(\bar{B} \cdot \bar{C}) = (\bar{B} \cdot \bar{C})\bar{A} = (\bar{C} \cdot \bar{B})\bar{A}\,.$$

The last two identities suggest that we can define a transpose of $\bar{\bar{D}}$, denoted by $\tilde{\bar{\bar{D}}}$, meaning

(4) $$\tilde{\bar{\bar{D}}} = \bar{B}\,\bar{A}$$

so that

(5) $$\bar{\bar{D}} \cdot \bar{C} = \bar{C} \cdot \tilde{\bar{\bar{D}}}\,.$$

The definition of a scalar product between a dyadic function and a vector function implies that the resultant is a vector. And it is this property which makes a dyadic function a very useful quantity in vector field theory. In addition to two scalar products, there are two vector products between a dyadic function and a vector function. They are defined as

(6) $$\bar{\bar{D}} \times \bar{C} = \bar{A}(\bar{B} \times \bar{C})$$

and

(7) $$\bar{C} \times \bar{\bar{D}} = (\bar{C} \times \bar{A})\bar{B}\,.$$

These products are, of course, dyadic functions by definition. If we decompose $\bar{A}$ and $\bar{B}$ into the component form in the rectangular coordinate system, (1) can be written in the form

(8) $$\bar{\bar{D}} = \begin{pmatrix} A_x B_x \hat{x}\hat{x} + A_x B_y \hat{x}\hat{y} + A_x B_z \hat{x}\hat{z} \\ + A_y B_x \hat{y}\hat{x} + A_y B_y \hat{y}\hat{y} + A_y B_z \hat{y}\hat{z} \\ + A_z B_x \hat{z}\hat{x} + A_z B_y \hat{z}\hat{y} + A_z B_z \hat{z}\hat{z} \end{pmatrix}.$$

Thus a dyadic function in general, has nine components each of which is distinct. An alternative form of writing $\bar{\bar{D}}$ is to condense the nine anterior elements into three vector functions, so that

(9) $$\bar{\bar{D}} = \bar{D}^{(x)}\hat{x} + \bar{D}^{(y)}\hat{y} + \bar{D}^{(z)}\hat{z}$$

where

$$\begin{cases} \bar{D}^{(x)} = A_x B_x \hat{x} + A_y B_x \hat{y} + A_z B_x \hat{z} = B_x \bar{A} \\ \bar{D}^{(y)} = A_x B_y \hat{x} + A_y B_y \hat{y} + A_z B_y \hat{z} = B_y \bar{A} \\ \bar{D}^{(z)} = A_x B_z \hat{x} + A_y B_z \hat{y} + A_z B_z \hat{z} = B_z \bar{A}\,. \end{cases}$$

It is obvious that

$$\bar{\bar{D}} \cdot \hat{x} = \bar{D}^{(x)}, \qquad \bar{\bar{D}} \cdot \hat{y} = \bar{D}^{(y)}, \qquad \bar{\bar{D}} \cdot \hat{z} = \bar{D}^{(z)}.$$

Similarly, we may introduce three vectors ${}^{(x)}\bar{D}$, ${}^{(y)}\bar{D}$, and ${}^{(z)}\bar{D}$ such that

$$\bar{\bar{D}} = \hat{x}\,{}^{(x)}\bar{D} + \hat{y}\,{}^{(y)}\bar{D} + \hat{z}\,{}^{(z)}\bar{D}\,. \tag{10}$$

In our work, form (9) is used more often. In view of (9) and (10) a dyadic function can be considered as a composite function which has the capacity, in the language of the computing age, to store three vector functions. And this is precisely the reason it was introduced in field theories whenever there are three distinct vector functions to be presented or analyzed.

A dyadic function can be operated on by the divergence and the curl. Thus the divergence of a dyadic function is defined by

$$\nabla \cdot \bar{\bar{D}} = [\nabla \cdot \bar{D}^{(x)}]\hat{x} + [\nabla \cdot \bar{D}^{(y)}]\hat{y} + [\nabla \cdot \bar{D}^{(z)}]\hat{z} \tag{11}$$

and the curl of $\bar{\bar{D}}$ produces another dyadic function defined by

$$\nabla \times \bar{\bar{D}} = [\nabla \times \bar{D}^{(x)}]\hat{x} + [\nabla \times \bar{D}^{(y)}]\hat{y} + [\nabla \times \bar{D}^{(z)}]\hat{z}\,. \tag{12}$$

One unique dyad, called the *idem factor* or unit dyad, is defined by

$$\bar{\bar{I}} = \hat{x}\hat{x} + \hat{y}\hat{y} + \hat{z}\hat{z}\,. \tag{13}$$

The characteristic feature of the idem factor is manifested in the following identities:

$$\bar{A} \cdot \bar{\bar{I}} = \bar{\bar{I}} \cdot \bar{A} = \bar{A} \tag{14}$$

and

$$\nabla \cdot (\bar{\bar{I}}\psi) = \nabla\psi \tag{15}$$

where ψ represents a scalar function of position. The above introduction covers all the essential operations which we have to know concerning the algebra of a dyadic function.

14. FREE-SPACE VECTOR GREEN'S FUNCTIONS AS THE CONSTITUENTS OF THE FREE-SPACE DYADIC GREEN'S FUNCTION

To introduce the dyadic Green's function, it is most convenient to commence with a problem which is already familiar to us. We consider the field due to a harmonically oscillating current in free space in the absence of any diffracting body. In Sec. 8 we reviewed the formulation and the solution to that problem based on the method of potentials. For clarity, some of the basic formulas are repeated here. The two equations to be solved are

(1) $$\nabla \times \nabla \times \bar{E}(\bar{R}) - k^2 \bar{E}(\bar{R}) = i\omega\mu_0 \bar{J}(\bar{R})$$

(2) $$\nabla \times \nabla \times \bar{H}(\bar{R}) - k^2 \bar{H}(\bar{R}) = \nabla \times \bar{J}(\bar{R}).$$

The solutions are

(3) $$\bar{E}(\bar{R}) = i\omega\left[1 + \frac{1}{k^2}\nabla\nabla\cdot\right]\bar{A}(\bar{R})$$

(4) $$\bar{H}(\bar{R}) = \frac{1}{\mu_0}\nabla \times \bar{A}(\bar{R})$$

where

(5) $$\bar{A}(\bar{R}) = \mu_0 \iiint G_0(\bar{R}|\bar{R}')\bar{J}(\bar{R}')\, dV$$

with

(6) $$G_0(\bar{R}|\bar{R}') = \frac{e^{ik|\bar{R} - \bar{R}'|}}{4\pi|\bar{R} - \bar{R}'|}.$$

It is obvious that the fields $\bar{E}(\bar{R})$ and $\bar{H}(\bar{R})$, in general, depend upon the distribution of the source current. We now consider an infinitesimal source located at $\bar{R}'$ with a current moment equal to $1/i\omega\mu_0$ pointed in the x-direction. This is described by

(7) $$\bar{J}(\bar{R}) = \frac{1}{i\omega\mu_0}\,\delta(\bar{R} - \bar{R}')\,\hat{x}$$

and

(8) $$\iiint \bar{J}(\bar{R})\, dV = \frac{1}{i\omega\mu_0}\,\hat{x}\,.$$

Substituting (7) and (8) into (5), we obtain

$$\bar{A}(\bar{R}) = \frac{1}{i\omega}\, G_0(\bar{R}|\bar{R}')\,\hat{x}\,.$$

If we denote the electric field produced by this elementary source by $\bar{G}_0^{(x)}(\bar{R}|\bar{R}')$, then

(9) $$\bar{G}_0^{(x)}(\bar{R}|\bar{R}') = \left(1 + \frac{1}{k^2}\nabla\nabla\cdot\right) G_0(\bar{R}|\bar{R}')\,\hat{x}$$

which is a solution of the equation

(10) $$\nabla \times \nabla \times \bar{G}_0^{(x)}(\bar{R}|\bar{R}') - k^2 \bar{G}_0^{(x)}(\bar{R}|\bar{R}') = \delta(\bar{R} - \bar{R}')\,\hat{x}\,.$$

In view of (8-18) the function $\bar{G}_0^{(x)}(\bar{R}|\bar{R}')$ satisfies the radiation condition as $R\rightarrow\infty$, or

$$\lim_{R\rightarrow\infty} R\,[\nabla\times\bar{G}_0^{(x)}(\bar{R}|\bar{R}')-ik\hat{R}\times\bar{G}_0^{(x)}(\bar{R}|\bar{R}')]=0\,. \tag{11}$$

We designate $\bar{G}_0^{(x)}(\bar{R}|\bar{R}')$ as a free-space vector Green's function with a source pointed in the x-direction, hence, the superscript notation (x). The function plays a similar role as $G_0(\bar{R}|\bar{R}')$ does with respect to a scalar wave equation. Since a vector source can point in the y-direction or in the z-direction as well, we have two more free-space vector Green's functions, denoted, respectively, by $\bar{G}_0^{(y)}(\bar{R}|\bar{R}')$ and $\bar{G}_0^{(z)}(\bar{R}|\bar{R}')$. The equations which are satisfied by these two functions and their solutions are given by

$$\nabla\times\nabla\times\bar{G}_0^{(y)}(\bar{R}|\bar{R}')-k^2\bar{G}_0^{(y)}(\bar{R}|\bar{R}')=\delta\,(\bar{R}-\bar{R}')\,\hat{y} \tag{12}$$

$$\nabla\times\nabla\times\bar{G}_0^{(z)}(\bar{R}|\bar{R}')-k^2\bar{G}_0^{(z)}(\bar{R}|\bar{R}')=\delta\,(\bar{R}-\bar{R}')\,\hat{z} \tag{13}$$

and

$$\bar{G}_0^{(y)}(\bar{R}|\bar{R}')=(1+\frac{1}{k^2}\nabla\nabla\cdot)\,G_0(\bar{R}|\bar{R}')\,\hat{y} \tag{14}$$

$$\bar{G}_0^{(z)}(\bar{R}|\bar{R}')=(1+\frac{1}{k^2}\nabla\nabla\cdot)\,G_0(\bar{R}|\bar{R}')\,\hat{z}\,. \tag{15}$$

These two functions obey the same radiation condition as described by (11).

As discussed in the previous section, a dyadic function can store three vector functions. It is now quite natural that a dyadic function can be introduced which would store the three distinct vector Green's functions. We call this a *free-space dyadic Green's function*, and it is defined by

$$\bar{\bar{G}}_0(\bar{R}|\bar{R}')=\bar{G}_0^{(x)}(\bar{R}|\bar{R}')\,\hat{x}+\bar{G}_0^{(y)}(\bar{R}|\bar{R}')\,\hat{y}+\bar{G}_0^{(z)}(\bar{R}|\bar{R}')\,\hat{z}\,. \tag{16}$$

By placing a unit vector $\hat{x}$, $\hat{y}$ and $\hat{z}$ respectively at the posterior position of (10), (12), and (13), and adding the three resultant equations, we can form a dyadic differential equation for $\bar{\bar{G}}_0(\bar{R}|\bar{R}')$, which reads

$$\nabla\times\nabla\times\bar{\bar{G}}_0(\bar{R}|\bar{R}')-k^2\bar{\bar{G}}_0(\bar{R}|\bar{R}')=\bar{\bar{I}}\,\delta\,(\bar{R}-\bar{R}') \tag{17}$$

where $\bar{\bar{I}}$ denotes the idem factor defined by (13-13). Upon substituting (9), (14) and (15) into (16), we obtain, in view of (13-15),

$$\bar{\bar{G}}_0(\bar{R}|\bar{R}')=\left(1+\frac{1}{k^2}\nabla\nabla\cdot\right)\,G_0(\bar{R}|\bar{R}')\,\bar{\bar{I}}=\left(\bar{\bar{I}}+\frac{1}{k^2}\nabla\nabla\right)\,G_0(\bar{R}|\bar{R}')\,. \tag{18}$$

Equations 17 and 18 contain all the information which is needed to describe the three separate differential equations and their solutions for three vector

Green's functions. In a similar manner, the three separate radiation conditions for the vector Green's functions can be combined to read

$$\lim_{R\to\infty} R[\nabla \times \bar{\bar{G}}_0(\bar{R}|\bar{R}') - ik\hat{R} \times \bar{\bar{G}}_0(\bar{R}|\bar{R}')] = 0 . \tag{19}$$

It is expected that once we know $\bar{\bar{G}}_0(\bar{R}|\bar{R}')$ the field due to any current distribution can be found by a summation process or by an integration. The integration formula can readily be derived by an application of the vector Green's theorem (2-5). Thus we let

$$\bar{P} = \bar{E}(\bar{R}), \quad \bar{Q} = \bar{\bar{G}}_0(\bar{R}|\bar{R}') \cdot \bar{a}$$

where $\bar{a}$ denotes a constant but arbitrary vector. Upon substituting them into (2-5), we have

$$\begin{aligned}\iiint_V \{\bar{E}(\bar{R}) \cdot \nabla \times \nabla \times \bar{\bar{G}}_0(\bar{R}|\bar{R}') \cdot \bar{a} - [\nabla \times \nabla \times \bar{E}(\bar{R})] \cdot \bar{\bar{G}}_0(\bar{R}|\bar{R}') \cdot \bar{a}\}\, dV \\ = -\oiint_S \{[\nabla \times \bar{E}(\bar{R})] \times \bar{\bar{G}}_0(\bar{R}|\bar{R}') \cdot \bar{a} + \bar{E}(\bar{R}) \times \nabla \times \bar{\bar{G}}_0(\bar{R}|\bar{R}') \cdot \bar{a}\} \cdot d\bar{S} \\ = -\oiint_S \{[\hat{n} \times \nabla \times \bar{E}(\bar{R})] \cdot \bar{\bar{G}}_0(\bar{R}|\bar{R}') \cdot \bar{a} + [\hat{n} \times \bar{E}(\bar{R})] \cdot \nabla \times \bar{\bar{G}}_0(\bar{R}|\bar{R}') \cdot \bar{a}\}\, dS\end{aligned} \tag{20}$$

where $\hat{n}$ denotes an outward unit vector normal to the surface S. The terms $\nabla \times \nabla \times \bar{E}(\bar{R})$ and $\nabla \times \nabla \times \bar{\bar{G}}_0(\bar{R}|\bar{R}')$ contained in the volume integral of (20) can be eliminated as a result of (1) and (17). Since

$$\iiint_V \bar{E}(\bar{R}) \cdot \bar{\bar{I}}\, \delta(\bar{R} - \bar{R}')\, dV = \bar{E}(\bar{R}')$$

(20) can be written as

$$\begin{aligned}\bar{E}(\bar{R}') \cdot \bar{a} = i\omega\mu_0 \iiint_V \bar{J}(\bar{R}) \cdot \bar{\bar{G}}_0(\bar{R}|\bar{R}') \cdot \bar{a}\, dV \\ - \iint_S \{[\hat{n} \times \nabla \times \bar{E}(\bar{R})] \cdot \bar{\bar{G}}_0(\bar{R}|\bar{R}') \cdot \bar{a} + [\hat{n} \times \bar{E}(\bar{R})] \cdot \nabla \times \bar{\bar{G}}_0(\bar{R}|\bar{R}') \cdot \bar{a}\}\, dS .\end{aligned}$$

By replacing $\nabla \times \bar{E}$ by $i\omega\mu_0 \bar{H}$ and interchanging the primed and the unprimed variables and deleting $(\cdot\, \bar{a})$ from the above equation, since $\bar{a}$ is a constant vector we have

$$\begin{aligned}\bar{E}(\bar{R}) = i\omega\mu_0 \iiint_V \bar{J}(\bar{R}') \cdot \bar{\bar{G}}_0(\bar{R}'|\bar{R})\, dV' - \oiint_S \{i\omega\mu_0 [\hat{n} \times \bar{H}(\bar{R}')] \cdot \bar{\bar{G}}_0(\bar{R}'|\bar{R}) \\ + [\hat{n} \times \bar{E}(\bar{R}')] \cdot \nabla' \times \bar{\bar{G}}_0(\bar{R}'|\bar{R})\}\, dS'\end{aligned} \tag{21}$$

where ($\nabla' \times$) means that the curl operation has to be performed in the primed coordinate system. If we consider the volume to be of infinite extent so that S recedes to infinity, then as a result of the radiation condition described by (8-18), or

$$\lim_{R\to\infty} R[i\omega\mu_0 \bar{H}(\bar{R}) - ik\hat{n} \times \bar{E}(\bar{R})] = 0 \tag{22}$$

the surface integral in (21) can be written in the form

$$\oiint_{S\to\infty} [\hat{n} \times \bar{E}(\bar{R}')] \cdot [\nabla' \times \bar{\bar{G}}_0(\bar{R}'|\bar{R}) - ik\hat{n} \times \bar{\bar{G}}_0(\bar{R}'|\bar{R})]\, dS'.$$

Because of the radiation condition described by (19) the surface integral vanishes as S recedes to infinity. Hence

$$\bar{E}(\bar{R}) = i\omega\mu_0 \iiint_V \bar{J}(\bar{R}') \cdot \bar{\bar{G}}_0(\bar{R}'|\bar{R})\, dV' = i\omega\mu_0 \iiint_V \tilde{\bar{\bar{G}}}_0(\bar{R}'\bar{R}) \cdot \bar{J}(\bar{R}')\, dV' \tag{23}$$

$$= i\omega\mu_0 \iiint_V \bar{\bar{G}}_0(\bar{R}|\bar{R}') \cdot \bar{J}(\bar{R}')\, dV'.$$

The last expression results from the identity that

$$\tilde{\bar{\bar{G}}}_0(\bar{R}'|\bar{R}) = \bar{\bar{G}}_0(\bar{R}|\bar{R}')\,. \tag{24}$$

This symmetrical property of the free-space Green's function will be proved later in Sec. 17.

Another formula which can be derived from (21) deals with the mathematical formulation of Huygen's principle. If we let V be bounded by two surfaces S_1 and S_2 as shown in Fig. 14-1 where the current source is assumed to be confined within S_1, then the volume integral is zero. And if we let S_2 recede to infinity (21) reduces to

(25)

$$\bar{E}(\bar{R}) = -\oiint_{S_1} \{i\omega\mu_0[\hat{n} \times \bar{H}(\bar{R}')] \cdot \bar{\bar{G}}_0(\bar{R}'|\bar{R}) + [\hat{n} \times \bar{E}(\bar{R}')] \cdot \nabla' \times \bar{\bar{G}}_0(\bar{R}'|\bar{R})\}\, dS'$$

or

(26)

$$\bar{E}(\bar{R}) = -\oiint_{S_1} \{i\omega\mu_0 \bar{\bar{G}}_0(\bar{R}|\bar{R}') \cdot [\hat{n} \times \bar{H}(\bar{R}')] + \nabla \times \bar{\bar{G}}_0(\bar{R}|\bar{R}') \cdot [\hat{n} \times \bar{E}(\bar{R}')]\}\, dS'.$$

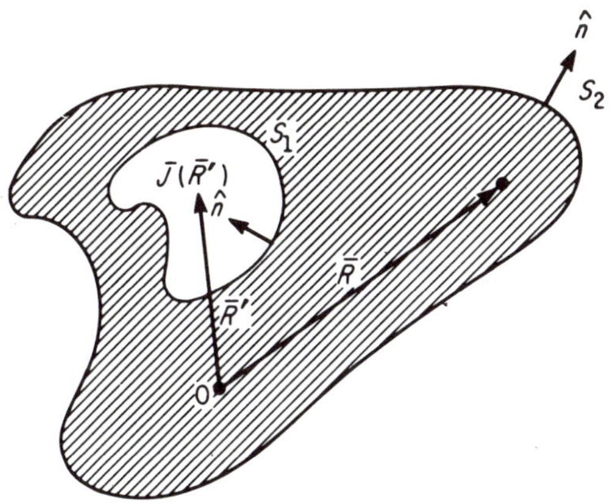

FIG. 14-1. A volume bounded by two surfaces, S_1 and S_2, with the current source confined within S_1.

The conversion from (25) to (26) results from the fact that we have

$$\tilde{\nabla}' \times \bar{\bar{G}}_0(\bar{R}'|\bar{R}) = \nabla \times \bar{\bar{G}}_0(\bar{R}|\bar{R}')^* \tag{27}$$

which will be proved later together with (24). Equation (25) or (26) is a precise mathematical statement of the Huygen's principle for an electromagnetic field. It means that whenever the tangential components of $\bar{E}$ and $\bar{H}$ on a surface which encloses a radiation source are known, the field elsewhere outside the surface can be determined with the aid of the free-space Green's function.

In a similar manner, if we apply the vector Green's theorem to integrate the equation for $\bar{H}(\bar{R})$, (2), the following expression can be derived.

(28)

$$\bar{H}(\bar{R}) = \iiint_V \bar{\bar{G}}_0(\bar{R}|\bar{R}') \cdot \nabla' \times \bar{J}(\bar{R}')\, dV' + \oint\!\!\!\oint_S \{i\omega\varepsilon_0 \bar{\bar{G}}_0(\bar{R}|\bar{R}') \cdot [\hat{n} \times \bar{E}(\bar{R}')]$$
$$- \nabla \times \bar{\bar{G}}_0(\bar{R}|\bar{R}') \cdot [\hat{n} \times \bar{H}(\bar{R}')]\}\, dS'.$$

In the special case, when V and S correspond to that of Fig. 14-1, we have

(29)

$$\bar{H}(\bar{R}) = \iint_{S_1} \{i\omega\varepsilon_0 \bar{\bar{G}}_0(\bar{R}|\bar{R}') \cdot [\hat{n} \times \bar{E}(\bar{R}')] - \nabla \times \bar{\bar{G}}_0(\bar{R}|\bar{R}') \cdot [\hat{n} \times \bar{H}(\bar{R}')]\}\, dS'.$$

* The transpose of the curl of a dyadic function is indicated by a tilde over the nabla sign: $\tilde{\nabla} \times \bar{\bar{G}} \equiv (\nabla \times \bar{\bar{G}}))_{transpose}$.

A compact formulation that embraces both (26) and (20) will be described in Sec. 16. Alternative expressions to describe Huygen's principle using other kinds of dyadic Green's functions will be described later, after these functions are introduced.

Finally, it should be mentioned that Huygen's principle for electromagnetic theory was previously formulated by Stratton and Chu (Stratton, pp. 464–468) based on the technique of vector Green's function. Because of their choice of the vector Green's function, the surface integral involves not only $\hat{n} \times \bar{E}$ and $\hat{n} \times \bar{H}$ but also $\hat{n} \cdot \bar{E}$. The term involving $\hat{n} \cdot \bar{E}$ can be eliminated if a proper vector Green's function is chosen. An exercise is assigned in Appendix D to reformulate this problem with a revised vector Green's function.

15. THE OPERATIONAL METHOD OF LEVINE AND SCHWINGER

The expression for $\bar{\bar{G}}_0(\bar{R}|\bar{R}')$ has so far been constructed using the known solutions for the vector Green's functions which are obtained by the method of potentials. Actually, there are two direct methods by which one can find $\bar{\bar{G}}_0(\bar{R}|\bar{R}')$ without taking advantage of the previous knowledge on this problem. One of them is the Fourier transform method. This is similar to the technique which we discussed in Chapter 3 in finding $g_0(x|x')$ for the transmission line. The main difference is that a three-dimensional Fourier transform is needed for the present problem and the evaluation of the inverse Fourier transform is rather long and tedious. Another method, due to Levine and Schwinger (1950), is short and elegant. However, it involves some operational steps with which we have to bear. This method is described in the following.

In view of the definition of $\nabla^2 \bar{F}$ as given by (10) of Appendix A, (14-17) can be written in the form

$$-\nabla^2 \bar{\bar{G}}_0(\bar{R}|\bar{R}') + \nabla\nabla \cdot \bar{\bar{G}}_0(\bar{R}|\bar{R}') - k^2 \bar{\bar{G}}_0(\bar{R}|\bar{R}') = \bar{\bar{I}}\delta(\bar{R} - \bar{R}') . \tag{1}$$

By taking the divergence of (14-17), we obtain

$$\nabla \cdot \bar{\bar{G}}_0(\bar{R}|\bar{R}') = -\frac{1}{k^2}\nabla \cdot [\bar{\bar{I}}\delta(\bar{R} - \bar{R}')] = -\frac{1}{k^2}\nabla\, [\delta(\bar{R} - \bar{R}')] . \tag{2}$$

Upon substituting (2) into (1) and rearranging the terms, we have

$$\nabla^2 \bar{\bar{G}}_0(\bar{R}|\bar{R}') + k^2 \bar{\bar{G}}_0(\bar{R}|\bar{R}') = -\left(\bar{\bar{I}} + \frac{1}{k^2}\nabla\nabla\right)\delta(\bar{R} - \bar{R}') . \tag{3}$$

To solve for $\bar{\bar{G}}_0(\bar{R}|\bar{R}')$, we let

$$\bar{\bar{G}}_0(\bar{R}|\bar{R}') = \left(\bar{\bar{I}} + \frac{1}{k^2}\nabla\nabla\right)\psi(\bar{R}|\bar{R}').$$

The above expression would be a solution of (3) provided that $\psi(\bar{R}|\bar{R}')$ satisfies the scalar wave equation

$$\nabla^2\psi(\bar{R}|\bar{R}') + k^2\psi(\bar{R}|\bar{R}') = -\delta(\bar{R}-\bar{R}').$$

In an infinite domain, the solution for $\psi(\bar{R}|\bar{R}')$ which fulfills the radiation condition, according to Sec. 12 is the free-space Green's function $G_0(\bar{R}|\bar{R}')$, hence

$$\bar{\bar{G}}_0(\bar{R}|\bar{R}') = \left(\bar{\bar{I}} + \frac{1}{k^2}\nabla\nabla\right) G_0(\bar{R}|\bar{R}') \tag{4}$$

which is identical to (14-18). The ingenuity of this method needs no elaboration. In Chapter 12, the same method will be applied to derive the dyadic Green's function for a moving medium.

16. A COMPACT FORMULATION OF HUYGENS' PRINCIPLE

In Sec. 14 we have derived separately the formulas for $\bar{E}$ and $\bar{H}$ involved in Huygens' principle. The double tasks can be avoided if we adopt the concept of a combined field vector* as first introduced by Bateman (1955, p. 4) and used effectively by Itoh (1959). By applying the dyadic Green's function technique to the equation for the combined field vector, we can present Huygens' principle in a very compact manner. The formulation does not add anything new, but it is an interesting exercise.

We start with the two independent Maxwell's equations in their definite form for an air-medium:

$$\nabla \times \bar{E} = i\omega\mu_0\bar{H} \tag{1}$$

$$\nabla \times \bar{H} = \bar{J} - i\omega\varepsilon_0\bar{E}. \tag{2}$$

A combined field vector is now introduced; it is defined by

$$\bar{F} = \bar{E} + \Gamma\bar{H} \tag{3}$$

where

* The term "unified field vector" was used previously by Itoh. Such a designation, however, is in conflict with a similar term encountered in relativistic theory. The new name "combined field vector" is therefore suggested here.

$$\Gamma = \pm i\eta_0 = hi\eta_0, \qquad \eta_0 = (\mu_0/\varepsilon_0)^{1/2}.$$

The ambiguity sign ± 1, denoted by h, will be used as a separation operator similar to $\sqrt{-1}$ or i in complex number theory. We adopt the rule that $h^2 = 1$. By multiplying (2) by Γ and adding it to (1) we obtain

(4) $$\nabla \times \bar{F} = q\bar{F} + \Gamma\bar{J}$$

where

$$q = -i\omega\varepsilon_0\Gamma = hk, \text{ and } k = \omega(\varepsilon_0\eta_0)^{1/2} = 2\pi/\lambda.$$

Equations (1) and (2) can be recovered from (4) by separating the parts with and without h. A function such as

$$f = A + hB$$

therefore plays a similar role as the function $u + iv$ in complex variable theory. For convenience, we call A the transversal part and B the longitudinal part of f. To integrate (4), we first take the curl of that equation, which yields

(5) $$\nabla \times \nabla \times \bar{F} - k^2\bar{F} = \bar{f}$$

where

$$\bar{f} = \Gamma(q\bar{J} + \nabla \times \bar{J}).$$

Applying now the vector Green theorem to (5) and (14-17) with $\bar{P} = \bar{F}$ and $\bar{Q} = \bar{\bar{G}}_0 \cdot \bar{a}$ as we did before for $\bar{E}$ and $\bar{H}$ and simplifying the result, we obtain

$$\bar{F}(\bar{R}') = \iiint \bar{f}(\bar{R}) \cdot \bar{\bar{G}}_0(\bar{R}|\bar{R}')\, dV + \oiint (\bar{F} \times \hat{n}) \cdot [\nabla \times \bar{\bar{G}}_0(\bar{R}|\bar{R}') + q\bar{\bar{G}}_0(\bar{R}|\bar{R}')]\, dS$$

or equivalently

(6)
$$\bar{F}(\bar{R}) = \iiint \bar{\bar{G}}_0(\bar{R}|\bar{R}') \cdot \bar{f}(\bar{R}')\, dV' + \oiint [\nabla \times \bar{\bar{G}}_0(\bar{R}|\bar{R}') + q\bar{\bar{G}}_0(\bar{R}|\bar{R}') \cdot (\bar{F} \times \hat{n})\, dS',$$

where we have interchanged the primed and unprimed variables and have made use of the symmetrical property of $\bar{\bar{G}}_0$ and $\nabla \times \bar{\bar{G}}_0$ as quoted previously. Equation (6) represents, perhaps, the most compact expression in describing Huygens' principle for a harmonically oscillating field. By taking the transversal part and the longitudinal part of that equation, using h as the separation operator, the expressions for $\bar{E}$ and $\bar{H}$ as derived previously by two separate manipulations can be recovered.

17. CLASSIFICATION OF VARIOUS KINDS OF DYADIC GREEN'S FUNCTIONS AND THEIR SYMMETRICAL PROPERTIES

Similar to the terminology used in the transmission-line theory, we classify various kinds of dyadic Green's functions according to one of the essential boundary conditions which they must satisfy on certain surfaces or across a surface of discontinuity. In general, we will use the notation $\bar{\bar{G}}(\bar{R}|\bar{R}')$ to denote a dyadic Green's function which is a solution of the equation

$$\nabla \times \nabla \times \bar{\bar{G}}(\bar{R}|\bar{R}') - k^2 \bar{\bar{G}}(\bar{R}|\bar{R}') = \bar{\bar{I}}\delta(\bar{R} - \bar{R}'). \tag{1}$$

Different kinds of $\bar{\bar{G}}$ will be distinguished by a subscript. Thus $\bar{\bar{G}}_0$ has been used to designate the free-space Green's function which satisfies the radiation condition at infinity, and that is all. Other kinds of dyadic Green's function will now be introduced together with the origin of these functions.

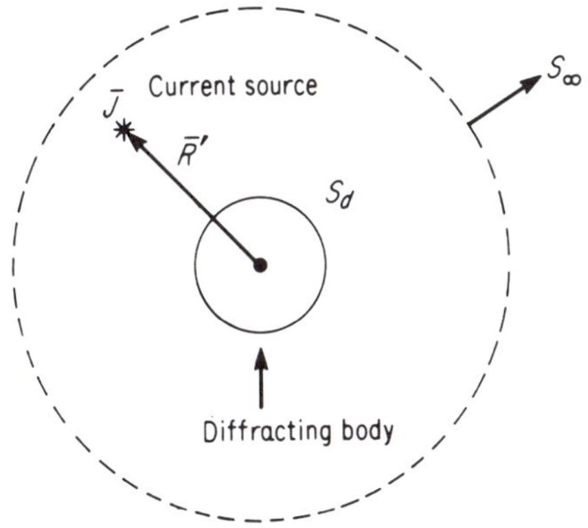

FIG. 17-1. A current source in the presence of a conducting body.

Let us consider the field fue to a current source in the presence of a diffracting body which is assumed to be perfectly conducting as shown schematically in Fig. 17-1.

We seek the solution for (1) which satisfies the boundary condition that on the surface of the body $\hat{n} \times \bar{E}(\bar{R}) = 0$. To integrate (1), we let

$$\bar{P} = \bar{E}(\bar{R})$$

$$\bar{Q} = \bar{\bar{G}}(\bar{R}|\bar{R}') \cdot \bar{a}.$$

Applying the vector Green's theorem to these two functions through a volume bounded by S_d and S_∞ we obtain, as a result of (14-1) and (1), an expression for $\bar{E}(\bar{R})$ which is similar to (14-20) except that the function $\bar{\bar{G}}_0$ is now replaced by $\bar{\bar{G}}$, that is

$$\text{(2)} \qquad \bar{E}(\bar{R}') = i\omega\mu_0 \iiint \bar{J}(\bar{R}) \cdot \bar{\bar{G}}(\bar{R}|\bar{R}')\, dV$$
$$- \oiint_{S_d, S_\infty} \{[\hat{n} \times \bar{E}(\bar{R})] \cdot \nabla \times \bar{\bar{G}}(\bar{R}|\bar{R}') + [\hat{n} \times \nabla \times \bar{E}(\bar{R})] \cdot \bar{\bar{G}}(\bar{R}|\bar{R}')\}\, dS$$

We now exchange the primed and the unprimed variables and postulate that the symmetrical property for $\bar{\bar{G}}_0$ as described by (14-24) and (14-27) also holds true for $\bar{\bar{G}}$. A proof of these relations will be shown at the end of this section. Equation 2 with $\nabla \times \bar{E}$ replaced by $i\omega\mu_0\bar{H}$ can therefore be written in the form

$$\text{(3)} \qquad \bar{E}(\bar{R}) = i\omega\mu_0 \iiint \bar{\bar{G}}(\bar{R}|\bar{R}') \cdot \bar{J}(\bar{R}')\, dV'$$
$$- \iint_{S_d, S_\infty} \{\nabla \times \bar{\bar{G}}(\bar{R}|\bar{R}') \cdot [\hat{n} \times \bar{E}(\bar{R}')] - \hat{n} \times \bar{\bar{G}}(\bar{R}|\bar{R}') \cdot [i\omega\mu_0 \bar{H}(\bar{R}')]\}\, dS'$$

At infinity, we impose the radiation condition that

$$\text{(4)} \qquad \lim_{R\to\infty} R[\nabla \times \bar{\bar{G}}(\bar{R}|\bar{R}') - ik\hat{R} \times \bar{\bar{G}}(\bar{R}|\bar{R}')] = 0$$

and

$$\text{(5)} \qquad \lim_{R\to\infty} R[\nabla \times \bar{E}(\bar{R}) - ik\hat{R} \times \bar{E}(\bar{R})] = 0$$

or

$$\text{(6)} \qquad \lim_{R\to\infty} R[i\omega\mu_0\bar{H}(\bar{R}) - ik\hat{R} \times \bar{E}(\bar{R})] = 0\,.$$

The surface integral evaluated on S_∞ therefore approaches zero. Only the one on S_d remains. On S_d, $\hat{n} \times \bar{E}(\bar{R}) = 0$ as we are dealing with a perfectly conducting body. Thus, if we can find a function denoted by $\bar{\bar{G}}_1$ such that

$$\text{(7)} \qquad \hat{n} \times \bar{\bar{G}}_1(\bar{R}|\bar{R}') = 0 \;\text{ on }\; S_d$$

the entire surface integral would disappear. We call this function the dyadic Green's function of the first kind. In terms of $\bar{\bar{G}}_1$, the electric field is given by

$$\text{(8)} \qquad \bar{E}(\bar{R}) = i\omega\mu_0 \iiint \bar{\bar{G}}_1(\bar{R}|\bar{R}') \cdot \bar{J}(\bar{R}')\, dV'.$$

Likewise, if we integrate the equation for $\bar{H}$, (14-2), using the same technique and construct a function such that

(9) $$\hat{n} \times \nabla \times \bar{\bar{G}}_2(\bar{R}|\bar{R}') = 0 \quad \text{on} \quad S_d$$

then we obtain

(10) $$\bar{H}(\bar{R}) = \iiint \bar{\bar{G}}_2(\bar{R}|\bar{R}') \cdot \nabla' \times \bar{J}(\bar{R}') \, dV'.$$

We will show later that there is an intimate relationship between $\bar{\bar{G}}_1$ and $\bar{\bar{G}}_2$, and (10) actually can be derived from (8) without a separate formulation. Besides $\bar{\bar{G}}_1$ and $\bar{\bar{G}}_2$, we will encounter a function of the third kind or fourth kind which satisfies a mixed boundary condition on a surface of discontinuity between two different media.

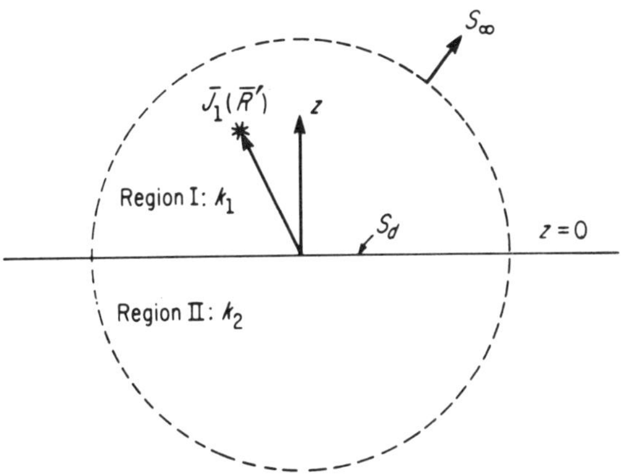

FIG. 17-2. Two half-spaces made of different media.

We consider, for example, two half-spaces made of different isotropic media as shown in Fig. 17-2 with a current source located in Region I. In this case, we have two differential equations for the electric fields in the two regions. They are

(11) $$\nabla \times \nabla \times \bar{E}_1 - k_1^2 \bar{E}_1 = i\omega\mu_1 \bar{J}_1, \qquad z \geqslant 0$$

and

(12) $$\nabla \times \nabla \times \bar{E}_2 - k_2^2 \bar{E}_2 = 0, \qquad z \leqslant 0.$$

The boundary conditions which must be satisfied by the fields at S_d are

(13) $$\hat{n} \times \bar{E}_1 = \hat{n} \times \bar{E}_2$$

(14) $$\hat{n} \times \bar{H}_1 = \hat{n} \times \bar{H}_2$$

or

$$\frac{1}{\varepsilon_1}\hat{n}\times\nabla\times\bar{H}_1=\frac{1}{\varepsilon_2}\hat{n}\times\nabla\times\bar{H}_2 \tag{15}$$

$$\frac{1}{\mu_1}\hat{n}\times\nabla\times\bar{E}_1=\frac{1}{\mu_2}\hat{n}\times\nabla\times\bar{E}_2 \tag{16}$$

where the permittivity $\varepsilon_1, \varepsilon_2$ and the permeability μ_1, μ_2 of the media can be complex. To integrate (11) we introduce a Green's function $\bar{\bar{G}}_3$ consisting of two parts; denoted, respectively, by $\bar{\bar{G}}_3^{(11)}$ and $\bar{\bar{G}}_3^{(21)}$, such that

$$\nabla\times\nabla\times\bar{\bar{G}}_3^{(11)}-k_1^2\bar{\bar{G}}_3^{(11)}=\bar{\bar{I}}\delta(\bar{R}-\bar{R}') \qquad z\geqslant 0 \tag{17}$$

and

$$\nabla\times\nabla\times\bar{\bar{G}}_3^{(21)}-k_2^2\bar{\bar{G}}_3^{(21)}=0, \qquad z\leqslant 0. \tag{18}$$

The reason for using double superscript is that if the current source is located in Region II, we would introduce a $\bar{\bar{G}}_3$ whose two parts would satisfy the equations

$$\nabla\times\nabla\times\bar{\bar{G}}_3^{(12)}-k_1^2\bar{\bar{G}}_3^{(12)}=0, \qquad z\geqslant 0 \tag{19}$$

and

$$\nabla\times\nabla\times\bar{\bar{G}}_3^{(22)}-k_2^2\bar{\bar{G}}_3^{(22)}=\bar{\bar{I}}\delta(\bar{R}-\bar{R}'), \quad z\leqslant 0. \tag{20}$$

To integrate (11), we let

$$\bar{P}_1=\bar{E}_1(\bar{R})$$

and

$$\bar{Q}_1=\bar{\bar{G}}_3^{(11)}(\bar{R}|\bar{R}')\cdot\bar{a}.$$

Applying the vector Green's theorem to these two functions through Region I, we obtain, after interchanging the primed and the unprimed variable,

$$\bar{E}_1(\bar{R})=i\omega\mu_1\iiint_{V_1}\bar{J}_1(\bar{R}')\cdot\bar{\bar{G}}_3^{(11)}(\bar{R}'|\bar{R})\,dV' \tag{21}$$
$$-\iint_{S_d}\{[\hat{n}\times\bar{E}_1(\bar{R}')]\cdot\nabla'\times\bar{\bar{G}}_3^{(11)}(\bar{R}'|\bar{R})-i\omega\mu_1\bar{H}_1(\bar{R}')\cdot[\hat{n}\times\bar{\bar{G}}_3^{(11)}(\bar{R}'|\bar{R})]\}\,dS'$$

where we have already imposed the radiation condition on $\bar{\bar{G}}_3^{(11)}$ to eliminate the surface integral at infinity. We now define another pair of vector functions for Region II, such that

$$\bar{P}_2 = \bar{E}_2(\bar{R})$$
$$\bar{Q}_2 = \bar{\bar{G}}_3^{(21)}(\bar{R}|\bar{R}') \cdot \bar{a}\,.$$

In this case an application of the vector Green's theorem to these two functions yields

(22)
$$\iint_{S_d} \{[\hat{n} \times \bar{E}_2(\bar{R}')] \cdot \nabla' \times \bar{\bar{G}}_3^{(21)}(\bar{R}'|\bar{R}) - i\omega\mu_2 \bar{H}_2(\bar{R}') \cdot [\hat{n} \times \bar{\bar{G}}_3^{(21)}(\bar{R}'|\bar{R})]\} \; dS' = 0$$

On account of the boundary conditions for $\bar{E}$ and $\bar{H}$ in (13) and (14), if we can construct a $\bar{\bar{G}}_3$ such that on S_d

$$\hat{n} \times \bar{\bar{G}}_3^{(11)} = \hat{n} \times \bar{\bar{G}}_3^{(21)} \tag{23}$$

$$\frac{1}{\mu_1} \hat{n} \times \nabla \times \bar{\bar{G}}_3^{(11)} = \frac{1}{\mu_2} \hat{n} \times \nabla \times \bar{\bar{G}}_3^{(21)} \tag{24}$$

then, the surface integral contained in (21) would be equal to zero as a result of (22). Thus we obtain

$$\bar{E}_1(\bar{R}) = i\omega\mu_1 \iiint \bar{J}_1(\bar{R}') \cdot \bar{\bar{G}}_3^{(11)}(\bar{R}'|\bar{R}) \; dV' \tag{25}$$

It will be shown in the following that $\bar{\bar{G}}_3^{(11)}$ also has the symmetrical property that

$$\tilde{\bar{\bar{G}}}_3^{(11)}(\bar{R}'|\bar{R}) = \bar{\bar{G}}_3^{(11)}(\bar{R}|\bar{R}')$$

hence (25) can be written in the form

$$\bar{E}_1(\bar{R}) = i\omega\mu_1 \iiint \bar{\bar{G}}_3^{(11)}(\bar{R}|\bar{R}') \cdot \bar{J}_1(\bar{R}') \; dV'. \tag{26}$$

By using (11) with (19) and (12) with (20) and the symmetrical property between $\bar{\bar{G}}_3^{(12)}$ and $\bar{\bar{G}}_3^{(21)}$, to be shown below, it can be shown that

$$\bar{E}_2(\bar{R}) = i\omega\mu_1 \iiint \bar{\bar{G}}_3^{(21)}(\bar{R}|\bar{R}') \cdot \bar{J}_1(\bar{R}') \; dV'. \tag{27}$$

The dyadic function of the fourth kind actually has the similar composition as the third kind. It is introduced to integrate the equation for $\bar{H}$. We give it a separate designation in order to avoid a possible confusion. In terms of $\bar{\bar{G}}_4$, we have

$$\bar{H}_1(\bar{R}) = \iiint \bar{\bar{G}}_4^{(11)}(\bar{R}|\bar{R}') \cdot \nabla' \times \bar{J}_1(\bar{R}') \, dV' \tag{28}$$

and

$$\bar{H}_2(\bar{R}) = \iiint \bar{\bar{G}}_4^{(21)}(\bar{R}|\bar{R}') \cdot \nabla' \times \bar{J}_1(\bar{R}')\, dV'. \tag{29}$$

The function $\bar{\bar{G}}_4$ satisfies the boundary condition that on S_d

$$\hat{n} \times \bar{\bar{G}}_4^{(11)} = \hat{n} \times \bar{\bar{G}}_4^{(21)} \tag{30}$$

and

$$\frac{1}{\varepsilon_1} \hat{n} \times \nabla \times \bar{\bar{G}}_4^{(11)} = \frac{1}{\varepsilon_2} \hat{n} \times \nabla \times \bar{\bar{G}}_4^{(21)}. \tag{31}$$

They differ from (23) and (24) in that the permeability constants have been replaced by the permittivity constants. An alternative nomenclature would be to call $\bar{\bar{G}}_3$ an electric dyadic function of the third kind and $\bar{\bar{G}}_4$ a magnetic dyadic function of the third kind.

In the remaining part of this section, we will demonstrate the symmetrical properties of the various kinds of the dyadic Green's functions which we have postulated previously. Let us denote, for the time being, any of the following functions

$$\bar{\bar{G}}_0,\ \bar{\bar{G}}_1,\ \bar{\bar{G}}_2,\ \bar{\bar{G}}_3^{(11)},\ \bar{\bar{G}}_3^{(22)},\ \bar{\bar{G}}_4^{(11)} \text{ and } \bar{\bar{G}}_4^{(22)}, \tag{32}$$

by $\bar{\bar{G}}$. For the third and fourth kinds, it is implied that the source point and the field point are located in the same region or medium. We now define two vector functions $\bar{P}$ and $\bar{Q}$ using a $\bar{\bar{G}}$ of the same species but with two different source locations $\bar{R}_a$ and $\bar{R}_b$ such that

$$\bar{P} = \bar{\bar{G}}(\bar{R}|\bar{R}_a) \cdot \bar{a}$$

$$\bar{Q} = \bar{\bar{G}}(\bar{R}|\bar{R}_b) \cdot \bar{b}$$

where $\bar{a}$ and $\bar{b}$ denote two constant arbitrary vectors. Applying the vector Green's theorem to these two functions, we obtain

$$\bar{a} \cdot \bar{\bar{G}}(\bar{R}_a|\bar{R}_b) \cdot \bar{b} = \bar{b} \cdot \bar{\bar{G}}(\bar{R}_b|\bar{R}_a) \cdot \bar{a}\,. \tag{33}$$

The surface integral vanishes either because of the radiation condition or because of the specific boundary condition which $\bar{\bar{G}}$ must satisfy on S_d. Equation (33) implies that

$$\tilde{\bar{\bar{G}}}(\bar{R}_a|\bar{R}_b) = \bar{\bar{G}}(\bar{R}_b|\bar{R}_a)\,. \tag{34}$$

This relation applies to all the functions listed in (32). For the third and fourth kinds, if $\bar{R}_a$ and $\bar{R}_b$ are located in two different regions, then we are seeking the symmetrical relations between $\bar{\bar{G}}_3^{(12)}$ and $\bar{\bar{G}}_3^{(21)}$ or between $\bar{\bar{G}}_4^{(12)}$

and $\bar{\bar{G}}_4^{(21)}$. The proof of these relations is more involved. Let us consider, for example, the third kind. We start with the following system of equations:

$$\nabla\times\nabla\times\bar{\bar{G}}_3^{(11)}(\bar{R}|\bar{R}_a)-k_1^2\bar{\bar{G}}_3^{(11)}(\bar{R}|\bar{R}_a)=\bar{\bar{I}}\delta(\bar{R}-\bar{R}_a) \tag{35}$$

$$\nabla\times\nabla\times\bar{\bar{G}}_3^{(21)}(\bar{R}|\bar{R}_a)-k_2^2\bar{\bar{G}}_3^{(21)}(\bar{R}|\bar{R}_a)=0 \tag{36}$$

$$\nabla\times\nabla\times\bar{\bar{G}}_3^{(22)}(\bar{R}|\bar{R}_b)-k_2^2\bar{\bar{G}}_3^{(22)}(\bar{R}|\bar{R}_b)=\bar{\bar{I}}\delta(\bar{R}-\bar{R}_b) \tag{37}$$

$$\nabla\times\nabla\times\bar{\bar{G}}_3^{(12)}(\bar{R}|\bar{R}_b)-k_1^2\bar{\bar{G}}_3^{(12)}(\bar{R}|\bar{R}_b)=0 \tag{38}$$

where $\bar{R}_a$ is assumed to be located in Region I or medium I and $\bar{R}_b$ is in Region II or medium II. At the interface, S_d, these functions satisfy the boundary condition that

$$\hat{n}\times\bar{\bar{G}}_3^{(11)}=\hat{n}\times\bar{\bar{G}}_3^{(21)} \tag{39}$$

$$\frac{1}{\mu_1}\hat{n}\times\nabla\times\bar{\bar{G}}_3^{(11)}=\frac{1}{\mu_2}\hat{n}\times\nabla\times\bar{\bar{G}}_3^{(21)} \tag{40}$$

$$\hat{n}\times\bar{\bar{G}}_3^{(22)}=\hat{n}\times\bar{\bar{G}}_3^{(12)} \tag{41}$$

$$\frac{1}{\mu_2}\hat{n}\times\nabla\times\bar{\bar{G}}_3^{(22)})=\frac{1}{\mu_1}\hat{n}\times\nabla\times\bar{\bar{G}}_3^{(12)}. \tag{42}$$

We form two vector functions defined by

$$\bar{P}_1=\bar{\bar{G}}_3^{(11)}(\bar{R}|\bar{R}_a)\cdot\bar{a}$$

$$\bar{Q}_1=\bar{\bar{G}}_3^{(12)}(\bar{R}|\bar{R}_b)\cdot\bar{b}\,.$$

An application of the vector Green's theorem to these two functions in Region I yields

$$\begin{aligned}\bar{a}\cdot\bar{\bar{G}}_3^{(12)}(\bar{R}_a|\bar{R}_b)\cdot\bar{b}&\\ =\iint_{S_d}[\bar{a}\cdot\tilde{\bar{\bar{G}}}_3^{(11)}(\bar{R}|\bar{R}_a)\times\nabla\times\bar{\bar{G}}_3^{(12)}(\bar{R}|\bar{R}_b)\cdot\bar{b}&\\ -\bar{b}\cdot\tilde{\bar{\bar{G}}}_3^{(12)}(\bar{R}|\bar{R}_b)\times\nabla\times\bar{\bar{G}}_3^{(11)}(\bar{R}|\bar{R}_a)\cdot\bar{a}]\,dS&\end{aligned} \tag{43}$$

The surface integral evaluated on S_∞ vanishes because of the radiation condition. Similarly, by letting

$$\bar{P}_2=\bar{\bar{G}}_3^{(22)}(\bar{R}|\bar{R}_b)\cdot\bar{b}$$

$$\bar{Q}_2=\bar{\bar{G}}_3^{(21)}(\bar{R}|\bar{R}_a)\cdot\bar{a}$$

and applying the vector Green's theorem in Region II we obtain

$$\bar{b} \cdot \bar{\bar{G}}_3^{(21)}(\bar{R}_b|\bar{R}_a) \cdot \bar{a} = \iint_{S_d} [\bar{b} \cdot \tilde{\bar{\bar{G}}}_3^{(22)}(\bar{R}|\bar{R}_b) \times \nabla \times \bar{\bar{G}}_3^{(21)}(\bar{R}|\bar{R}_a) \cdot \bar{a} - \bar{a} \cdot \tilde{\bar{\bar{G}}}_3^{(21)}(\bar{R}|\bar{R}_a) \times \nabla \times \bar{\bar{G}}_3^{(22)}(\bar{R}|\bar{R}_b) \cdot \bar{b}]\, dS \,. \tag{44}$$

By comparing the expressions for the two surface integrals in (43) and (44), and making use of (39) through (42) we find that

$$\frac{1}{\mu_1} \tilde{\bar{\bar{G}}}_3^{(12)}(\bar{R}_a|\bar{R}_b) = \frac{1}{\mu_2} \bar{\bar{G}}_3^{(21)}(\bar{R}_b|\bar{R}_a) \,. \tag{45}$$

This symmetrical relation differs slightly in form as compared to (34). It is analogous to (11-10) found in the transmission-line theory. If the same technique is applied to the fourth kind, we would obtain

$$\frac{1}{\varepsilon_1} \tilde{\bar{\bar{G}}}_4^{(12)}(\bar{R}_a|\bar{R}_b) = \frac{1}{\varepsilon_2} \bar{\bar{G}}_4^{(21)}(\bar{R}_b|\bar{R}_a) \tag{46}$$

Two other symmetrical relations deal with the curl of the functions. The two relations which we use often are

$$\nabla \times \bar{\bar{G}}_0(\bar{R}_a|\bar{R}_b) = \tilde{\nabla} \times \bar{\bar{G}}_0(\bar{R}_b|\bar{R}_a) \tag{47}$$

and

$$\nabla \times \bar{\bar{G}}_1(\bar{R}_a|\bar{R}_b) = \tilde{\nabla} \times \bar{\bar{G}}_2(\bar{R}_b|\bar{R}_a) \,. \tag{48}$$

To prove these relations it is convenient to use an alternative version of the vector Green's theorem, sometimes called *Green's first identity*. Thus if we let

$$\bar{A} = \bar{Q} \times \nabla \times \bar{P}$$

and substitute it into Gauss' theorem, we obtain

$$\iiint_V [(\nabla \times \bar{P}) \cdot (\nabla \times \bar{Q}) - \bar{Q} \cdot \nabla \times \nabla \times \bar{P}]\, dV = \oiint_S (\bar{Q} \times \nabla \times \bar{P}) \cdot d\bar{S} \qquad \text{(Green's first identity)} \tag{49}$$

To prove (48), we let

$$\bar{P} = \bar{\bar{G}}_2(\bar{R}|\bar{R}_b) \cdot \bar{b}$$

$$\bar{Q} = \nabla \times \bar{\bar{G}}_1(\bar{R}|\bar{R}_a) \cdot \bar{a} \,.$$

Upon substituting them into (49) we have

(50)

$$\bar{b}\cdot\nabla\times\bar{\bar{G}}_2(\bar{R}_a|\bar{R}_b)\cdot\bar{a}-\bar{b}\cdot\nabla\times\bar{\bar{G}}_1(\bar{R}_b|\bar{R}_a)\cdot\bar{a}+k^2\iiint\{[\nabla\times\bar{\bar{G}}_2(\bar{R}|\bar{R}_b)\cdot\bar{b}]$$

$$\cdot[\bar{\bar{G}}_1(\bar{R}|\bar{R}_a)\cdot\bar{a}]-[\nabla\times\bar{\bar{G}}_1(\bar{R}|\bar{R}_a)\cdot\bar{a}]\cdot[\bar{\bar{G}}_2(\bar{R}|\bar{R}_b)\cdot\bar{b}]\}\,dV$$

$$=\oint\!\!\oint\{[\nabla\times\bar{\bar{G}}_1(\bar{R}|\bar{R}_a)\cdot\bar{a}]\times[\nabla\times\bar{\bar{G}}_2(\bar{R}|\bar{R}_b)\cdot\bar{b}]\}\cdot d\bar{S}\,.$$

The volume integral in (50) can be changed into a surface integral by means of the Gauss theorem since it is equal to

$$k^2\iiint\nabla\cdot\{[\bar{\bar{G}}_2(\bar{R}|\bar{R}_b)\cdot\bar{b}]\times[\bar{\bar{G}}_1(\bar{R}|\bar{R}_a)\cdot\bar{a}]\}\,dV\,.$$

As a result of the radiation condition at infinity and the Dirichlet or Neumann condition for $\bar{\bar{G}}_1$ and $\bar{\bar{G}}_2$ on a surface of discontinuity, the resultant surface integral vanishes, thus, we have

(51) $$\widetilde{\nabla\times\bar{\bar{G}}_2(\bar{R}_a|\bar{R}_b)}=\nabla\times\bar{\bar{G}}_1(\bar{R}_b|\bar{R}_a)\,.$$

The proof for (42) is the same. We summarize in Table 17-1 the symmetrical relations which we have proved so far, whereby we have replaced $\bar{R}_a$ by $\bar{R}'$ and $\bar{R}_b$ by $\bar{R}$.

TABLE 17-1
Symmetrical Relations of Dyadic Green's Functions

Relation	Functions
$\tilde{\bar{\bar{G}}}(\bar{R}'\|\bar{R})=\bar{\bar{G}}(\bar{R}\|\bar{R}')$	$\bar{\bar{G}}_0,\bar{\bar{G}}_1,\bar{\bar{G}}_2,\bar{\bar{G}}_3^{(11)},\bar{\bar{G}}_3^{(22)}$ $\bar{\bar{G}}_4^{(11)},\bar{\bar{G}}_4^{(22)}$
$\frac{1}{\mu_1}\tilde{\bar{\bar{G}}}_3^{(12)}(\bar{R}'\|\bar{R})=\frac{1}{\mu_2}\bar{\bar{G}}_3^{(21)}(\bar{R}\|\bar{R}')$	
$\frac{1}{\varepsilon_1}\tilde{\bar{\bar{G}}}_4^{(12)}(\bar{R}'\|\bar{R})=\frac{1}{\varepsilon_2}\bar{\bar{G}}_4^{(21)}(\bar{R}\|\bar{R}')$	
$\widetilde{\nabla'\times\bar{\bar{G}}_0(\bar{R}'\|\bar{R})}=\nabla\times\bar{\bar{G}}_0(\bar{R}\|\bar{R}')$	
$\widetilde{\nabla'\times\bar{\bar{G}}_2(\bar{R}'\|\bar{R})}=\nabla\times\bar{\bar{G}}_1(\bar{R}\|\bar{R}')$	
$\widetilde{\nabla'\times\bar{\bar{G}}_1(\bar{R}'\|\bar{R})}=\nabla\times\bar{\bar{G}}_2(\bar{R}\|\bar{R}')$	

18. DYADIC GREEN'S FUNCTION FOR A HALF-SPACE

In addition to $\overline{\overline{G}}_0$, the dyadic Green's function for a half-space, either of the first kind or of the second kind, has a relatively simple and closed form. It can be derived by applying the method of images. In the following, we give a derivation for $\overline{\overline{G}}_1$. The derivation for $\overline{\overline{G}}_2$ is similar. According to the method of images, the field due to certain distribution of currents in a half-space with an infinitely conducting ground plane is the same as the one produced by the original current and its image in a whole space. The equivalence is illustrated in Fig. 18-1. By applying the method of potentials to the configuration shown in Fig. 18-1b, we obtain

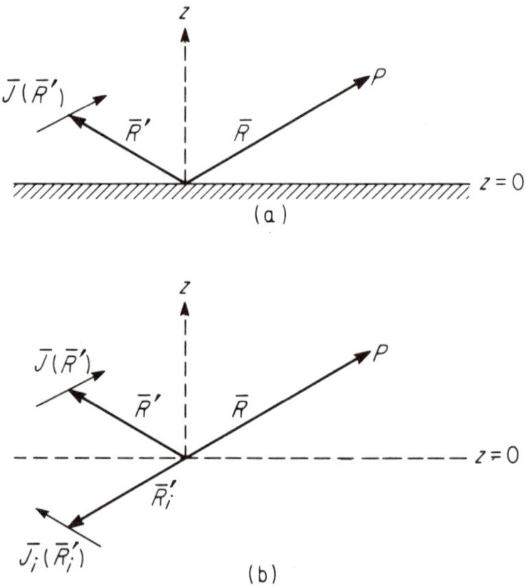

FIG. 18-1. The method of images. (a) Half-space with an infinite conducting ground plane. (b) Whole-space with the original current and its image.

$$\bar{A}(\bar{R}) = \mu_0 \iiint G_0(\bar{R}|\bar{R}')\bar{J}(\bar{R}')\, dV' + \mu_0 \iiint G_0(\bar{R}|\bar{R}_i')\bar{J}_i(\bar{R}_i')\, dV_i' \,, \tag{1}$$

where

$$G_0(\bar{R}|\bar{R}') = \frac{e^{ik|\bar{R}-\bar{R}'|}}{4\pi|\bar{R}-\bar{R}'|} \tag{2}$$

and $G_0(\bar{R}|\bar{R}_i')$ is obtained by replacing $\bar{R}'$ by $\bar{R}_i'$, where

$$|\bar{R}-\bar{R}'| = [(x-x')^2 + (y-y')^2 + (z-z')^2]^{1/2}$$

and

$$|\bar{R}-\bar{R}_i'| = [(x-x')^2+(y-y')^2+(z+z')^2]^{1/2}.$$

The corresponding electric field at $\bar{R}$ is given by

$$\bar{E}(\bar{R}) = i\omega[\bar{A}(\bar{R}) + \frac{1}{k^2}\nabla\nabla\cdot\bar{A}(\bar{R})] .$$

If we let

$$\bar{J}(\bar{R}) = \frac{1}{i\omega\mu_0}\,\bar{a}\delta(\bar{R}-\bar{R}')$$

where $\bar{a}$ denotes an arbitrary constant vector that consists of a transversal part and a longitudinal part with respect to the z-axis, that is,

(3) $$\bar{a} = a_t\hat{t} + a_z\hat{z} ,$$

then

$$\bar{A}(\bar{R}) = \frac{1}{i\omega}\,[\bar{a}G_0(\bar{R}|\bar{R}') + \bar{a}_iG_0(\bar{R}|\bar{R}_i')]$$

and

(4) $$\bar{E}(\bar{R}) = \left(1+\frac{1}{k^2}\nabla\nabla\cdot\right)[\bar{a}G_0(\bar{R}|\bar{R}') + \bar{a}_iG_0(\bar{R}|\bar{R}_i')]$$

where

(5) $$\bar{a}_i = -a_t\hat{t} + a_z\hat{z} .$$

Using the nomenclature introduced in the previous section, we can write

(6) $$\bar{E}(\bar{R}) = i\omega\mu_0\iiint\bar{\bar{G}}_1(\bar{R}|\bar{R}')\cdot\bar{J}(\bar{R}')\,dV' = \bar{\bar{G}}_1(\bar{R}|\bar{R}')\cdot\bar{a}\cdot$$

To find $\bar{\bar{G}}_1$ it is just a matter of putting (4) in a dyadic form with $\bar{a}$ as its posterior multiplier. This can be done by performing the following transformations. First, we write

(7) $$\begin{aligned}\bar{a}G_0(\bar{R}|\bar{R}') + \bar{a}_iG_0(\bar{R}|\bar{R}_i') &= \bar{a}[G_0(\bar{R}|\bar{R}') - G_0(\bar{R}|\bar{R}_i')] + 2a_z\hat{z}G_0(\bar{R}|\bar{R}_i')\\ &= \{\bar{\bar{I}}[G_0(\bar{R}|\bar{R}') - G_0(\bar{R}|\bar{R}_i')] + 2\hat{z}\hat{z}G_0(\bar{R}|\bar{R}_i')\}\cdot\bar{a}\end{aligned}$$

then

$$\nabla\cdot[\bar{a}_iG_0(\bar{R}|\bar{R}_i')] = \left(-a_x\frac{\partial}{\partial x} - a_y\frac{\partial}{\partial y} + a_z\frac{\partial}{\partial z}\right)G_0(\bar{R}|\bar{R}_i')$$

$$= \left(a_x \frac{\partial}{\partial x'} + a_y \frac{\partial}{\partial y'} + a_z \frac{\partial}{\partial z'} \right) G_0(\bar{R}|\bar{R}_i') = \nabla' \cdot [\bar{a} G_0(\bar{R}|\bar{R}')] ,$$

and

$$\nabla \cdot [\bar{a} G_0(\bar{R}|\bar{R}')] = -\nabla' \cdot [\bar{a} G_0(\bar{R}|\bar{R}')]$$

hence

$$(8) \quad \nabla\nabla \cdot [\bar{a} G_0(\bar{R}|\bar{R}') + \bar{a}_i G_0(\bar{R}|\bar{R}_i')] = -\nabla\nabla' \cdot [\bar{a} G_0(\bar{R}|\bar{R}') - \bar{a} G_0(\bar{R}|\bar{R}_i')]$$

$$= -\nabla\nabla' [G_0(\bar{R}|\bar{R}') - G_0(\bar{R}|\bar{R}_i')] \cdot \bar{a}$$

Upon substituting (7) and (8) into (4), and comparing the resultant expression with (6), we obtain

$$(9) \quad \bar{\bar{G}}_1(\bar{R}|\bar{R}') = \left(\bar{\bar{I}} - \frac{1}{k^2} \nabla\nabla' \right) [G_0(\bar{R}|\bar{R}') - G_0(\bar{R}|\bar{R}_i')] + 2\hat{z}\hat{z}\, G_0(\bar{R}|\bar{R}_i').$$

It can be verified that $\hat{n} \times \bar{\bar{G}}_1$ indeed vanishes at $z = 0$. Furthermore, if we introduce a complementary unit dyad defined by

$$\bar{\bar{I}}_i = \hat{x}\hat{x} + \hat{y}\hat{y} - \hat{z}\hat{z}$$

then

$$(10) \quad \nabla \times \bar{\bar{G}}_1(\bar{R}|\bar{R}') = \nabla G_0(\bar{R}|\bar{R}') \times \bar{\bar{I}} - \nabla G_0(\bar{R}|\bar{R}_i') \times \bar{\bar{I}}_i .$$

We leave readers to show that $\bar{\bar{G}}_1(\bar{R}|\bar{R}')$ as expressed by (9) does indeed satisfy the radiation condition at infinity.

chapter **5**

Rectangular Waveguides

In this chapter the expressions for the dyadic Green's functions of the first and second kinds pertaining to a rectangular waveguide will be derived. The method and the general procedure would apply equally well to all other bodies treated in the remaining chapters. The readers should therefore grasp the concept and follow the key steps in a firm manner to assure a smooth passage to the rest of the book.

19. RECTANGULAR VECTOR WAVE FUNCTIONS

The vector wave functions are the building blocks of the eigenfunction expansions of various kinds of dyadic Green's functions. These functions were first introduced by Hansen (1935, 1936, 1937) in formulating certain electromagnetic problems. The effectiveness of these functions was recognized by Stratton (1941) who, for example, reformulated Mie's theory of the diffraction of a plane electromagnetic wave by a sphere using the spherical vector wave function. In his original work Hansen introduced three kinds of vector wave functions, denoted by $\bar{L}$, $\bar{M}$, and $\bar{N}$, which are solutions of the homogeneous vector Helmholtz equation. Such a presentation was followed by Stratton (1941) and by Morse and Feshbach (1953). To derive the eigenfunction expansion of the dyadic Green's functions that satisfy with the vector wave equation, the $\bar{L}$ functions are not needed. The elimination of the third set not only simplifies considerably the presentation but also makes a complete orthogonal relationship between the remaining two sets possible. When we speak of the vector wave functions in this book, the $\bar{L}$ functions are not included nor even introduced.

A vector wave function, by definition, is an eigenfunction or a characteristic function which is a solution of the homogeneous vector wave equation

(1) $$\nabla \times \nabla \times \bar{F} - \kappa^2 \bar{F} = 0$$

where κ, designated as the propagation constant, is so far arbitrary. There are two independent sets of vector wave functions which can be constructed using the characteristic function pertaining to a scalar wave equation as the generating functions. One kind of vector wave function, called the *Cartesian vector wave function*, is formed if we let

(2) $$\bar{F} = \nabla \times (\psi_1 \bar{c})$$

where ψ_1 denotes a characteristic function which satisfies the scalar wave equation

(3) $$\nabla^2 \psi + \kappa^2 \psi = 0$$

and $\bar{c}$ denotes a constant vector, such as $\hat{x}$, $\hat{y}$, or $\hat{z}$. For convenience we shall designate $\bar{c}$ as the piloting vector and ψ as the generating function. Another kind, designated as the *spherical vector wave function*, will be introduced in Sec. 39, whereby the piloting vector is identified as the spherical radial vector $\bar{R}$. Except for spherical problems, we are always dealing with the Cartesian vector wave functions. When (2) is substituted into (1), we obtain

$$\nabla \times [\nabla \times \nabla \times (\psi_1 \bar{c}) - \kappa^2 (\psi_1 \bar{c})] = 0 .$$

In view of identity (10) of Appendix A, this is equivalent to

$$\nabla \times [\bar{c}(\nabla^2 \psi_1 + \kappa^2 \psi_1)] = 0$$

hence (2) is a solution for (1) if ψ_1 is a solution for (3). The set of functions so obtained will be denoted by the letter $\bar{M}$, that is,

(4) $$\bar{M}_1 = \nabla \times (\psi_1 \bar{c}) .$$

Another set of vector wave functions, denoted by $\bar{N}$, is obtained by letting

(5) $$\bar{N}_2 = \frac{1}{\kappa} \nabla \times \nabla \times (\psi_2 \bar{c})$$

where ψ_2 denotes a characteristic function which also satisfies (3), but may be different from the function used to define $\bar{M}$. By substituting (5) into (1), we obtain

$$\nabla \times \nabla \times [\bar{c}(\nabla^2 \psi_2 + \kappa^2 \psi_2)] = 0$$

hence $\bar{N}_2$ is a solution for (1), if ψ_2 satisfies (3). In the case in which an identical scalar wave function is used for both $\bar{M}$ and $\bar{N}$ we have the following symmetrical relations between these two types of functions:

(6) $$\bar{N} = \frac{1}{\kappa} \nabla \times \bar{M}$$

(7) $$\bar{M} = \frac{1}{\kappa} \nabla \times \bar{N} .$$

Equation (6) follows directly from (4) and (5), and (7) is obtained if we take the curl of (6), which yields

$$\nabla \times \bar{N} = \frac{1}{\kappa} \nabla \times \nabla \times \bar{M} = \kappa \bar{M}.$$

Equations (6) and (7) show why the constant κ is introduced in the definition for the N-functions, otherwise the relations between the two sets will not be perfectly symmetrical.

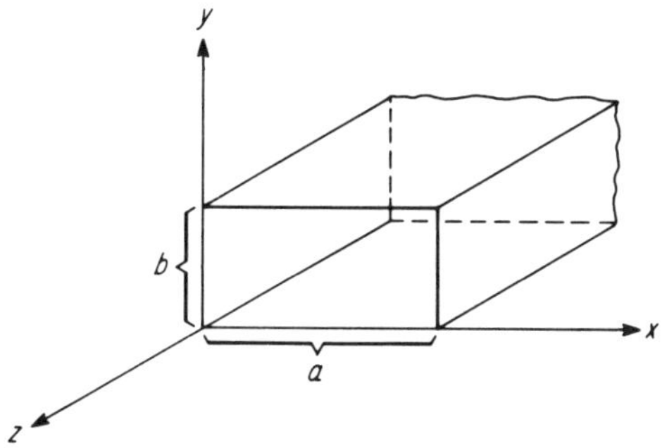

FIG. 19-1. A rectangular waveguide.

The exact expressions for the two sets of vector wave functions depend, of course, not only on the specific expression for the scalar wave function which is being used but also on the choice of the piloting vector $\bar{c}$. For the rectangular waveguide problems to be discussed in this chapter, we will use the configuration shown in Fig. 19-1 for the orientation of the guide with respect to the rectangular coordinate system, and we will choose the unit vector $\hat{z}$ to represent the piloting vector $\bar{c}$. By doing so, the two sets of vector wave functions thus constructed would provide us the TE and TM modes described in the theory of rectangular wave guides.

The scalar wave equation (3), when solved by the method of separation of variables, yields

(8) $$\psi = (A \cos k_x x + B \sin k_x x)\,(C \cos k_y y + D \cos k_y y) e^{ihz}$$

where $k_x^2 + k_y^2 + h^2 = \kappa^2$.

Now if we demand that the vector wave functions, both $\bar{M}$ and $\bar{N}$, so generated would satisfy the boundary condition that on the walls of the guide

(9) $$\hat{n} \times \bar{M} = 0$$

and

(10) $$\hat{n} \times \bar{N} = 0$$

then the constants k_x and k_y must have certain characteristic values or eigenvalues. The boundary condition specified by (9) or (10) corresponds to the one satisfied by the electric field on a perfectly conducting surface. Using (4), with $\bar{c}$ replaced by $\hat{z}$, it is not difficult to show that the only allowed functions in this case are the cosine functions or the even functions and the constants k_x and k_y should have the following characteristic values,

(11) $$k_x = \frac{m\pi}{a}, \quad m = 0,1, \ldots$$

(12) $$k_y = \frac{n\pi}{b}, \quad n = 0,1, \ldots$$

The complete expression and the notation for the set of $\bar{M}$-functions which satisfy the vector Dirichlet condition are

(13) $$\bar{M}_{emn}(h) = \nabla \times [\psi_{emn}(h)\hat{z}] = \nabla \times \left(\cos\frac{m\pi x}{a}\cos\frac{n\pi y}{b}\, e^{ihz}\,\hat{z}\right)$$

$$= \left(-\frac{n\pi}{b}\cos\frac{m\pi x}{a}\sin\frac{n\pi y}{b}\,\hat{x} + \frac{m\pi}{a}\sin\frac{m\pi x}{a}\cos\frac{n\pi y}{b}\,\hat{y}\right) e^{ihz}$$

where

(14) $$\left(\frac{m\pi}{a}\right)^2 + \left(\frac{n\pi}{b}\right)^2 + h^2 = \kappa^2.$$

The subscript e is an abbreviation for the word "even." We might put h as an additional subscript, but that would make the notation too long and awkward. In a similar manner, the set of $\bar{N}$-functions which satisfy the vector Dirichlet condition are given by

(14)
$$\bar{N}_{omn}(h) = \frac{1}{\kappa}\nabla \times \nabla \times [\psi_{omn}(h)\hat{z}] = \frac{1}{\kappa}\nabla \times \nabla \times \left(\sin\frac{m\pi x}{a}\sin\frac{n\pi y}{b}\, e^{ihz}\,\hat{z}\right)$$

$$= \frac{1}{\kappa}\left[ih\frac{m\pi}{a}\cos\frac{m\pi x}{a}\sin\frac{n\pi y}{b}\,\hat{x} + ih\frac{n\pi}{b}\sin\frac{m\pi x}{a}\cos\frac{n\pi y}{b}\,\hat{y}\right.$$

$$\left. + \left(\frac{m^2\pi^2}{a^2} + \frac{n^2\pi^2}{b^2}\right)\sin\frac{m\pi x}{a}\sin\frac{n\pi y}{b}\,\hat{z}\right] e^{ihz}$$

The subscript o is an abbreviation for the word "odd," and we can still set $m=0, 1, 2 \cdots$; $n=0, 1, 2, \cdots$ with the modes $m=0$ or $n=0$ treated as null modes. It is obvious that $\bar{M}_{emn}(h)$ represents the electric field of the TE_{mn} mode while $\bar{N}_{omn}(h)$ represents that of the TM_{mn} mode. In view of (6) and (7) we have

$$\bar{M}_{omn}(h) = \frac{1}{\kappa} \nabla \times \bar{N}_{omn}(h) = \nabla \times [\psi_{omn}(h)\, \hat{z}] \tag{15}$$

and

$$\bar{N}_{emn}(h) = \frac{1}{\kappa} \nabla \times \bar{M}_{emn}(h) = \frac{1}{\kappa} \nabla \times \nabla \times [\psi_{emn}(h)\, \hat{z}]\,. \tag{16}$$

The vector functions $\bar{M}_{omn}(h)$ and $\bar{N}_{emn}(h)$ are the proper functions to represent the H field in a rectangular waveguide and they satisfy the vector Neumann condition on the boundary, namely

$$\hat{n} \times \left\{ \begin{matrix} \nabla \times \bar{M}_{omn}(h) \\ \nabla \times \bar{N}_{emn}(h) \end{matrix} \right\} = 0\,. \tag{17}$$

In summary, the vector wave functions which can be used to represent the electromagnetic field inside a rectangular waveguide are of the form

$$\bar{M}_{\substack{e\\o}mn}(h) = \nabla \times [\psi_{\substack{e\\o}mn}(h)\, \hat{z}] \tag{18}$$

$$\bar{N}_{\substack{e\\o}mn}(h) = \frac{1}{\kappa} \nabla \times \nabla \times [\psi_{\substack{e\\o}mn}(h)\, \hat{z}] \tag{19}$$

where

$$\psi_{\substack{e\\o}mn}(h) = \left\{ \begin{matrix} \cos \dfrac{m\pi x}{a} \cos \dfrac{n\pi y}{b} \\ \sin \dfrac{m\pi x}{a} \sin \dfrac{n\pi y}{b} \end{matrix} \right\} e^{ihz}$$

and

$$\left(\frac{m\pi}{a}\right)^2 + \left(\frac{n\pi}{b}\right)^2 + h^2 = \kappa^2$$

with

$$m, n = 0, 1, 2, \ldots$$

It is understood that the odd functions with m or $n=0$ are null modes. For convenience, we will call these functions the *rectangular vector wave functions*.

It should be pointed out that the above functions have been formed as a result of our proper choice of the piloting vector $\bar{c}$ in (4) and (5). If we had chosen another piloting vector for $\bar{c}$ such as $\bar{x}$, then we would generate a hybrid rectangular vector wave function. For example, the function defined by

$$\bar{M}^{(x)}_{mn}(h) = \nabla \times \left[\sin \frac{m\pi x}{a} \cos \frac{n\pi y}{b} e^{ihz} \hat{x} \right] \tag{20}$$

$$= \left(ih \sin \frac{m\pi x}{a} \cos \frac{n\pi y}{b} \hat{y} + \frac{n\pi}{b} \sin \frac{m\pi x}{a} \sin \frac{n\pi y}{b} \hat{z} \right) e^{ihz}$$

also satisfies the vector Dirichlet condition at $x=0$ and a, and $y=0$ and b, but it does not represent a pure TE or TM mode with respect to the z-axis. In fact, it can be shown that

$$\bar{M}^{(x)}_{mn}(h) = \frac{1}{\kappa^2 - h^2} \left[\kappa \frac{n\pi}{b} \bar{N}_{omn}(h) + ih \frac{m\pi}{a} \bar{M}_{emn}(h) \right]$$

thus, the function represents a superposition of the TE_{mn} mode and the TM_{mn} mode. In Appendix B, we list a number of these hybrid rectangular vector wave functions together with their relations to the $\bar{M}_{\substack{e\\o}mn}$ and $\bar{N}_{\substack{e\\o}mn}$ functions. The complete expressions for various vector wave functions are also tabulated there for convenient reference.

Having defined the rectangular vector wave functions, we can now discuss their orthogonal properties. By referring to (13), (14) and (1) and (2) of Appendix B, it is relatively simple to show that

$$\left\{ \begin{aligned} &\iiint \bar{M}_{\substack{e\\o}mn}(h) \cdot \bar{N}_{\substack{e\\o}m'n'}(-h')\, dV = 0 \\ &\iiint \bar{M}_{\substack{e\\o}mn}(h) \cdot \bar{N}_{\substack{o\\e}m'n'}(-h')\, dV = 0 \\ &\iiint \bar{M}_{\substack{e\\o}mn}(h) \cdot \bar{M}_{\substack{o\\e}m'n'}(-h')\, dV = 0 \\ &\iiint \bar{N}_{\substack{e\\o}mn}(h) \cdot \bar{N}_{\substack{o\\e}m'n'}(-h')\, dV = 0 \end{aligned} \right. \tag{21}$$

where the m,n,h and m',n',h' denote two sets of eigenvalues which may be distinct or the same and the volume of integration extends from $x=0$ to a, $y=0$ to b, and $z=-\infty$ to $+\infty$. Actually, these orthogonal relationships require only the integration with respect to x and y. However, we include the integration with respect to z for completeness because in the eigenfunction expansion for the dyadic Green's function we would encounter the volume integral instead of the surface integral. When $m \neq n'$ or $n \neq n'$, it is also quite simple to show that

$$
\begin{aligned}
&\iiint \bar{M}_{emn}(h) \cdot \bar{M}_{em'n'}(-h')\, dV = 0 \\
&\iiint \bar{M}_{omn}(h) \cdot \bar{M}_{om'n'}(-h')\, dV = 0 \\
&\iiint \bar{N}_{emn}(h) \cdot \bar{N}_{em'n'}(-h')\, dV = 0 \\
&\iiint \bar{N}_{omn}(h) \cdot \bar{N}_{om'n'}(-h')\, dV = 0
\end{aligned}
\tag{22}
$$

Thus all the rectangular vector wave functions are orthogonal to each other. What remains is the determination of the normalization factor for the case that $m = m'$ and $n = n'$. We will let h and h' be different for good reason.

We consider first the integral

$$
\begin{aligned}
&\int_0^a \int_0^b \int_{-\infty}^{\infty} \bar{M}_{emn}(h) \cdot \bar{M}_{emn}(-h')\, dx\, dy\, dz \\
&= (1+\delta_0)\,\frac{ab}{4}\left[\left(\frac{m\pi}{a}\right)^2 + \left(\frac{n\pi}{b}\right)^2\right]\int_{-\infty}^{\infty} e^{i(h-h')z}\, dz \qquad (23) \\
&= (1+\delta_0)\,\frac{\pi ab}{2}\left[\left(\frac{m\pi}{a}\right)^2 + \left(\frac{n\pi}{b}\right)^2\right]\delta(h-h')
\end{aligned}
$$

where the delta function $\delta(h-h')$ results from the identity given by (3-24), and δ_0 denotes the Kronecker delta function defined as follows:

$$
\delta_0 = \begin{cases} 1 & m \text{ or } n = 0 \\ 0 & m \text{ and } n \neq 0 \end{cases}
$$

Similarly, we have

$$
\begin{aligned}
&\int_0^a \int_0^b \int_{-\infty}^{\infty} \bar{M}_{omn}(h) \cdot \bar{M}_{omn}(-h')\, dx\, dy\, dz \qquad (24) \\
&= \frac{\pi ab}{2}\left[\left(\frac{m\pi}{a}\right)^2 + \left(\frac{n\pi}{b}\right)^2\right]\delta(h-h'), \\
&\qquad\qquad\qquad\qquad m \neq 0,\ n \neq 0\,.
\end{aligned}
$$

It is recalled that when m or n is equal to zero, the function $\bar{M}_{omn}$ is a null function. The normalization factors for the $\bar{N}$-functions are more complicated. By carrying out the integration we obtain

$$\int_0^a \int_0^b \int_{-\infty}^{\infty} \bar{N}_{emn}(h) \cdot \bar{N}_{emn}(-h')\, dx\, dy\, dz \tag{25}$$

$$= (1+\delta_0)\, \frac{\pi ab}{2\kappa\kappa'} \left[\left(\frac{m\pi}{a}\right)^2 + \left(\frac{n\pi}{b}\right)^2\right] \left[\left(\frac{m\pi}{a}\right)^2 + \left(\frac{n\pi}{b}\right)^2 + hh'\right] \delta(h-h')$$

where δ_0 has been defined previously, and

$$\int_0^a \int_0^b \int_{-\infty}^{\infty} \bar{N}_{omn}(h) \cdot \bar{N}_{omn}(-h')\, dx\, dy\, dz \tag{26}$$

$$= \frac{\pi ab}{2\kappa\kappa'} \left[\left(\frac{m\pi}{a}\right)^2 + \left(\frac{n\pi}{b}\right)^2\right] \left[\left(\frac{m\pi}{a}\right)^2 + \left(\frac{n\pi}{b}\right)^2 + hh'\right] \delta(h-h'),$$

$m \neq 0,\ n \neq 0$.

Equations (22) to (26) are the basic relations which will be needed in finding the orthogonal expansion of the dyadic Green's functions inside a rectangular waveguide. We shall take full advantage of the delta function to determine the unknown coefficients in such an expansion.

20. DYADIC GREEN'S FUNCTIONS OF THE FIRST AND SECOND KIND FOR A RECTANGULAR WAVEGUIDE

We shall now apply the Ohm-Rayleigh method to derive the dyadic Green's function of the first kind for a rectangular waveguide. The function $\bar{\bar{G}}_1(\bar{R}|\bar{R}')$ satisfies the equation

$$\nabla \times \nabla \times \bar{\bar{G}}_1(\bar{R}|\bar{R}') - k^2 \bar{\bar{G}}_1(\bar{R}|\bar{R}') = \bar{\bar{I}}\, \delta(\bar{R}-\bar{R}') \tag{1}$$

defined for the domain $a \geqslant x \geqslant 0$, $b \geqslant y \geqslant 0$, $-\infty < z < \infty$, and the boundary condition that

$$\hat{n} \times \bar{\bar{G}}_1(\bar{R}|\bar{R}') = 0 \text{ at } x=0 \text{ and } a, \text{ and } y=0 \text{ and } b. \tag{2}$$

At $z = \pm\infty$, it satisfies the radiation condition that

$$\lim_{z \to \pm\infty} [\nabla \times \bar{\bar{G}}_1(\bar{R}|\bar{R}') \mp ik\hat{z} \times \bar{\bar{G}}_1(\bar{R}|\bar{R}')] = 0\,. \tag{3}$$

The radiation condition specified here differs slightly in form from the one described by (17-4) because we are dealing with a guided wave instead of a spherical wave propagating outward in an infinite domain. The propagation constant $k(=2\pi/\lambda)$ in (1) is assumed to be given. According to the Ohm-Rayleigh method, we first seek an eigenfunction expansion for the source function $\bar{\bar{I}}\delta(\bar{R}-\bar{R}')$ using the vector wave functions introduced in the previous sections. The proper functions to be used are $\bar{M}_{emn}$ and $\bar{N}_{omn}$, since they satisfy the boundary condition specified by (2). Thus we let

$$\text{(4)}\quad \bar{\bar{I}}\delta(\bar{R}-\bar{R}')=\int_{-\infty}^{\infty} dh \sum_{m=0}^{\infty} \sum_{n=0}^{\infty} [\bar{M}_{emn}(h)\,\bar{A}_{emn}(h)+\bar{N}_{omn}(h)\,\bar{B}_{omn}(h)]$$

where $\bar{A}_{emn}(h)$ and $\bar{B}_{omn}(h)$ are two unknown functions (coefficients) to be determined. These two unknown functions are found by the same method as the one described in Sec. 10 for the infinitely long transmission line, except that we are dealing with a dyadic delta function, and the eigenfunctions used are the vector wave functions. By taking the anterior scalar product of (4) with a vector function $\bar{M}_{em'n'}(-h')$ with certain fixed eigenvalues m', n', and h' and integrating the resultant equation through the entire volume of the guide, we obtain

$$\iiint_V \bar{M}_{em'n'}(-h')\cdot\bar{\bar{I}}\delta(\bar{R}-\bar{R}')\,dV=\iiint_V dV \int_{-\infty}^{\infty} dh \sum_m \sum_n \bar{M}_{em'n'}(-h') \cdot[\bar{M}_{emn}(h)\bar{A}_{emn}(h)+\bar{N}_{omn}(h)\bar{B}_{omn}(h)]\,.$$

On account of the dyadic delta function $\bar{\bar{I}}\delta(\bar{R}-\bar{R}')$ and the orthogonal relations between the rectangular vector wave functions, (19-22) and (19-23), we have

$$\begin{aligned}\bar{M}'_{em'n'}(-h') &= \int_{-\infty}^{\infty} dh(1+\delta_0)\,\frac{\pi ab}{2}\left[\left(\frac{m'\pi}{a}\right)^2+\left(\frac{n'\pi}{b}\right)^2\right]\bar{A}_{em'n'}(h)\,\delta(h-h')\\ &=(1+\delta_0)\,\frac{\pi ab}{2}\left[\left(\frac{m'\pi}{a}\right)^2+\left(\frac{n'\pi}{b}\right)^2\right]\bar{A}_{em'n'}(h'),\end{aligned}$$

where the primed function $\bar{M}'$ signifies that the function is defined with respect to the primed coordinates x', y', z', corresponding to the location of $\bar{R}'$. The function $\bar{A}_{emn}(h)$ therefore, is given by

$$\text{(5)}\qquad \bar{A}_{emn}(h)=\frac{2-\delta_0}{\pi abk_c^2}\,\bar{M}'_{emn}(-h)$$

where $k_c^2=\left(\frac{m\pi}{a}\right)^2+\left(\frac{n\pi}{b}\right)^2$.

In a similar manner, by taking the anterior scalar product of (4) with $\bar{N}_{om'n'}(-h')$ and making use of (19-22) and (19-26), we obtain

(6) $$\bar{B}_{omn}(h) = \frac{2-\delta_0}{\pi abk_c^2}\,\bar{N}'_{omn}(-h)$$

Actually in this case neither m nor n is equal to zero so the factor $2-\delta_0$ is always equal to 2. We keep this factor in (6) to make it identical to the coefficient appearing in (5). Substituting (5) and (6) into (4), we have the complete eigenfunction expansion of the dyadic delta function, that is,

(7) $$\bar{\bar{I}}\delta(\bar{R}-\bar{R}') = \int_{-\infty}^{\infty} dh \sum_m \sum_n \frac{2-\delta_0}{\pi abk_c^2} \cdot [\bar{M}_{emn}(h)\bar{M}'_{emn}(-h) + \bar{N}_{omn}(h)\bar{N}'_{omn}(-h)]\,.$$

To find $\bar{\bar{G}}_1(\bar{R}|\bar{R}')$ we let

$$\bar{\bar{G}}_1(\bar{R}|\bar{R}') = \int_{-\infty}^{\infty} dh \sum_m \sum_n \frac{2-\delta_0}{\pi abk_c^2} \cdot [a(h)\bar{M}_{emn}(h)\bar{M}'_{emn}(-h) + b(h)\bar{N}_{omn}(h)\bar{N}'_{omn}(-h)]\,,$$

where the coefficients $a(h)$ and $b(h)$ are to be determined. By substituting it into (1), and making use of the identity that

$$\nabla\times\nabla\times \begin{Bmatrix}\bar{M}_{emn}\\ \bar{N}_{omn}\end{Bmatrix} = \kappa^2 \begin{Bmatrix}\bar{M}_{emn}\\ \bar{N}_{omn}\end{Bmatrix}$$

we obtain, as a result of (7), the following equations

$$(\kappa^2 - k^2)a(h) = 1$$

$$(\kappa^2 - k^2)b(h) = 1$$

where $\kappa^2 = k_c^2 + h^2$,

hence

$$a(h) = b(h) = \frac{1}{h^2 + k_c^2 - k^2}$$

thus

$$\bar{\bar{G}}_1(\bar{R}|\bar{R}') = \int_{-\infty}^{\infty} \frac{dh}{h^2+k_c^2-k^2} \sum_m \sum_n \frac{2-\delta_0}{\pi abk_c^2} \cdot [\bar{M}_{emn}(h)\bar{M}'_{emn}(-h) + \bar{N}_{omn}(h)\bar{N}'_{omn}(-h)]\,.$$

Since the dyadic functions $\bar{M}\bar{M}'$ and $\bar{N}\bar{N}'$ contain the exponential function $e^{ih(z-z')}$, the integration with respect to h can be evaluated in a closed form by the method of contour integration in view of the fact that the integrand has a pair of poles located at $h = \pm\sqrt{k^2 - k_c^2}$. The method and the details are exactly the same as the ones described in Sec. 10. Without repeating them, we find that

$$\bar{\bar{G}}_1(\bar{R}|\bar{R}') = \frac{i}{ab} \sum_m \sum_n \frac{2-\delta_0}{k_c^2 k_g}$$

(8) $$\cdot \begin{cases} \bar{M}_{emn}(k_g)\bar{M}'_{emn}(-k_g) + \bar{N}_{omn}(k_g)\bar{N}'_{omn}(-k_g), & z > z' \\ \bar{M}_{emn}(-k_g)\bar{M}'_{emn}(k_g) + \bar{N}_{omn}(-k_g)\bar{N}'_{omn}(k_g), & z < z', \end{cases}$$

where

(9) $$k_g = (k^2 - k_c^2)^{1/2}$$

and we adopt the rule that when k_g is real

$$k_g = |(k^2 - k_c^2)^{1/2}|$$

and when k_g is imaginary $(k < k_c)$

$$k_g = i|(k_c^2 - k^2)^{1/2}|$$

so that the radiation condition at $z = \pm\infty$ is satisfied. For real k_g, corresponding to a propagating mode, (9) can be written in the form

$$\frac{1}{\lambda_g} = \left(\frac{1}{\lambda^2} - \frac{1}{\lambda_c^2}\right)^{1/2}$$

where λ_g = guided wavelength

λ_c = cutoff wavelength for the TE_{mn} or TM_{mn} mode

$$= \left[\left(\frac{m}{2a}\right)^2 + \left(\frac{n}{2b}\right)^2\right]^{-1/2}$$

Equation (8) gives the complete eigenfunction expansion of $\bar{\bar{G}}_1(\bar{R}|\bar{R}')$ for an infinite rectangular waveguide. To find $\bar{\bar{G}}_2(\bar{R}|\bar{R}')$ the procedure is exactly the same, except that we would use $\bar{M}_{omn}$ and $\bar{N}_{emn}$ as the proper eigenfunctions. However, because of the symmetrical property between $\bar{\bar{G}}_1$ and $\bar{\bar{G}}_2$ as described by (17-51), the expression for $\bar{\bar{G}}_2$ can be obtained by inspection without another lengthy analysis, that is,

(10) $$\bar{\bar{G}}_2(\bar{R}|\bar{R}') = \frac{i}{ab} \sum_m \sum_n \frac{(2-\delta_0)}{k_c^2 k_g}$$

$$\cdot \begin{cases} \bar{M}_{omn}(k_g)\bar{M}'_{omn}(-k_g) + \bar{N}_{emn}(k_g)\bar{N}'_{emn}(-k_g), & z > z', \\ \bar{M}_{omn}(-k_g)\bar{M}'_{omn}(k_g) + \bar{N}_{emn}(-k_g)\bar{N}'_{emn}(k_g), & z < z'. \end{cases}$$

This completes our derivation for $\overline{\overline{G}}_1$ and $\overline{\overline{G}}_2$. The derivation of $\overline{\overline{G}}_3$ and $\overline{\overline{G}}_4$ and some problems illustrating the application of these functions are given as exercises in Appendix D.

chapter **6**

Cylindrical Waveguides

The derivation of the dyadic Green's functions for a cylindrical waveguide follows the same procedures as that for a rectangular waveguide. The only difference is that cylindrical vector wave functions are used in the eigenfunction expansion. Once the orthogonal relations between these vector wave functions are known, the remaining steps are exactly the same as in the rectangular case.

21. CYLINDRICAL VECTOR WAVE FUNCTIONS WITH DISCRETE EIGENVALUES

The cylindrical vector wave functions with discrete eigenvalues can be used to describe the electromagnetic field inside a cylindrical waveguide with circular cross section, the geometry of which is shown in Fig. 21-1.

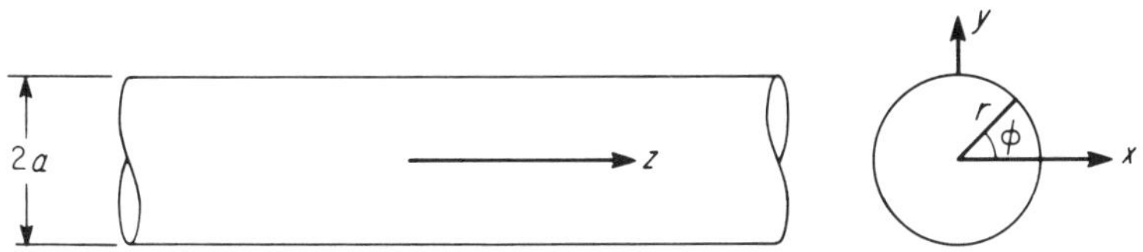

FIG. 21-1. A cylindrical waveguide with circular cross section.

Before we define these functions, we shall review the roots of the Bessel functions with integer order, $J_n(x)$, and the roots of their derivatives. We will designate the roots of the equation

$$J_n(x) = 0$$

by p_{nm}. Thus p_{23} represents the third root of the Bessel function of second order. The roots of the equation

$$\frac{dJ_n(x)}{dx} = 0$$

will be designated by q_{nm}. Tables 21-1 and 21-2 list a number of these roots in an ascending order.

TABLE 21-1

Roots of $J_n(x) = 0$: p_{nm}
(TM_{nm} modes)

$n \backslash m$	1	2	3
0	2.405	5.520	8.654
1	3.832	7.106	10.173
2	5.136	8.417	11.620
3	6.380	9.761	13.015

TABLE 21-2

Roots of $\dfrac{dJ_n(x)}{dx} = 0$: q_{nm}
(TE_{nm} modes)

$n \backslash m$	1	2	3
0	0	3.832	7.016
1	1.841	5.331	8.536
2	3.054	6.706	9.969
3	4.201	8.015	11.346

With such a designation, the values of p_{nm} and q_{nm} increase with either n or m in an orderly manner as shown more vividly in Fig. 21-2. This figure can also be used to interpolate the roots of the Bessel function of fractional order or the roots of their derivatives. We now define two types of cylindrical vector wave functions both of which satisfy the vector Dirichlet boundary condition at $r = a$, corresponding to the site of the wall of the cylindrical waveguide shown in Fig. 21-1. They are

(1) $$\bar{M}_{\substack{e\\o}n\mu}(h) = \nabla \times \left[J_n(\mu r) \begin{matrix}\cos\\ \sin\end{matrix}\, n\phi\, e^{ihz}\hat{z}\right]$$

with $\mu = q_{nm}/a$

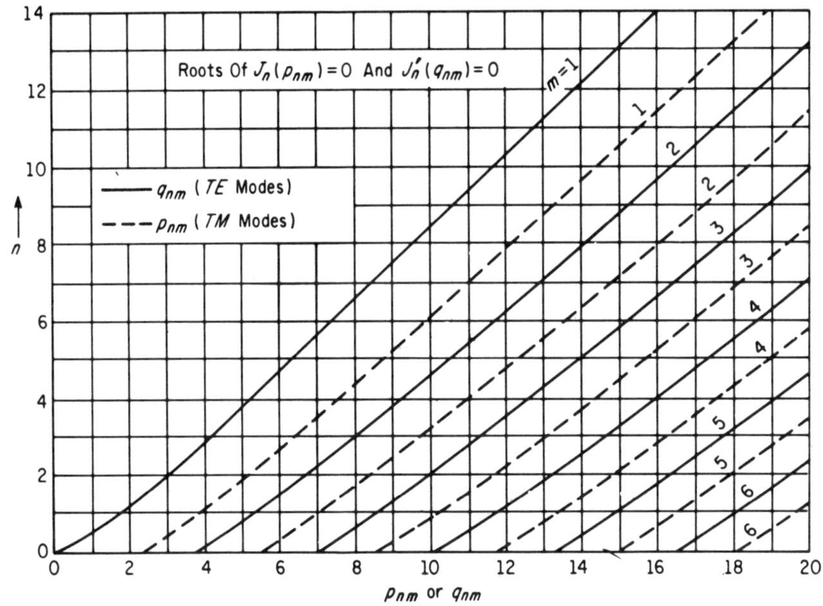

FIG. 21-2

and

(2)
$$\bar{N}_{\substack{e\\o}n\lambda}(h)=\frac{1}{\kappa_\lambda}\nabla\times\nabla\times\left[J_n(\lambda r)\begin{matrix}\cos\\ \sin\end{matrix}\, n\phi\, e^{ihz}\hat{z}\right]$$

with $\lambda=p_{nm}/a$, $\kappa_\lambda^2=\lambda^2+h^2$.

It is obvious that these two sets of functions are generated in the same manner as the rectangular vector wave functions except that the scalar wave equation is now solved, by the method of separation of variables, in cylindrical coordinate system, and the eigenvalues λ and μ are so chosen that the functions would satisfy the vector Dirichlet boundary condition at $r=a$. The complete expressions of these two sets are

(3)
$$\bar{M}_{\substack{e\\o}n\mu}(h)=\left[\mp\frac{nJ_n(\mu r)}{r}\begin{matrix}\sin\\ \cos\end{matrix}\, n\phi\hat{r}-\frac{\partial J_n(\mu r)}{\partial r}\begin{matrix}\cos\\ \sin\end{matrix}\, n\phi\hat{\phi}\right]e^{ihz}$$

(4)
$$\bar{N}_{\substack{e\\o}n\lambda}(h)=\frac{1}{\kappa_\lambda}\left[ih\frac{\partial J_n(\lambda r)}{\partial r}\begin{matrix}\cos\\ \sin\end{matrix}\, n\phi\hat{r}\right.$$
$$\left.\mp\frac{ihn}{r}J_n(\lambda r)\begin{matrix}\sin\\ \cos\end{matrix}\, n\phi\hat{\phi}+\lambda^2J_n(\lambda r)\begin{matrix}\cos\\ \sin\end{matrix}\, n\phi\hat{z}\right]e^{ihz}$$

where $\kappa_\lambda^2=\lambda^2+h^2$.

It should be noted that both λ and μ are denumerable numbers depending on n and m as defined previously. The functions $\bar{M}_{\substack{e\\o}n\mu}(h)$ can be used to describe the electric field of the TE_{nm} mode in a cylindrical waveguide with radius equal to a, and the function $\bar{N}_{\substack{e\\o}n\lambda}(h)$ for the TM_{nm} mode. According to our nomenclature, the mode TE_{01} is a null mode because $\mu_{01}=0$ and the dominant mode is TE_{11}. Such a nomenclature appears to be more logical than the one used in the current literature. This view is shared by Ramo, Whinnery and Van Duzer (p. 432). To describe the magnetic field in the guide, the proper vector wave functions to be used are

$$\bar{M}_{\substack{e\\o}n\lambda}(h)=\nabla\times\left[J_n(\lambda r)\substack{\cos\\\sin}\,n\phi e^{ihz}\hat{z}\right] \tag{5}$$

$$=\left[\mp\frac{nJ_n(\lambda r)}{r}\substack{\sin\\\cos}\,n\phi\hat{r}-\frac{\partial J_n(\lambda r)}{\partial r}\substack{\cos\\\sin}\,n\phi\hat{\phi}\right]e^{ihz}$$

$$\bar{N}_{\substack{e\\o}n\mu}(h)=\frac{1}{\kappa_\mu}\nabla\times\nabla\times\left[J_n(\mu r)\substack{\cos\\\sin}\,n\phi e^{ihz}\hat{z}\right]=\frac{1}{\kappa_\mu}\left[ih\frac{\partial J_n(\mu r)}{\partial r}\substack{\cos\\\sin}\,n\phi\hat{r}\right. \tag{6}$$

$$\left.\mp\frac{ihn}{r}J_n(\mu r)\substack{\sin\\\cos}\,n\phi\hat{\phi}+\mu^2J_n(\mu r)\substack{\cos\\\sin}\,n\phi\hat{z}\right]e^{ihz}$$

where

$$\kappa_\mu^2=\mu^2+h^2,\qquad \kappa_\lambda^2=\lambda^2+h^2,$$

$$\bar{N}_{\substack{e\\o}n\mu}(h)=\frac{1}{\kappa_\mu}\nabla\times\bar{M}_{\substack{e\\o}n\mu}(h)$$

and

$$\bar{N}_{\substack{e\\o}n\lambda}(h)=\frac{1}{\kappa_\lambda}\nabla\times\bar{M}_{\substack{e\\o}n\lambda}(h),$$

and vice versa. The $\bar{M}_{\substack{e\\o}n\lambda}(h)$ and $\bar{N}_{\substack{e\\o}n\mu}(h)$ functions obviously satisfy the vector Neumann boundary condition at $r=a$.

Before we prove the orthogonality between the various vector wave functions, the orthogonal properties of $J_n(\lambda r)$ and $J_n(\mu r)$ should be discussed first. We consider two Bessel functions $J_n(\alpha r)$ and $J_n(\beta r)$ which satisfy, respectively, the differential equations

$$\frac{1}{r}\frac{d}{dr}\left[r\frac{dJ_n(\alpha r)}{dr}\right]+\left(\alpha^2-\frac{n^2}{r^2}\right)J_n(\alpha r)=0 \tag{7}$$

and

$$\frac{1}{r}\frac{d}{dr}\left[r\frac{dJ_n(\beta r)}{dr}\right]+\left(\beta^2-\frac{n^2}{r^2}\right)J_n(\beta r)=0. \tag{8}$$

Multiplying (7) by $J_n(\beta r)$ and (8) by $J_n(\alpha r)$ and taking the differences we obtain

$$(\alpha^2-\beta^2)J_n(\alpha r)J_n(\beta r)=\frac{1}{r}\left[J_n(\alpha r)\frac{d}{dr}\left(r\,\frac{dJ_n(\beta r)}{dr}\right)-J_n(\beta r)\frac{d}{dr}\left(r\,\frac{dJ_n(\alpha r)}{dr}\right)\right].$$

An integration of the above equation with respect to $r\,dr$ from $r=0$ to a yields

$$(\alpha^2-\beta^2)\int_0^a J_n(\alpha r)J_n(\beta r)r\,dr=\left[rJ_n(\alpha r)\,\frac{dJ_n(\beta r)}{dr}-rJ_n(\beta r)\frac{dJ_n(\alpha r)}{dr}\right]_0^a.$$

If we let $\alpha=\lambda=p_{nm}/a$ and $\beta=\lambda'=p_{nm'}/a$,

(9) $$\int_0^a J_n(\lambda r)J_n(\lambda' r)r\,dr=0$$

since $J_n(p_{nm})=0$ and $J_n(p_{nm'})=0$. Similarly, if we let $\alpha=\mu=q_{nm}/a$ and $\beta=\mu'=q_{nm'}/a$, we obtain

(10) $$\int_0^a J_n(\mu r)J_n(\mu' r)r\,dr=0\,.$$

To determine the normalization factor when $\alpha=\beta=\lambda$ or $\alpha=\beta=\mu$, we let $\beta=\alpha+\triangle$. Then

$$I_\alpha=\int_0^a J_n^2(\alpha r)r\,dr=\lim_{\triangle\to 0}\int_0^a J_n(\alpha r)J_n[(\alpha+\Delta)r]\,r\,dr$$

$$=\frac{-1}{2\alpha}\lim_{\triangle\to 0}\frac{1}{\Delta}\left\{rJ_n(\alpha r)\,\frac{\partial J_n[(\alpha+\Delta)r]}{\partial r}-rJ_n[(\alpha+\Delta)r]\,\frac{\partial J_n(\alpha r)}{\partial r}\right\}_0^a$$

$$=\frac{-1}{2\alpha}\left[rJ_n(\alpha r)\,\frac{\partial^2 J_n(\alpha r)}{\partial\alpha\partial r}-r\,\frac{\partial J_n(\alpha r)}{\partial\alpha}\,\frac{\partial J_n(\alpha r)}{\partial r}\right]_0^a$$

$$=\frac{-a}{2\alpha}\left[\,J_n(\alpha r)\,\frac{\partial^2 J_n(\alpha r)}{\partial\alpha\partial r}-\frac{\partial J_n(\alpha r)}{\partial\alpha}\,\frac{\partial J_n(\alpha r)}{\partial r}\,\right]_{r=a}.$$

Now

$$\frac{\partial^2 J_n(\alpha r)}{\partial a\partial r}=\frac{\partial}{\partial\alpha}\left[\alpha\,\frac{dJ_n(\alpha r)}{d(\alpha r)}\right]=\frac{dJ_n(\alpha r)}{d(\alpha r)}+\alpha r\,\frac{d^2 J_n(\alpha r)}{d^2(\alpha r)}$$

$$=-\frac{r}{\alpha}\left(\alpha^2-\frac{n^2}{r^2}\right)J_n(\alpha r)$$

and

$$\frac{\partial J_n(\alpha r)}{\partial \alpha} = \frac{r}{\alpha} \frac{\partial J_n(\alpha r)}{\partial r}$$

hence

$$I_\alpha = \frac{a^2}{2\alpha^2} \left\{ \left(\alpha^2 - \frac{n^2}{r^2} \right) J_n^2(\alpha r) + \left[\frac{\partial J_n(\alpha r)}{\partial r} \right]^2 \right\}_{r=a} .$$

Thus, when $\alpha = \lambda$, we obtain

$$(11) \qquad I_\lambda = \int_0^a J_n^2(\lambda r) r \, dr = \frac{a^2}{2\lambda^2} \left[\frac{\partial J_n(\lambda r)}{\partial r} \right]_{r=a}^2$$

because

$$J_n(\lambda a) = 0, \quad \lambda a = p_{nm} .$$

When $\alpha = \mu$, we have

$$(12) \qquad I_\mu = \int_0^a J_n^2(\mu r) r \, dr = \frac{a^2}{2\mu^2} \left(\mu^2 - \frac{n^2}{a^2} \right) J_n^2(\mu a)$$

because

$$\left. \frac{dJ_n(x)}{dx} \right|_{x=\mu a} = 0, \quad \mu a = q_{nm} .$$

The orthogonal relations (9) and (10) and the normalization factors, (11) and (12), are needed in the discussion of the orthogonal property of the vector wave functions.

The orthogonal relations between the various cylindrical vector wave functions can be stated as follows:

$$(13) \qquad \iiint \bar{M}_{\substack{e\\o} n\mu}(h) \cdot \bar{N}_{\substack{e\\o} n'\lambda}(-h') \, dV = 0$$

$$(14) \qquad \iiint \bar{M}_{\substack{e\\o} n\mu}(h) \cdot \bar{M}_{\substack{e\\o} n'\mu'}(-h') \, dV$$

$$= \iiint \bar{N}_{\substack{e\\o} n\mu}(h) \cdot \bar{N}_{\substack{e\\o} n'\mu'}(-h') \, dV$$

$$= \begin{cases} 0, & n \neq n', \quad \mu \neq \mu' \\ (1+\delta_0) 2\pi^2 \mu^2 I_\mu \delta(h-h') & n = n', \, \mu = \mu' \end{cases}$$

$$(15) \qquad \iiint \bar{M}_{\substack{e\\o} n\lambda}(h) \cdot \bar{M}_{\substack{e\\o} n'\lambda'}(-h') \, dV = \iiint \bar{N}_{\substack{e\\o} n\lambda}(h) \cdot \bar{N}_{\substack{e\\o} n'\lambda'}(-h') \, dV$$

$$= \begin{cases} 0, & n \neq n', \quad \lambda \neq \lambda' \\ (1+\delta_0) 2\pi^2 \lambda^2 I_\lambda \delta(h-h'), & n=n', \ \lambda=\lambda', \end{cases}$$

where the volume integral is extended through the entire volume of an infinitely long guide. Because of the trigonometric functions, all the cylindrical vector wave functions are orthogonal in the ϕ-domain when $n \neq n'$. It is, therefore, sufficient to discuss those cases where $n=n'$. We shall demonstrate only two of them, as the proofs of the remaining ones are similar. We consider for example the integral

$$I = \iiint \bar{M}_{en\mu}(h) \cdot \bar{N}_{on\lambda}(-h')\, dV, \quad n \neq 0.$$

In view of (3) and (4), we obtain

$$I = \int_0^a r\,dr \int_0^{2\pi} d\phi \int_{-\infty}^{\infty} dz \,\frac{1}{\kappa_\lambda}\left[ih'n \frac{J_n(\mu r)}{r} \frac{\partial J_n(\lambda r)}{\partial r} \sin^2 n\phi \right.$$

$$\left. + ih'n \frac{J_n(\lambda r)}{r} \frac{\partial J_n(\mu r)}{\partial r} \cos^2 n\phi \right] e^{i(h-h')z}$$

$$= \int_0^a r\,dr \int_{-\infty}^{\infty} dz \,\frac{ih'n\pi}{\kappa_\lambda} \left[\frac{J_n(\mu r)}{r} \frac{\partial J_n(\lambda r)}{\partial r} + \frac{J_n(\lambda r)}{r} \frac{\partial J_n(\mu r)}{\partial r} \right] e^{i(h-h')z}$$

$$= \int_0^a r\,dr \int_{-\infty}^{\infty} dz \,\frac{ih'n\pi}{\kappa_\lambda r} \frac{\partial}{\partial r} \left[J_n(\lambda r) J_n(\mu r) \right] e^{i(h-h')z}$$

$$= \frac{ih'n\pi}{\kappa_\lambda} J_n(\lambda a) J_n(\mu a) 2\pi \delta(h-h').$$

Since $J_n(\lambda a)=0$, I vanishes identically. The normalization factor for functions of the same species can be found as follows. Using the function $\bar{M}_{en\mu}$ as an example, we have

$$\iiint \bar{M}_{en\mu}(h) \cdot \bar{M}_{en\mu}(-h')\, dV$$

$$= \iiint \left\{ \frac{n^2 J_n^2(\mu r)}{r^2} \sin^2 n\phi + \left[\frac{\partial J_n(\mu r)}{\partial r} \right]^2 \cos^2 n\phi \right\} \cdot e^{i(h-h')z}\, dV$$

$$= (1+\delta_0) 2\pi^2 \delta(h-h') \int_0^a \left\{ \frac{n^2 J_n^2(\mu r)}{r^2} + \left[\frac{\partial J_n(\mu r)}{\partial r} \right]^2 \right\} r\,dr$$

where δ_0 denotes the Kronecker delta function defined with respect to n. The integral involving the square of the derivative can be simplified by integration by parts that cancels the term involving J_n^2/r^2 and yields

$$2(1+\delta_0)\pi^2\delta(h-h')\int_0^a \mu^2 J_n^2(\mu r)\, r\, dr\,.$$

In view of (12), the final results for the normalization factor is given by

$$2(1+\delta_0)\pi^2\mu^2 I_\mu \delta(h-h')\,.$$

The readers can verify other relations listed in (12) through (14) as exercises. Knowing these orthogonal relations we can find the eigenfunction expansion of the dyadic Green's functions by the Ohm-Rayleigh method.

22. DYADIC GREEN'S FUNCTIONS OF THE FIRST AND SECOND KIND FOR A CYLINDRICAL WAVEGUIDE

Based on the method described in Sec. 21 for the derivation of $\bar{\bar{G}}_1$ we let

$$\bar{\bar{I}}\delta(\bar{R}-\bar{R}') = \int_{-\infty}^{\infty} \sum_{n=0}^{\infty} \sum_{m=1}^{\infty} [\bar{M}_{{}^e_o n\mu}(h)\, \bar{A}_{{}^e_o n\mu}(h) + \bar{N}_{{}^e_o n\lambda}(h)\, \bar{B}_{{}^e_o n\lambda}(h)\,]\, dh\,. \tag{1}$$

The index m is used here to designate the ordinal number associated with

$$\lambda\left(=\frac{p_{nm}}{a}\right) \text{ and } \mu\left(=\frac{q_{nm}}{a}\right).$$

We adopt the simplified notation that

$$\bar{M}_{{}^e_o n\mu}\bar{A}_{{}^e_o n\mu} = \bar{M}_{en\mu}\bar{A}_{en\mu} + \bar{M}_{on\mu}\bar{A}_{on\mu}$$

Taking the anterior scalar products of (1) with $\bar{M}_{en\mu}(-h')$, $\bar{M}_{on\mu}(-h')$, $\bar{N}_{en\lambda}(-h')$ and $\bar{N}_{on\lambda}(-h')$ in turn and integrating the resultant equations through the entire volume of the guide, we can determine, as a consequence of the orthogonal relations, the unknown coefficients $\bar{A}_{{}^e_o n\mu}(h)$ and $\bar{B}_{{}^e_o n\mu}(h)$. They are

$$\bar{A}_{{}^e_o n\mu}(h) = \frac{2-\delta_0}{4\pi^2\mu^2 I_\mu}\, \bar{M}'_{{}^e_o n\mu}(-h) \tag{2}$$

$$\bar{B}_{{}^e_o n\lambda}(h) = \frac{2-\delta_0}{4\pi^2\lambda^2 I_\lambda}\, \bar{N}'_{{}^e_o n\lambda}(-h)\,. \tag{3}$$

The primed functions are defined with respect to the primed coordinates (r', ϕ', z'), corresponding to the location of $\bar{R}'$. To find $\bar{\bar{G}}_1(\bar{R}|\bar{R}')$ we let

(4) $$\bar{\bar{G}}_1(\bar{R}|\bar{R}') = \int_{-\infty}^{\infty} dh \sum_n \sum_m \frac{2-\delta_0}{4\pi^2} \cdot \left[\frac{a(h)}{\mu^2 I_\mu} \bar{M}_{\substack{e\\o}n\mu}(h)\bar{M}'_{\substack{e\\o}n\mu}(-h) + \frac{b(h)}{\lambda^2 I_\lambda} \bar{N}_{\substack{e\\o}n\lambda}(h)\bar{N}'_{\substack{e\\o}n\lambda}(-h) \right].$$

Substituting (1) and (4) into the differential equation for $\bar{\bar{G}}_1(\bar{R}|\bar{R}')$, we can determine the unknown coefficients $a(h)$ and $b(h)$. The results are

$$a(h) = \frac{1}{h^2 + \mu^2 - k^2}$$

$$b(h) = \frac{1}{h^2 + \lambda^2 - k^2}.$$

The remaining integration with respect to h can be evaluated by the method of contour integration by taking advantage of the poles at $h = \pm(k^2 - \mu^2)^{1/2}$ and $h = \pm(k^2 - \lambda^2)^{1/2}$. The final expression for $\bar{\bar{G}}_1(\bar{R}|\bar{R}')$ is given by

(5) $$\bar{\bar{G}}_1(\bar{R}|\bar{R}') = \frac{i}{4\pi} \sum_{n=0}^{\infty} \sum_{m=1}^{\infty} (2-\delta_0) \cdot \left[\frac{1}{\mu^2 I_\mu k_\mu} \bar{M}_{\substack{e\\o}n\mu}(k_\mu)\bar{M}'_{\substack{e\\o}n\mu}(-k_\mu) + \frac{1}{\lambda^2 I_\lambda k_\lambda} \bar{N}_{\substack{e\\o}n\lambda}(k_\lambda)\bar{N}'_{\substack{e\\o}n\lambda}(-k_\lambda) \right], \qquad z > z'$$

where

(6) $$k_\mu = (k^2 - \mu^2)^{1/2}$$

(7) $$k_\lambda = (k^2 - \lambda^2)^{1/2}$$

For the case $z < z'$, we replace the dyadic functions in (5) by $\bar{M}(-k_\mu)\bar{M}'(k_\mu)$ and $\bar{N}(-k_\lambda)\,\bar{N}'(k_\lambda)$ To satisfy the radiation condition we adopt the rule that when k_μ and k_λ are real:

$$k_\mu = |(k^2 - \mu^2)^{1/2}|$$

$$k_\lambda = |(k^2 - \lambda^2)^{1/2}|$$

When they are imaginary ($k^2 < \mu^2$, $k^2 < \lambda^2$),

$$k_\mu = i|(\mu^2 - k^2)^{1/2}|$$

$$k_\lambda = i|(\lambda^2 - k^2)^{1/2}|.$$

The function $\bar{\bar{G}}_2(\bar{R}|\bar{R}')$ can be found by making use of the symmetrical relation as described by (17-51).

For $z > z'$, it is given by

$$(8) \qquad \overline{\overline{G}}_2(\bar{R}|\bar{R}') = \frac{i}{4\pi}\sum_{n=0}^{\infty}\sum_{m=1}^{\infty}(2-\delta_0)\cdot\left[\frac{1}{\mu^2 I_\mu k_\mu}\bar{N}_{\substack{e\\o}n\mu}(k_\mu)\bar{N}'_{\substack{e\\o}n\mu}(-k_\mu) + \frac{1}{\lambda^2 I_\lambda k_\lambda}\bar{M}_{\substack{e\\o}n\lambda}(k_\lambda)\bar{M}'_{\substack{e\\o}n\lambda}(-k_\lambda)\right]$$

The similarity in structure between (5), (8) and (20-8), (20-10) is expected. It should be noted that the symbols "even" and "odd" used for the cylindrical functions apply only to the angular functions. The same symbols when used for the rectangular functions apply to the two trigonometric. functions or the entire scalar wave function. The even and odd functions for the rectangular waveguide actually play the same role as the λ and μ functions for the cylindrical waveguide.

chapter **7**

Circular Cylinder in Free Space

When we are dealing with radiation problems in the presence of an infinite cylinder as a diffracting body, the vector wave functions with continuous eigenvalues both in the h and λ domains are needed. The orthogonal properties of these functions are not quite the same as the ones with discrete eigenvalues. We will discuss these properties first, and then find the eigenfunction expansion of the free-space dyadic Green's function in terms of three functions. Once the latter is known, the functions of the other kinds can be found by the method of scattering superposition.

23. CYLINDRICAL VECTOR WAVE FUNCTIONS WITH CONTINUOUS EIGENVALUES

In dealing with an integral representation of fields in free space, the cylindrical vector wave functions to be used have continuous eigenvalues in the λ-domain as well as in the h-domain. The formulae for $\bar{M}$ and $\bar{N}$ as defined by (21-3) and (21-4), with $\mu=\lambda$, are still valid but no constraint is placed on λ or μ. Since λ and μ merge into one parameter, only two sets of functions,

$$\bar{M}_{\substack{e\\o}n\lambda} \quad \text{and} \quad \bar{N}_{\substack{e\\o}n\lambda}$$

are left to represent both the electric and magnetic field. These functions are now defined in the entire space, corresponding to $\infty > r \geqslant 0$, $2\pi \geqslant \phi \geqslant 0$, and $-\infty < z < \infty$. The orthogonal properties of these functions can be stated as follows:

$$(1) \quad \iiint \bar{M}_{\substack{e\\o}n\lambda}(h) \cdot \bar{N}_{\substack{e\\o}n'\lambda'}(-h')\, dV = 0$$

$$(2)\quad \iiint \overline{M}_{\substack{e\\o}n\lambda}(h)\cdot\overline{M}_{\substack{e\\o}n'\lambda'}(-h')\,dV=\begin{cases}0, & n\neq n'\\ 2(1+\delta_0)\pi^2\lambda\delta(\lambda-\lambda')\delta(h-h'), & n=n'\end{cases}$$

$$(3)\quad \iiint \overline{N}_{\substack{e\\o}n\lambda}(h)\cdot\overline{N}_{\substack{e\\o}n'\lambda'}(-h')\,dV=\begin{cases}0, & n\neq n'\\ 2(1+\delta_0)\pi^2\lambda\delta(\lambda-\lambda')\delta(h-h'), & n=n',\end{cases}$$

where the domain of integration encloses the entire space. The proofs of these formulae are slightly different from the ones for (21-13, 21-16). Because of the angular function, all these functions, either of the same species or of different species, are orthogonal when $n\neq n'$. It is therefore sufficient to discuss those cases when $n=n'$. To prove (1), let us consider, for example, the integral

$$I=\iiint \overline{M}_{en\lambda}(h)\cdot\overline{N}_{on\lambda'}(-h')\,dV$$

where (λ, h) and $(\lambda', -h')$ denote two distinct pairs of eigenvalues. Using (21-3) and (21-4), with $\mu=\lambda$, we obtain, after an integration with respect to ϕ,

$$I=\frac{(1+\delta_0)\pi}{\sqrt{\lambda'^2+h'^2}}\int_0^\infty r\,dr\int_{-\infty}^{\infty}dz$$

$$\cdot\frac{inh'}{r}\left[J_n(\lambda r)\frac{\partial J_n(\lambda' r)}{\partial r}+J_n(\lambda' r)\frac{\partial J_n(\lambda r)}{\partial r}\right]e^{i(h-h')z}$$

$$=\frac{inh'(1+\delta_0)\pi}{\sqrt{\lambda'^2+h'^2}}\cdot 2\pi\delta(h-h')\left[J_n(\lambda r)J_n(\lambda' r)\right]_0^\infty=0\,.$$

Actually, $\overline{M}_{en\lambda}$ and $\overline{N}_{on\lambda'}$ are orthogonal in the $(r$–$\phi)$ plane, the integration with respect to z is not necessary. We include it here for completeness because, in general, we always deal with the volume integral of these functions. To prove (2) or (3), let us consider the integral

$$I=\iiint \overline{N}_{en\lambda}(h)\cdot\overline{N}_{en\lambda'}(-h')\,dV=\frac{1}{\kappa\kappa'}\iiint\left[hh'\frac{\partial J_n(\lambda r)}{\partial r}\frac{\partial J_n(\lambda' r)}{\partial r}\cos^2 n\phi\right.$$

$$\left.+\frac{hh'n^2}{r^2}J_n(\lambda r)J_n(\lambda' r)\sin^2 n\phi+\lambda^2\lambda'^2J_n(\lambda r)J_n(\lambda' r)\cos^2 n\phi\right]e^{i(h-h')z}\,dV$$

where

$$\kappa=\sqrt{\lambda^2+h^2},\qquad \kappa'=\sqrt{\lambda'^2+h'^2}\,.$$

The integrations with respect to ϕ and z yield

$$I = \frac{(1+\delta_0)2\pi^2\delta(h-h')}{\kappa\kappa'}$$

$$\cdot \int_0^\infty \left[hh' \frac{\partial J_n(\lambda r)}{\partial r} \frac{\partial J_n(\lambda' r)}{\partial r} + \left(\frac{hh'n^2}{r^2} + \lambda^2\lambda'^2 \right) J_n(\lambda r) J_n(\lambda' r) \right] r\, dr\,.$$

Using the recurrence relations of the Bessel functions,

$$J_n(x) = \frac{x}{2n} \left[J_{n-1}(x) + J_{n+1}(x) \right] \tag{4}$$

$$\frac{dJ_n(x)}{dx} = \frac{1}{2} \left[J_{n-1}(x) - J_{n+1}(x) \right] \tag{5}$$

we can change the integral into the form

$$I = \frac{(1+\delta_0)2\pi^2\delta(h-h')}{\kappa\kappa'}$$

$$\cdot \int_0^\infty \left\{ \frac{1}{2} \lambda\lambda' hh' \left[J_{n-1}(\lambda r) J_{n-1}(\lambda' r) + J_{n+1}(\lambda r) J_{n+1}(\lambda' r) \right] + \lambda^2\lambda'^2 J_n(\lambda r) J_n(\lambda' r) \right\} r\, dr\,.$$

As a result of the integral representation of the weighted delta function given by (3-25), we obtain

$$\begin{aligned} I &= \frac{(1+\delta_0)2\pi^2\delta(h-h')}{\kappa\kappa'} \cdot (\lambda\lambda' hh' + \lambda^2\lambda'^2) \frac{\delta(\lambda-\lambda')}{\lambda} \\ &= 2(1+\delta_0)\pi^2\lambda\delta(\lambda-\lambda')\delta(h-h')\,, \end{aligned}$$

which is the normalization factor for two cylindrical vector wave functions of the same species. The proofs for the remaining combinations are practically the same. It should be mentioned that the orthogonal properties of these functions have previously been investigated by Stratton (1941, pp. 397–399). However, he used a mixed domain consisting of (r,ϕ,λ) which is partly spatial and partly in the eigenvalue domain. As a result the fine structure of these properties is not spelled out. Our presentation shows that the normalization factor in the spatial domain contains two delta functions in the eigenvalue domain. It is this important feature that facilitates greatly the eigenfunction expansion of a dyadic Green's function or any other vector functions. In regard to (1), it is clear that $\bar{M}_e$ and $\bar{N}_o$ or $\bar{M}_o$ and $\bar{N}_e$ are orthogonal in the r-ϕ plane but not in the ϕ-plane alone as asserted by Stratton (p. 398). This is presumably just an oversight.

24. EIGENFUNCTION EXPANSION OF THE FREE-SPACE DYADIC GREEN'S FUNCTION; FUNCTIONS OF THE FIRST AND THE THIRD KIND

The free space dyadic Green's function $\bar{\bar{G}}_0(\bar{R}|\bar{R}')$ introduced previously in Sec. 14, by definition, satisfies the equation

$$\nabla \times \nabla \times \bar{\bar{G}}_0(\bar{R}|\bar{R}') - k^2\bar{\bar{G}}_0(\bar{R}|\bar{R}') = \bar{\bar{I}}\delta(\bar{R} - \bar{R}') \tag{1}$$

and the radiation condition at infinity. According to the Ohm-Rayleigh method we let

$$\bar{\bar{I}}\delta(\bar{R} - \bar{R}') = \int_0^\infty d\lambda \int_{-\infty}^{\infty} dh \sum_{n=0}^{\infty} [\bar{M}_{\substack{e\\o}n\lambda}(h)\bar{A}_{\substack{e\\o}n\lambda}(h) + \bar{N}_{\substack{e\\o}n\lambda}(h)\bar{B}_{\substack{e\\o}n\lambda}(h)] \tag{2}$$

where the unknown posterior functions $\bar{A}$ and $\bar{B}$ are to be determined. This step is similar to the one described by (22-1) except that λ is now continuous. Only positive values of λ are included because $J_n(\lambda r)$ and $J_n(-\lambda r)$ are not independent functions. By taking the anterior product of (2) with $\bar{M}_{en'\lambda'}(-h')$ and integrating the resultant equation through the entire space, we obtain, as a consequence of the orthogonal solutions described by (23-1) through (23-3),

$$\bar{M}'_{en'\lambda'}(-h') = (1 + \delta_0)2\pi^2\lambda'\bar{A}_{en'\lambda'}(h')$$

or

$$\bar{A}_{en\lambda}(h) = \frac{2 - \delta_0}{4\pi^2\lambda}\bar{M}'_{en\lambda}(-h)$$

where the primed function, $\bar{M}'$, is defined with respect to the primed coordinates. Repeating the same with the odd $\bar{M}$ functions and the $\bar{N}$ functions, we have

$$\bar{A}_{\substack{e\\o}n\lambda}(h) = \frac{2 - \delta_0}{4\pi^2\lambda}\bar{M}'_{\substack{e\\o}n\lambda}(-h)$$

$$\bar{B}_{\substack{e\\o}n\lambda}(h) = \frac{2 - \delta_0}{4\pi^2\lambda}\bar{N}'_{\substack{e\\o}n\lambda}(-h)\,.$$

Once the eigenfunction expansion of the dyadic delta function is known, we can write down immediately the integral representation of $\bar{\bar{G}}_0(\bar{R}|\bar{R}')$ on account of (1). It is given by

(3) $$\overline{\overline{G}}_0(\bar{R}|\bar{R}') = \int_0^\infty d\lambda \int_{-\infty}^\infty dh \sum_{n=0}^\infty \frac{2-\delta_0}{4\pi^2\lambda(h^2+\lambda^2-k^2)}$$

$$\cdot \left[\bar{M}_{\substack{e\\o}n\lambda}(h)\bar{M}'_{\substack{e\\o}n\lambda}(-h) + \bar{N}_{\substack{e\\o}n\lambda}(h)\bar{N}'_{\substack{e\\o}n\lambda}(-h)\right].$$

Although (3) contains a double integral, one of them can be eliminated. The elimination depends on the nature of the problem. For the construction of the dyadic Green's functions associated with an infinite cylinder we want to eliminate the λ-integration. The reason for doing so will become clear later. In dealing with a flat ground, the integration with respect to h will be eliminated.

To perform the integration with respect to λ we can apply the integral relation (4-10) to (3) in an operational manner. Thus we write

$$\bar{M}_{en\lambda}(h)\bar{M}'_{en\lambda}(-h) = \overline{\overline{T}}_{en\lambda}\left[J_n(\lambda r)J_n(\lambda r')\right]$$

where $\overline{\overline{T}}_{en\lambda}$ represents a dyadic spatial operator. Its precise form can be written out if necessary but this is not required. Then

$$\int_0^\infty \frac{\bar{M}_{en\lambda}(h)\bar{M}'_{en\lambda}(-h)}{\lambda(\lambda^2+h^2-k^2)}\,d\lambda = \int_0^\infty \frac{\overline{\overline{T}}_{en\lambda}\left[J_n(\lambda r)J_n(\lambda r')\right]}{\lambda(\lambda^2+h^2-k^2)}\,d\lambda\,.$$

In view of (4-13), this is equal to

$$\frac{i\pi}{2\eta^2}\begin{cases}\overline{\overline{T}}_{en\eta}\left[H_n^{(1)}(\eta r)J_n(\eta r')\right], & r>r'\\ \overline{\overline{T}}_{en\eta}\left[J_n(\eta r)H_n^{(1)}(\eta r')\right], & r<r'\end{cases}$$

$$=\frac{i\pi}{2\eta^2}\begin{cases}\bar{M}^{(1)}_{en\eta}(h)\bar{M}'_{en\eta}(-h), & r>r'\\ \bar{M}_{en\eta}(h)\bar{M}'^{(1)}_{en\eta}(-h), & r<r'\end{cases}$$

where $\eta=\sqrt{k^2-h^2}$, and $H_n^{(1)}$ denotes the Hankel function of the first kind. The superscript (1) attached to $\bar{M}^{(1)}$ means that the vector function is defined with respect to the Hankel function of the first kind, i.e.,

(4) $$\bar{M}^{(1)}_{\substack{e\\o}n\eta}(h) = \nabla\times\left[H_n^{(1)}(\eta r)\,{\cos \atop \sin}\,n\phi\; e^{ihz}\,\hat{z}\right].$$

Our notation is getting more and more complicated. This is unavoidable. Repeating the same for the integration of $\bar{N}\bar{N}'$, we obtain finally an expression for $\overline{\overline{G}}_0$, involving only a Fourier integration with respect to h, which has the form

(5a)
(5b)
$$\overline{\overline{G}}_0(\overline{R}|\overline{R}') = \frac{i}{8\pi}\int_{-\infty}^{\infty} dh \sum_{n=0}^{\infty} \frac{2-\delta_0}{\eta^2}$$

$$\cdot \begin{cases} \overline{M}^{(1)}_{\substack{e\\o}n\eta}(h)\overline{M}'_{\substack{e\\o}n\eta}(-h) + \overline{N}^{(1)}_{\substack{e\\o}n\eta}(h)\overline{N}'_{\substack{e\\o}n\eta}(-h), & r > r' \\ \overline{M}_{\substack{e\\o}n\eta}(h)\overline{M}'^{(1)}_{\substack{e\\o}n\eta}(-h) + \overline{N}_{\substack{e\\o}n\eta}(h)\overline{N}'^{(1)}_{\substack{e\\o}n\eta}(-h), & r < r' \end{cases}$$

The main reason for us to have an eigenfunction expansion for $\overline{\overline{G}}_0(\overline{R}|\overline{R}')$ is for the construction of other kinds of dyadic Green's functions. Thus, the functions $\overline{\overline{G}}_1$ and $\overline{\overline{G}}_2$ are the pertinent ones associated with the field exterior to a perfectly conducting cylinder as shown schematically in

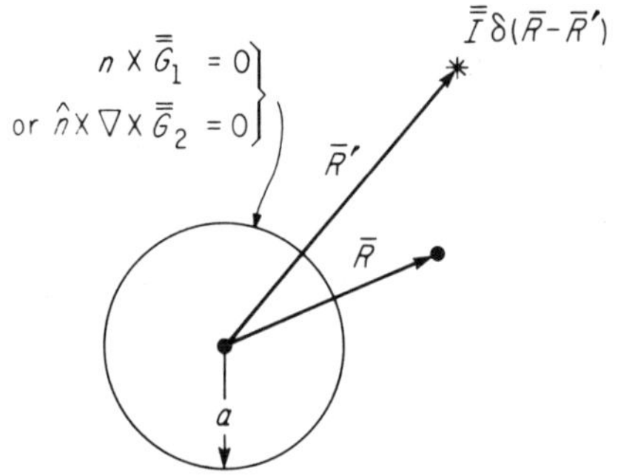

FIG. 24-1. Field exterior to a perfectly conducting cylinder.

Fig. 24-1. The simplest way of finding $\overline{\overline{G}}_1(\overline{R}|\overline{R}')$ is to apply the method of scattering superposition. Thus we let

(6)
$$\overline{\overline{G}}_1 = \overline{\overline{G}}_0 + \overline{\overline{G}}_{1s}$$

where $\overline{\overline{G}}_{1s}$ must have the form

(7)
$$\overline{\overline{G}}_{1s}(\overline{R}|\overline{R}') = \frac{i}{8\pi}\int_{-\infty}^{\infty} dh \sum_{n=0}^{\infty} \frac{2-\delta_0}{\eta^2}$$

$$\cdot \left[a_{\substack{e\\o}n}\overline{M}^{(1)}_{\substack{e\\o}n\eta}(h)\overline{M}'^{(1)}_{\substack{e\\o}n\eta}(-h) + b_{\substack{e\\o}n}\overline{N}^{(1)}_{\substack{e\\o}n\eta}(h)\overline{N}'^{(1)}_{\substack{e\\o}n\eta}(-h)\right]$$

The choice of $\overline{M}^{(1)}$ and $\overline{N}^{(1)}$ as the anterior elements is dictated by the radiation condition, and the choice of $\overline{M}'^{(1)}$ and $\overline{N}'^{(1)}$ as the posterior elements is guided by the expression for $\overline{\overline{G}}_0$ and the boundary condition that at $r = a$, $\hat{n} \times \overline{\overline{G}}_1(\overline{R}|\overline{R}') = 0$ which can be satisfied only if the posterior elements are the same. Substituting (5) and (7) into (6) and imposing the vector Dirichlet boundary condition at $r = a$, we have

$$\hat{r} \times [\bar{M}_{\substack{e\\o}n\eta}(h) + a_{\substack{e\\o}n} \bar{M}^{(1)}_{\substack{e\\o}n\eta}(h)]_{r=a} = 0$$

$$\hat{r} \times [\bar{N}_{\substack{e\\o}n\eta}(h) + b_{\substack{e\\o}n} \bar{N}^{(1)}_{\substack{e\\o}n\eta}(h)]_{r=a} = 0$$

hence

(8) $$a_{\substack{e\\o}n} = -\frac{\partial J_n(x)}{\partial x} \Big/ \frac{\partial H_n^{(1)}(x)}{\partial x}, \qquad x = \eta a$$

(9) $$b_{\substack{e\\o}n} = -J_n(x) \;/\; H_n^{(1)}(x), \qquad x = \eta a.$$

The expression for $\bar{\bar{G}}_2$ can be obtained either by taking advantage of the symmetrical property between $\bar{\bar{G}}_1$ and $\bar{\bar{G}}_2$ or by the method of scattering superposition and imposing the vector Neumann boundary condition at the surface of the cylinder. The net result is that if we interchange the role of $a_{\substack{e\\o}n}$ and $b_{\substack{e\\o}n}$ in $\bar{\bar{G}}_1$ we have the expression for $\bar{\bar{G}}_2$.

For a cylinder made of homogeneous isotropic material such as a dielectric cylinder, the pertinent functions which we are seeking would be the third or the fourth kind. For clarity, we shall adopt the following notations for similar parameters defined in the regions exterior and interior to the cylinder:

$$k_1 = \omega\sqrt{\mu_0 \varepsilon_0}, \qquad k_2 = \omega\sqrt{\mu\varepsilon}$$

$$\eta = \sqrt{k_1^2 - h^2}, \qquad \zeta = \sqrt{k_2^2 - h^2}$$

where μ and ε denote, respectively, the permeability and the permittivity of the cylinder which may be complex. To construct the Green's function of the third kind, we let

(10) $$\bar{\bar{G}}_3^{(11)} = \bar{\bar{G}}_0 + \bar{\bar{G}}_{3s}^{(11)}, \qquad r \geqslant a$$

(11) $$\bar{\bar{G}}_3^{(21)} = \bar{\bar{G}}_{3s}^{(21)}, \qquad r \leqslant a$$

where the significance of the superscript notation has been explained in Sec. 17. In order to satisfy the radiation and the boundary condition at the interface, the two scattering terms must have the following forms

(12) $$\bar{\bar{G}}_{3s}^{(11)}(\bar{R}|\bar{R}') = \frac{i}{8\pi}\int_{-\infty}^{\infty} dh \sum_{n=0}^{\infty} \frac{2-\delta_0}{\eta^2}$$

$$\cdot \{[A_{\substack{e\\o}n} \bar{M}^{(1)}_{\substack{e\\o}n\eta}(h) + B_{\substack{o\\e}n} \bar{N}^{(1)}_{\substack{o\\e}n\eta}(h)]\; \bar{M}'^{(1)}_{\substack{e\\o}n\eta}(-h)$$

$$+ [C_{\substack{e\\o}n} \bar{N}^{(1)}_{\substack{e\\o}n\eta}(h) + D_{\substack{o\\e}n} \bar{M}^{(1)}_{\substack{o\\e}n\eta}(h)]\; \bar{N}'^{(1)}_{\substack{e\\o}n\eta}(-h)\}$$

$$(13)\qquad \overline{\overline{G}}_{3s}^{(21)}(\overline{R}|\overline{R}')=\frac{i}{8\pi}\int_{-\infty}^{\infty}dh\sum_{n=0}^{\infty}\frac{2-\delta_0}{\eta^2}$$

$$\cdot\{[a_{\substack{e\\o}n}\overline{M}_{\substack{e\\o}n\zeta}(h)+b_{\substack{o\\e}n}\overline{N}_{\substack{o\\e}n\zeta}(h)]\,\overline{M}'^{(1)}_{\substack{e\\o}n\eta}(-h)$$

$$+[c_{\substack{e\\o}n}\overline{N}_{\substack{e\\o}n\zeta}(h)+d_{\substack{o\\e}n}\overline{M}_{\substack{o\\e}n\zeta}(h)]\,\overline{N}'^{(1)}_{\substack{e\\o}n\eta}(-h)\}\,.$$

We use the functions of the first kind as the anterior elements for $\overline{\overline{G}}_{3s}^{(11)}$ in order to satisfy the radiation condition at $r\to\infty$. $\overline{M}_{\substack{e\\o}n\zeta}$ and $\overline{N}_{\substack{e\\o}n\zeta}$ were chosen for $\overline{\overline{G}}_{3s}^{(21)}$ because they are solutions for the vector wave equation in the interior region and they satisfy the finiteness condition at $r=0$. The inclusion of the terms with coefficients B, D, b, and d are necessary in order to satisfy the boundary condition at the interface although these terms are absent in the free-space Green's function. Without these terms the two tangential components resulting from $\hat{n}\times\overline{N}$ and $\hat{n}\times\nabla\times\overline{M}$ cannot satisfy the boundary condition simultaneously. The fact that the even $\overline{M}$ functions play the similar roles as the odd $\overline{N}$ functions or vice versa is because they have the same angular functions. The boundary conditions at the interface, $r=a$, to be satisfied by the function of the third kind are, according to (17-23) and (17-24),

$$\hat{r}\times\overline{\overline{G}}_3^{(11)}=\hat{r}\times\overline{\overline{G}}_3^{(21)}$$

$$\frac{1}{\mu_0}\hat{r}\times\nabla\times\overline{\overline{G}}_3^{(11)}=\frac{1}{\mu}\hat{r}\times\nabla\times\overline{\overline{G}}_3^{(21)}$$

which enable us to determine the unknown coefficients A, B, C, D and a, b, c, d. The results are given below in the form of two sets of linear systems of equations, where these coefficients can be solved. It is understood that the function H_n stands for $H_n^{(1)}$ in these equations.

Knowing the explicit form of the dyadic Green's function of the third kind, the electric field due to a current source in the presence of a homogeneous cylinder can be obtained through the formula

$$\overline{E}_1(\overline{R})=i\omega\mu_0\iiint\overline{\overline{G}}_3^{(11)}(\overline{R}|\overline{R}')\cdot\overline{J}_1(\overline{R}')\,dV'$$

$$\overline{E}_2(\overline{R})=i\omega\mu_0\iiint\overline{\overline{G}}_3^{(21)}(\overline{R}|\overline{R}')\cdot\overline{J}_1(\overline{R}')\,dV'.$$

Of course, we still have the Fourier integral to be evaluated as far as the numerical result is concerned. If we are only interested in the far-zone field, the saddle-point method of integration can be applied to simplify

(14)

$$\begin{bmatrix} 0 & \dfrac{-\eta^2 H_n(\eta a)}{k_1} & 0 & \dfrac{\zeta^2 J_n(\zeta a)}{k_2} \\ \dfrac{-\partial H_n(\eta a)}{\partial a} & \dfrac{\pm ihnH_n(\eta a)}{k_1 a} & \dfrac{\partial J_n(\zeta a)}{\partial a} & \dfrac{\mp ihnJ_n(\zeta a)}{k_2 a} \\ \dfrac{-\eta^2 H_n(\eta a)}{\mu_0} & 0 & \dfrac{\zeta^2 J_n(\zeta a)}{\mu} & 0 \\ \dfrac{\mp ihnH_n(\eta a)}{\mu_0 a} & \dfrac{-k_1}{\mu_0}\dfrac{\partial H_n(\eta a)}{\partial a} & \dfrac{\pm ihnJ_n(\zeta a)}{\mu a} & \dfrac{k_2}{\mu}\dfrac{\partial J_n(\zeta a)}{\partial a} \end{bmatrix} \begin{bmatrix} A_{\substack{e\\o}n} \\ B_{\substack{o\\e}n} \\ a_{\substack{e\\o}n} \\ b_{\substack{o\\e}n} \end{bmatrix} = \begin{bmatrix} 0 \\ \dfrac{\partial J_n(\eta a)}{\partial a} \\ \dfrac{\eta^2 J_n(\eta a)}{\mu_0} \\ \dfrac{\pm ihnJ_n(\eta a)}{\mu_0 a} \end{bmatrix}$$

(15)

$$\begin{bmatrix} \dfrac{-\eta^2 H_n(\eta a)}{k_1} & 0 & \dfrac{\zeta^2 J_n(\zeta a)}{k_2} & 0 \\ \dfrac{\mp ihnH(\eta a)}{k_1 a} & \dfrac{-\partial H_n(\eta a)}{\partial a} & \dfrac{\pm ihnJ_n(\zeta a)}{k_2 a} & \dfrac{\partial J_n(\zeta a)}{\partial a} \\ 0 & \dfrac{-\eta^2 H_n(\eta a)}{\mu_0} & 0 & \dfrac{\zeta^2 J_n(\zeta a)}{\mu} \\ \dfrac{-k_1}{\mu_0}\dfrac{\partial H_n(\eta a)}{\partial a} & \dfrac{\pm ihnH_n(\eta a)}{\mu_0 a} & \dfrac{k_2}{\mu}\dfrac{\partial J_n(\zeta a)}{\partial a} & \dfrac{\mp ihnJ_n(\zeta a)}{\mu a} \end{bmatrix} \begin{bmatrix} C_{\substack{e\\o}n} \\ D_{\substack{o\\e}n} \\ c_{\substack{e\\o}n} \\ d_{\substack{o\\e}n} \end{bmatrix} = \begin{bmatrix} \dfrac{\eta^2 J_n(\eta a)}{k_1} \\ \dfrac{\pm ihnJ_n(\eta a)}{k_1 a} \\ 0 \\ \dfrac{k_1}{\mu_0}\dfrac{\partial J_n(\eta a)}{\partial a} \end{bmatrix}$$

the Fourier integral. This technique is illustrated in the following section for the function of the first kind. The same can be applied to the third kind, provided that the pole of the integrand is far away from the saddle point. Otherwise, a more refined saddle-point method of integration is required.

25. ASYMPTOTIC EXPRESSION

To find the far-zone field of a radiating source in the presence of a perfectly conducting cylinder or the far-zone field of an aperture antenna on the surface of the cylinder we need only the asymptotic expression for $\overline{\overline{G}}_1$ or $\overline{\overline{G}}_2$. This expression can be found by the method of saddle-point integration. Let us consider the function of the first kind. Assuming that ηr is large compared to unity, the Hankel function in $\overline{M}^{(1)}$ and $\overline{N}^{(1)}$ can be approximated by its asymptotic expression, that is,

$$H_n^{(1)}(\eta r) \simeq \sqrt{\frac{2}{\pi \eta r}}\,(-i)^{n+\frac{1}{2}}\, e^{i\eta r} .$$

The functions $\overline{M}^{(1)}$ and $\overline{N}^{(1)}$, therefore, become

$$\overline{M}^{(1)}_{\substack{e\\o}n\eta}(h) \simeq (-i)^{n+\frac{3}{2}}\,\eta \sqrt{\frac{2}{\pi\eta r}}\, e^{i(\eta r + hz)} \begin{matrix}\cos\\ \sin\end{matrix}\, n\phi\, \hat{\phi} \tag{1}$$

$$\overline{N}^{(1)}_{\substack{e\\o}n\eta}(h) \simeq (-i)^{n+\frac{1}{2}}\,\frac{\eta}{k} \sqrt{\frac{2}{\pi\eta r}}\, e^{i(\eta r + hz)} \begin{matrix}\cos\\ \sin\end{matrix}\, n\phi\, (-h\hat{r} + \eta\hat{z}) . \tag{2}$$

The approximate expression for $\overline{\overline{G}}_1$, using (24-5a) and (24-6) through (24-9) with the functions $\overline{M}^{(1)}$ and $\overline{N}^{(1)}$ replaced by (1) and (2), can be written in the form

$$\begin{aligned}\overline{\overline{G}}_1(\overline{R}|\overline{R}') \simeq \frac{i}{4\pi}\int_{-\infty}^{\infty} dh \sum_{n=0}^{\infty} \frac{2-\delta_0}{\eta}\, \frac{1}{\sqrt{2\pi\eta r}}\,(-i)^{n+\frac{1}{2}}\, e^{i(\eta r + hz)} \\ \cdot \Big\{ -i \begin{matrix}\cos\\ \sin\end{matrix} n\phi\,\hat{\phi}\,[\overline{M}'_{\substack{e\\o}n\eta}(-h) + a_{\substack{e\\o}n}\overline{M}'^{(1)}_{\substack{e\\o}n\eta}(-h)] \\ + \frac{1}{k}\begin{matrix}\cos\\ \sin\end{matrix} n\phi(-h\hat{r} + \eta\hat{z})[\overline{N}'_{\substack{e\\o}n\eta}(-h) + b_{\substack{e\\o}n}\overline{N}'^{(1)}_{\substack{e\\o}n\eta}(-h)]\Big\} ,\end{aligned} \tag{3}$$

where terms of the order equal to and higher than $(\eta r)^{-\frac{3}{2}}$ have been neglected.

We now change the cylindrical variables into spherical variables by letting

$$\eta = k \sin\beta, \qquad h = k\cos\beta$$

$$r = R\sin\theta, \qquad z = R\cos\theta$$

hence

$$\eta r + hz = kR\cos(\theta - \beta)$$

$$-h\hat{r} + \eta\hat{z} = k(-\hat{r}\cos\beta + \hat{z}\sin\beta)$$

$$dh = -k\sin\beta d\beta.$$

Applying the integral formula (4-9) to (3), we obtain

$$\overline{\overline{G}}_1(\overline{R}|\overline{R}') \simeq \frac{e^{ikR}}{4\pi kR\sin\theta} \sum_{n=0}^{\infty} (2-\delta_0)(-i)^{n+1} \begin{matrix}\cos\\ \sin\end{matrix} n\phi \cdot \{\hat{\phi}[\overline{M}'_{\substack{e\\o}ns}(-k\cos\theta) + a_{\substack{e\\o}n}\overline{M}'^{(1)}_{\substack{e\\o}ns}(-k\cos\theta)] - i\hat{\theta}[\overline{N}'_{\substack{e\\o}ns}(-k\cos\theta) + b_n\overline{N}'^{(1)}_{\substack{e\\o}ns}(-k\cos\theta)]\} \tag{4}$$

where $s = k\sin\theta$, corresponding to the values of η evaluated at $\beta = \theta$. Equation (4), of course, is valid only when $kR \gg 1$. The singularity at $\theta = 0$, in practice, does not exist because θ never goes to zero due to the finite radius of the cylinder. Conversely, when $kR' \gg 1$, corresponding to a remote source, we can find the asymptotic expression for $\overline{\overline{G}}_1$ using (24-5b) and (24-6) through (24-9). The result would provide us the implicit solution of the field due to a plane wave incident on a conducting cylinder. We leave this problem as an exercise in Appendix D where the readers are requested to recover some of the classical formulas based on the present technique. As far as the actual field is concerned, it is of course necessary to use the formula

$$\overline{E}(\overline{R}) = i\omega\mu_0 \iiint \overline{\overline{G}}_1(\overline{R}|\overline{R}') \cdot \overline{J}(\overline{R}')\, dV' \tag{5}$$

to find the field for a current source, or

$$\overline{E}(\overline{R}) = \iint \nabla \times \overline{\overline{G}}_1(\overline{R}|\overline{R}') \cdot [\hat{n} \times \overline{E}(\overline{R}')]\, dS' \tag{6}$$

for an aperture field source, where $\hat{n}$ denotes the outward normal to the cylindrical surface.

chapter **8**

Flat Earth

The cylindrical vector wave functions introduced previously can also be used in the eigenfuntion expansion of the dyadic Green's functions associated with a flat earth. The free-space Green's function will be transformed into an integral form suitable for the construction of the function of the third or the fourth kind. The present formulation will be compared with Sommerfeld's classical formulation, and certain unique features of the dyadic Green's function technique will be pointed out. Finally, the asymptotic expression for the dyadic Green's function of the third kind will be derived.

26. DYADIC GREEN'S FUNCTION OF THE THIRD KIND FOR A FLAT EARTH

When a space is partitioned into two halves, one of which is filled with air and the other half with a homogeneous lossy dielectric medium as shown in Fig. 26-1, the geometry corresponds to that of a flat earth. We assume that the earth is characterized by the constitutive constants ε, μ_0, and σ. For convenience, we will designate the propagation constants in the two media, respectively by

$$k_1 = \omega\sqrt{\mu_0\varepsilon_0} \qquad \text{and} \qquad k_2 = \omega\sqrt{\mu_0\varepsilon(1 + i\sigma/\omega\varepsilon)}.$$

The very nature of the composition suggests that the pertinent dyadic Green's function under consideration is either the third kind or the fourth kind. To find these functions, we shall first transform the free-space dyadic Green's function into an integral form that enables us to construct these functions by the method of scattering superposition. For this reason we would like to eliminate the Fourier integration with respect to h in (24-3).

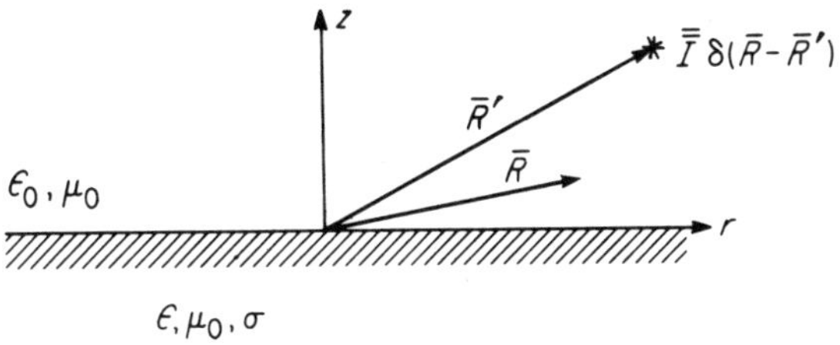

FIG. 26-1. Flat earth.

Applying the method of contour integration in the h-plane, we obtain, as a result of the residue theorem,

$$\bar{\bar{G}}_0(\bar{R}|\bar{R}') = \frac{i}{4\pi}\int_0^\infty d\lambda \sum_{n=0}^{\infty} \frac{2-\delta_0}{\lambda h_1} \tag{1}$$

$$\begin{cases} \bar{M}_{\substack{e\\o}n\lambda}(h_1)\bar{M}'_{\substack{e\\o}n\lambda}(-h_1) + \bar{N}_{\substack{e\\o}n\lambda}(h_1)\bar{N}'_{\substack{e\\o}n\lambda}(-h_1), & z \geqslant z' \\ \bar{M}_{\substack{e\\o}n\lambda}(-h_1)\bar{M}'_{\substack{e\\o}n\lambda}(h_1) + \bar{N}_{\substack{e\\o}n\lambda}(-h_1)\bar{N}'_{\substack{e\\o}n\lambda}(h_1), & z \leqslant z' \end{cases}$$

where

$$h_1 = \sqrt{k_1^2 - \lambda^2}.$$

To find the dyadic Green's function of the third kind, we let

$$\bar{\bar{G}}_3^{(11)}(\bar{R}|\bar{R}') = \bar{\bar{G}}_0(\bar{R}|\bar{R}') + \bar{\bar{G}}_{3s}^{(11)}(\bar{R}|\bar{R}'), \qquad z \geqslant 0 \tag{2}$$

$$\bar{\bar{G}}_3^{(21)}(\bar{R}|\bar{R}') = \bar{\bar{G}}_{3s}^{(21)}(\bar{R}|\bar{R}'), \qquad z \leqslant 0. \tag{3}$$

In order to satisfy the radiation condition and the boundary conditions at the interface the two scattering terms must have the following forms:

$$\bar{\bar{G}}_{3s}^{(11)}(\bar{R}|\bar{R}') = \frac{i}{4\pi}\int_0^\infty d\lambda \sum_{n=0}^{\infty} \frac{2-\delta_0}{\lambda h_1} \tag{4}$$

$$\cdot \left[a\bar{M}_{\substack{e\\o}n\lambda}(h_1)\bar{M}'_{\substack{e\\o}n\lambda}(h_1) + b\bar{N}_{\substack{e\\o}n\lambda}(h_1)\bar{N}'_{\substack{e\\o}n\lambda}(h_1)\right]$$

$$\bar{\bar{G}}_{3s}^{(21)}(\bar{R}|\bar{R}') = \frac{i}{4\pi}\int_0^\infty d\lambda \sum_{n=0}^{\infty} \frac{2-\delta_0}{\lambda h_1} \tag{5}$$

$$\cdot \left[c\bar{M}_{\substack{e\\o}n\lambda}(-h_2)\bar{M}'_{\substack{e\\o}n\lambda}(h_1) + d\bar{N}_{\substack{e\\o}n\lambda}(-h_2)\bar{N}'_{\substack{e\\o}n\lambda}(h_1)\right]$$

where $h_2 = \sqrt{k_2^2 - \lambda^2}$.

We use $\bar{M}(h_1)$ and $\bar{N}(h_1)$ as the anterior elements for $\bar{\bar{G}}_{3s}^{(11)}$ in order to satisfy the radiation condition at $z \to \infty$. $\bar{M}(-h_2)$ and $\bar{N}(-h_2)$ were chosen for $\bar{\bar{G}}_{3s}^{(21)}$ because they are solutions for the vector wave equation in the earth region and satisfy the radiation condition at $z \to -\infty$. The posterior elements of these functions must be the same as those of $\bar{\bar{G}}_0$ defined for $z < z'$ so that the boundary conditions at the interface can be satisfied for any excitation. According to (17-39) and (17-40), the boundary conditions required for $\bar{\bar{G}}_3$ at $z=0$ under the condition that $\mu_1 = \mu_2 = \mu_0$ as assumed here are:

(6) $$\hat{z} \times \bar{\bar{G}}_3^{(11)} = \hat{z} \times \bar{\bar{G}}_3^{(21)}$$

(7) $$\hat{z} \times \nabla \times \bar{\bar{G}}_3^{(11)} = \hat{z} \times \nabla \times \bar{\bar{G}}_3^{(21)}$$

which yield

$$1 + a = c$$

$$h_1(1-a) = h_2 c$$

$$\frac{h_1}{k_1}(1-b) = \frac{h_2}{k_2} d$$

$$k_1(1+b) = k_2 d$$

hence

(8) $$a = \frac{h_1 - h_2}{h_1 + h_2}$$

(9) $$b = \frac{k_2^2 h_1 - k_1^2 h_2}{k_2^2 h_1 + k_1^2 h_2} = \frac{n^2 h_1 - h_2}{n^2 h_1 + h_2}$$

where n denotes the complex index of refraction of the earth medium,

(10) $$c = \frac{2h_1}{h_1 + h_2}$$

(11) $$d = \frac{2k_1 k_2 h_1}{k_2^2 h_1 + k_1^2 h_2} = \frac{2n h_1}{n^2 h_1 + h_2}$$

recalling that

$$h_1 = \sqrt{k_1^2 - \lambda^2} \qquad \text{and} \qquad h_2 = \sqrt{k_2^2 - \lambda^2}.$$

We have thus obtained an integral representation of the dyadic Green's function of the third kind for the flat earth. The complete expressions for $\bar{\bar{G}}_3^{(11)}$ and $\bar{\bar{G}}_3^{(21)}$ are given below:

(12) $$\bar{\bar{G}}_3^{(11)}(\bar{R}|\bar{R}') = \frac{i}{4\pi}\int_0^\infty \frac{d\lambda}{\lambda h_1} \sum_{n=0}^{\infty} (2-\delta_0)$$

$$\begin{cases} \bar{M}_{\substack{e\\o}n\lambda}(h_1)\,[\bar{M}'_{\substack{e\\o}n\lambda}(-h_1) + a\bar{M}'_{\substack{e\\o}n\lambda}(h_1)] + \bar{N}_{\substack{e\\o}n\lambda}(h_1)\,[\bar{N}'_{\substack{e\\o}n\lambda}(-h_1) + b\bar{N}'_{\substack{e\\o}n\lambda}(h_1)] & z \geqslant z', \\ [\bar{M}_{\substack{e\\o}n\lambda}(-h_1) + a\bar{M}_{\substack{e\\o}n\lambda}(h_1)]\,\bar{M}'_{\substack{e\\o}n\lambda}(h_1) + [\bar{N}_{\substack{e\\o}n\lambda}(-h_1) + b\bar{N}_{\substack{e\\o}n\lambda}(h_1)]\,\bar{N}'_{\substack{e\\o}n\lambda}(h_1) & z' \geqslant z \geqslant 0, \end{cases}$$

(13) $$\bar{\bar{G}}_3^{(21)} = \frac{i}{4\pi}\int_0^\infty \frac{d\lambda}{\lambda h_1} \sum_{n=0}^{\infty} (2-\delta_0)$$

$$\cdot\{c\bar{M}_{\substack{e\\o}n\lambda}(-h_2)\bar{M}'_{\substack{e\\o}n\lambda}(h_1) + d\bar{N}_{\substack{e\\o}n\lambda}(-h_2)\bar{N}'_{\substack{e\\o}n\lambda}(h_1)\}, \qquad z \leqslant 0$$

where a,b,c and d are given by (8–11).

The derivation of the fourth kind is practially the same except that the boundary conditions at the interface should be replaced by (17-30) and (17-31). Knowing $\bar{\bar{G}}_3$, we can find the electric field due to a given current distribution through the formulas

(14) $$\bar{E}_1(\bar{R}) = i\omega\mu_0 \iiint \bar{\bar{G}}_3^{(11)}(\bar{R}|\bar{R}')\cdot \bar{J}_1(\bar{R}')\,dV'$$

(15) $$\bar{E}_2(\bar{R}) = i\omega\mu_0 \iiint \bar{\bar{G}}_3^{(21)}(\bar{R}|\bar{R}')\cdot \bar{J}_1(\bar{R}')\,dV'$$

where we have assumed that $\bar{J}_1$ is located in the air. If the current source is located in the earth what we need would be the functions $\bar{\bar{G}}_3^{(22)}$ and $\bar{\bar{G}}_3^{(12)}$. These can be obtained by interchanging the roles of k_1 and k_2. Two simple examples are given in the following section to illustrate the resultant field.

27. RADIATION FROM SHORT VERTICAL AND HORIZONTAL ELECTRIC DIPOLES; COMPARISON WITH SOMMERFELD'S FORMULATION

For a short (infinitesimal) vertical electric dipole with current moment $c\hat{z}$ located at $(0,0,z_0)$ in air, we write

$$\bar{J}_1(\bar{R}') = c\hat{z}\delta(x-0)\delta(y-0)\delta(z-z_0).$$

In order to compare with Sommerfeld's result later, we choose c to be numerically equal to $4\pi k_1^2/i\omega\mu_0$. Then according to (26-14), we obtain

$$\bar{E}_1(\bar{R}) = 4\pi k_1^2 \bar{\bar{G}}_3^{(11)}(\bar{R}|\bar{R}') \cdot \hat{z}|_{\bar{R}'=(0,0,z_0)} \tag{1}$$

Using the expression for $\bar{\bar{G}}_3^{(11)}$ found in the previous section and the expressions for $\bar{M}'_{\substack{e\\o}n\lambda}$ and $\bar{N}'_{\substack{e\\o}n\lambda}$ defined by (21-5) and (21-6) but in the primed variables, we find that only $\bar{N}'_{e0\lambda} \cdot \hat{z}$ survives when we let $\bar{R}' = (0, 0, z_0)$. For simplicity, let us consider just the case corresponding to $z > z_0$, then

$$\bar{E}_1(\bar{R}) = ik_1 \int_0^\infty \frac{\lambda}{h_1} \bar{N}_{e0\lambda}(h_1)\,[e^{-ih_1z_0} + be^{ih_1z_0}]\,d\lambda \tag{2}$$

where

$$b = \frac{n^2h_1 - h_2}{n^2h_1 + h_2}.$$

For a short horizontal electric dipole with the same current moment pointed in the x-direction, we have

$$\bar{E}_1(\bar{R}) = 4\pi k_1^2 \bar{\bar{G}}_3(\bar{R}|\bar{R}') \cdot \hat{x}|_{\bar{R}'=(0,0,z_0)} \tag{3}$$

In this case, only $\bar{M}_{o1\lambda}$ and $\bar{N}_{e1\lambda}$ survive, and we obtain

$$\bar{E}_1(\bar{R}) = ik_1 \int_0^\infty \left\{ \frac{k_1}{h_1} \bar{M}_{o1\lambda}(h_1)\,[e^{-ih_1z_0} + ae^{ih_1z_0}] \right. \\ \left. - i\bar{N}_{e1\lambda}(h_1)\,[e^{-ih_1z_0} - be^{ih_1z_0}] \right\} d\lambda \tag{4}$$

where b is the same as the one defined previously and

$$a = \frac{h_1 - h_2}{h_1 + h_2}.$$

We shall show now that the expressions for the electric field as given by (2) and (3) are equivalent to the ones obtained by Sommerfeld in his famous work as treated very completely in Chapter 6 of his book on Partial Differential Equations (1949, pp. 236–261).

Sommerfeld formulated these problems by the method of potentials. He used an electric Hertzian potential alone, so that

$$\bar{E}_1 = k_1^2\bar{\pi} + \nabla\nabla \cdot \bar{\pi}. \tag{5}$$

For $\bar{R} \neq \bar{R}'$, (5) is equivalent to

(6) $$\bar{E}_1(\bar{R}) = \nabla \times \nabla \times \bar{\pi}\,.$$

In the case of a vertical dipole he showed that a z-component of $\bar{\pi}$ alone is sufficient to formulate and to solve the problem. The resultant expression for π_z, including both the primary and the scattered field, for $z > z_0$, is given by

(7) $$\pi_z = i\int_0^\infty \frac{\lambda}{h_1} J_0(\lambda r) e^{ih_1 z} (e^{-ih_1 z_0} + be^{ih_1 z_0})\, d\lambda\,.$$

We have written all his parameters in our notation. Substituting (7) into (6), we obtain

(8) $$\bar{E}_1(\bar{R}) = ik_1 \int_0^\infty \frac{\lambda}{h_1} \bar{N}_{e0\lambda}(h_1) (e^{-ih_1 z_0} + be^{ih_1 z_0})\, d\lambda$$

which is identical to (2).

In the case of a horizontal dipole, he showed that two components of $\bar{\pi}$, corresponding to π_x and π_z, are required. For $z > z_0$, they are

(9) $$\pi_x = i\int_0^\infty \frac{\lambda}{h_1} J_0(\lambda r) e^{ih_1 z} (e^{-ih_1 z_0} + ae^{ih_1 z_0})\, d\lambda$$

and

(10) $$\pi_z = -2\cos\phi \int_0^\infty \frac{\lambda^2 (h_1 - h_2)}{k_2^2 h_1 + k_1^2 k_2} J_1(\lambda r) e^{ih_1(z+z_0)}\, d\lambda\,.$$

It is obvious that π_z and the part of π_x containing the coefficient a represent the scattered field. Substituting (9) and (10) into (6), we obtain

(11) $$\bar{E}_1(\bar{R}) = ik_1 \int_0^\infty \frac{\lambda}{h_1} \bar{N}^{(x)}_{e0\lambda}(h_1) (e^{-ih_1 z_0} + ae^{ih_1 z_0})\, d\lambda$$

$$- 2k_1 \int_0^\infty \frac{\lambda^2 (h_1 - h_2)}{k_2^2 h_1 + h_1^2 h_2} \bar{N}_{e1\lambda}(h_1) e^{ih_1 z_0}\, d\lambda$$

where $\bar{N}^{(x)}_{e0\lambda}$ represents a cylindrical vector wave function of the x-type defined by

$$\bar{N}^{(x)}_{e0\lambda}(h_1) = \frac{1}{k_1} \nabla \times \nabla \times [\, J_0(\lambda r) e^{ih_1 z} \hat{x}]$$

As shown in Appendix B, this type of function is related to the functions of the z-type which we have used exclusively in the construction of the dyadic

Green's function. In particular, when $n=0$, Equation (18) of Appendix B becomes

$$\bar{N}^{(x)}_{e0\lambda}(h_1)=\frac{1}{\lambda}\left[k_1\,\bar{M}_{o1\lambda}(h_1)-ih_1\,\bar{N}_{e1\lambda}(h_1)\right].$$

Substituting it into (11) and making use of the identity that

$$\frac{2\lambda^2(h_1-h_2)}{k_2^2h_1+k_1^2h_2}=a+b$$

where

$$a=\frac{h_1-h_2}{h_1+h_2},\quad b=\frac{k_2^2h_1-k_1^2h_2}{k_2^2h_1+k_1^2h_2},\quad h_1=\sqrt{k_1^2-\lambda^2},\quad \text{and } h_2=\sqrt{k_2^2-\lambda^2}$$

we find that (11), indeed, is identical to (4).

In retrospect, we see that Sommerfeld's original formulation used only the electric Hertzian potential. The formulation discussed here, from the point of view of potential theory, uses both the electric Hertzian potential and the magnetic Hertzian potential. The former generates the $\bar{N}$ functions and the latter generates the $\bar{M}$ functions. Whereas Sommerfeld's treatment is very ingenious, ours is more or less methodical. One unique feature of the dyadic Green's function formulation is that for any other current distributions the same technique applies. In doing so we have bypassed the need of determining specifically the potential functions.

Finally, we would like to mention the fine book written by Baños (1966) on dipole radiation in the presence of a flat earth. He has systematically reduced the various integrals into some basic ones for the convenience of numerical computation as well as for correlating different problems. Undoubtedly the integrals resulting from the present formulation can all be expressed in terms of Baños' basic functions. It is beyond the scope of this book to discuss the numerical aspect of these problems. An extension of Baños' treatment to multipole radiation based on the present formulation may be of interest to students working in the field.

28. ASYMPTOTIC EXPRESSION

Under certain conditions to be stated later, we can find the asymptotic expression for the dyadic Green's function of the third kind by the method of saddle-point integration. To derive this expression we shall first transform the integral representation of $\bar{\bar{G}}_3^{(11)}$ as given by (26-12) from a semi-infinite path to an infinite path in the λ-plane.

We consider an integral of the type

$$F = \int_0^\infty f_n(\lambda) J_n(\lambda r)\, d\lambda \tag{1}$$

which can be written in the form

$$F = \frac{1}{2} \int_0^\infty f_n(\lambda)\, [H_n^{(1)}(\lambda r) + H_n^{(2)}(\lambda r)]\, d\lambda$$

where $H_n^{(1)}$ and $H_n^{(2)}$ denote, respectively, the Hankel functions of the first and the second kind. The integral involving the second kind in (2) can be transformed into the first kind with a different path as follows:

$$\frac{1}{2} \int_0^\infty f_n(\lambda) H_n^{(2)}(\lambda r)\, d\lambda$$

$$= \frac{1}{2} \int_{-\infty}^0 f_n(-\lambda) H_n^{(2)}(-\lambda r)\, d\lambda = -\frac{1}{2} \int_{-\infty}^0 f_n(-\lambda) e^{in\pi} H_n^{(1)}(\lambda r)\, d\lambda$$

where we have made use of the half-circuit relation between the two kinds of Hankel functions (Sommerfeld, 1949, p. 315). If $f_n(\lambda)$ satisfies the relation that

$$f_n(\lambda) = -e^{in\pi} f_n(-\lambda) \tag{2}$$

then the original integral becomes

$$F = \int_{-\infty}^\infty f_n(\lambda) H_n^{(1)}(\lambda r)\, dr \tag{3}$$

The transform from (1) to (3) can be extended to a dyadic function by an operational method. Thus, if we have

$$F = \int_0^\infty \bar{\bar{f}}_n(\lambda) [J_n(\lambda r)]\, d\lambda \tag{4}$$

where $\bar{\bar{f}}_n$ denotes a dyadic operator which satisfies the relation that

$$\bar{\bar{f}}_n(\lambda) = -e^{in\pi} \bar{\bar{f}}_n(-\lambda) \tag{5}$$

then

$$\bar{\bar{F}} = \int_{-\infty}^\infty \bar{\bar{f}}_n(\lambda) [H_n^{(1)}(\lambda r)]\, d\lambda. \tag{6}$$

Now we consider a typical term of $\bar{\bar{G}}_3^{(11)}$ described by (26-12) for $z \geqslant z_0$, such as

$$\overline{\overline{g}} = \frac{i}{4\pi} \int_0^\infty \frac{1}{\lambda h_1} \bar{M}_{en\lambda}(h_1) \bar{M}'_{en\lambda}(-h_1)\, d\lambda. \tag{7}$$

Since $\bar{M}_{en\lambda}(h_1)$ contains $J_n(\lambda r)$ and $\bar{M}'_{en\lambda}(-h_1)$ contains $J_n(\lambda r')$, (7) can be written in an operational form as (4) and the operator does satisfy the relation as described by (5). Consequently, in view of (6), (7) can be transformed to

$$\overline{\overline{g}} = \frac{i}{4\pi} \int_{-\infty}^\infty \frac{1}{\lambda h_1} \bar{M}^{(1)}_{en\lambda}(h_1) \bar{M}'_{en\lambda}(-h_1)\, d\lambda \tag{8}$$

where $\bar{M}^{(1)}$ is defined with respect to the Hankel function of the first kind. The operational method avoids the long and tedious task of applying (3) repeatedly to each individual term contained in the dyadic function $\bar{M}\bar{M}'$. In view of (8) and similar transformations for the other terms, (26-12) can be written in the form

$$\begin{aligned} \overline{\overline{G}}_3(\bar{R}|\bar{R}') = \frac{i}{4\pi} \int_{-\infty}^\infty \frac{d\lambda}{\lambda h_1} \sum_{n=0}^\infty (2-\delta_0) \\ \cdot \{\bar{M}^{(1)}_{\substack{e\\o}n\lambda}(h_1)\, [\bar{M}'_{\substack{e\\o}n\lambda}(-h_1) + a\bar{M}'_{\substack{e\\o}n\lambda}(h_1)] + \bar{N}^{(1)}_{\substack{e\\o}n\lambda}(h_1)\, [\bar{N}'_{\substack{e\\o}n\lambda}(-h_1) + b\bar{N}'_{\substack{e\\o}n\lambda}(h_1)]\}, \\ z \geqslant z'. \end{aligned} \tag{9}$$

Having changed the contour of integration into an infinite path, we can apply the method of saddle-point integration to (9) provided that $kR \gg 1$. Following the same procedure as described in Sec. 25, we found that the first-order solution for the far-zone field is given by the following expression

$$\begin{aligned} \overline{\overline{G}}{}^{(11)}_3(\bar{R}|\bar{R}') \simeq \frac{e^{ikR}}{4\pi kR \sin\theta} \sum_{n=0}^\infty (2-\delta_0)(-i)^{n+1} \begin{matrix}\cos\\ \sin\end{matrix} n\phi \\ \cdot \{\hat{\phi}[\bar{M}'_{\substack{e\\o}ns}(-k_1\cos\theta) + a(\theta)\bar{N}'_{\substack{e\\o}ns}(k_1\cos\theta)] \\ - i\hat{\theta}\,[\bar{N}'_{\substack{e\\o}ns}(-k_1\cos\theta) + b(\theta)\bar{N}'_{\substack{e\\o}ns}(k_1\cos\theta)]\} \end{aligned} \tag{10}$$

where

$$s = k_1 \sin\theta$$

$$a(\theta) = \frac{k_1\cos\theta - \sqrt{k_2^2 - k_1^2\sin^2\theta}}{k_1\cos\theta + \sqrt{k_2^2 - k_1^2\sin^2\theta}}$$

$$b(\theta) = \frac{n^2 k_1\cos\theta - \sqrt{k_2^2 - k_1^2\sin^2\theta}}{n^2 k_1\cos\theta + \sqrt{k_2^2 - k_1^2\sin^2\theta}}$$

$\theta =$ angle between $\hat{R}$ and $\hat{z}$.

It is recognized that the coefficients $a(\theta)$ and $b(\theta)$ play the same role as the plane-wave reflection coefficients for an incident E-field either perpendicular or parallel to the plane of incidence. Being the first-order solution, it does not take into consideration the proximity effect when the saddle point is very close to the poles of the coefficients a and b lying in the λ-plane. In practice, (10) only represents the contribution due to the so-called *space wave*. It is a valid approximation when the point of observation is not near the ground. For a detailed discussion of the high-order solutions, readers are referred to Baños' book quoted previously, and the detailed discussion by Feynberg (1961) on the field near the ground, or the so-called *ground wave*.

chapter **9**

Perfectly Conducting Elliptical Cylinder

Vector wave functions in an elliptical cylinder coordinate system are generated when the elliptical cylinder scalar wave functions are used. Once the orthogonal properties of these functions are known, we can find the eigenfunction expansion of the free-space dyadic Green's function without any difficulty. The function of the first and second kind can then be constructed by the method of scattering superposition. Because both the angular functions and the radial functions depend on the wave number it is impossible to find an orthogonal expansion for the functions of the third and fourth kind. Problems involving a dielectric elliptical cylinder, therefore, cannot be formulated in the same manner as those involving a dielectric circular cylinder.

29. VECTOR WAVE FUNCTIONS IN AN ELLIPTICAL CYLINDER COORDINATE SYSTEM

The scalar wave equation in an elliptical cylinder coordinate system can be written in the form

$$\frac{1}{\beta^2}\left[\frac{\partial^2\psi}{\partial u^2}+\frac{\partial^2\psi}{\partial v^2}\right]+\frac{\partial^2\psi}{\partial z^2}+\kappa^2\psi=0 \tag{1}$$

where

$$\beta = c\,(\cos\text{h}^2\,u-\cos^2 v)^{1/2}$$

The variables u, v, z and the parameter c are defined in Sec. 7 and illustrated in Fig. 1-3. Equation (1) is derived by using (28), (29), and (33) of Appendix A. The eigenfunctions associated with (1), called the elliptical cylinder wave functions, have been discussed very thoroughly by Stratton (1941, pp. 375–387). We shall follow very closely Stratton's presentation with, however, some minor changes of notation in order to conform with our previous notation. These changes will be obvious to the readers when the two texts are compared. In order to give a full treatment of the orthogonal property of the vector wave functions in elliptical cylinder coordinate systems a brief review of the scalar wave functions is necessary. According to Stratton, the eigenfunctions pertaining to (1) can be written in the form

$$\psi_{{}^e_o m\lambda}(h) = S_{{}^e_o m\lambda}(v) R_{{}^e_o m\lambda}(u) e^{ihz} \tag{2}$$

where

$$h^2 + \lambda^2 = \kappa^2 .$$

The angular functions $S_{{}^e_o m\lambda}$ and the radial functions $R_{{}^e_o m\lambda}$ satisfy, respectively, the Mathieu equations

$$\frac{d^2 S_{{}^e_o m\lambda}(v)}{dv^2} + (b_{{}^e_o m} - c^2\lambda^2 \cos^2 v) S_{{}^e_o m\lambda}(v) = 0 \tag{3}$$

$$\frac{d^2 R_{{}^e_o m\lambda}(u)}{d\, u^2} - (b_{{}^e_o m} - c^2\lambda^2 \cosh^2 u) R_{{}^e_o m\lambda}(u) = 0 . \tag{4}$$

The eigenvalues b_{em} or b_{om} form a denumerable set such that the corresponding angular functions will be periodic functions of v. We require the angular functions to be periodic with respect to v so that the field represented by these functions would be a single-valued function of position. These periodic angular fuctions can be represented by either the cosine series in the case of even functions or the sine series in the case of odd functions. They are

$$S_{em\lambda}(v) = \sum_n{}' D_n^m \cos nv, \qquad (m = 1,2,3 \cdots) \tag{5}$$

and

$$S_{om\lambda}(v) = \sum_n{}' F_n^m \sin nv, \qquad (m = 1,2,3 \cdots) \tag{6}$$

where the primed summation is to be extended over even values of n if m is even and odd values of n if m is odd. The coefficients D_n^m and F_n^m are so normalized that

$$\sum_n{}' D_n^m = 1 \;\text{ or }\; S_{em\lambda}(0) = 1 \tag{7}$$

$$\sum_n{}' nF_n^m = 1 \;\text{ or }\; \frac{d}{dv} S_{om\lambda}(v)\big|_{v=0} = 1 . \tag{8}$$

Using (3) and (4), we can show that S_{em} and S_{om} form a complete orthogonal set, that is,

$$\int_0^{2\pi} S_{em\lambda}(v)S_{om'\lambda}(v)\,dv = 0 \tag{9}$$

$$\int_0^{2\pi} S_{em\lambda}(v)S_{em'\lambda}(v)\,dv = \begin{cases} 0, & m \neq m' \\ I_{em\lambda} & m = m' \end{cases} \tag{10}$$

with

$$I_{em\lambda} = \pi \sum_n{}' (1+\delta_0)(D_n^m)^2$$

$$\int_0^{2\pi} S_{om\lambda}(v)S_{om'\lambda}(v)\,dv = \begin{cases} 0, & m \neq m' \\ I_{om\lambda}, & m = m' \end{cases} \tag{11}$$

with

$$I_{om\lambda} = \pi \sum_n{}' (F_n^m)^2.$$

The radial functions which are solutions of (4) and finite at the origin can be written in the form of the series of Bessel functions. They are

$$R_{em\lambda}(u) = \sqrt{\frac{\pi}{2}} \sum_n{}' (i)^{m-n} D_n^m J_n(c\lambda \cosh u) \tag{12}$$

$$R_{om\lambda}(u) = \sqrt{\frac{\pi}{2}} \tanh u \sum_n{}' (i)^{n-m} F_n^m J_n(c\lambda \cosh u). \tag{13}$$

We have omitted the superscript 1 as used by Stratton in the designation of these functions. Two other radial functions which would represent outgoing waves in the $e^{-i\omega t}$ system involve the Hankel function of the first kind. They are given by the series

$$R_{em\lambda}^{(1)}(u) = \sqrt{\frac{\pi}{2}} \sum_n{}' (i)^{m-n} D_n^m H_n^{(1)}(c\lambda \cosh u) \tag{14}$$

$$R_{om\lambda}^{(1)}(u) = \sqrt{\frac{\pi}{2}} \tanh u \sum_n{}' (i)^{n-m} F_n^m H_n^{(1)}(c\lambda \cosh u). \tag{15}$$

In Stratton's original work, these functions are designated by $R_{\substack{e\\o} m\lambda}^{(3)}$. It is hoped that these minor changes of notation will not cause any inconvenience to readers. For the determination of the normalization constant for the vector wave functions we need the expansion formulae of the elliptical wave functions in terms of the circular wave functions. They are given by (Stratton, p. 386).

(16) $$R_{em\lambda}(u)S_{em\lambda}(v) = \sqrt{\frac{\pi}{2}}\sum_n{}' (i)^{n-m}D_n^m \cos n\phi\, J_n(\lambda r)$$

and

(17) $$R_{om\lambda}(u)S_{om\lambda}(v) = \sqrt{\frac{\pi}{2}}\sum_n{}' (i)^{n-m}F_n^m \sin n\phi\, J_n(\lambda r).$$

The above brief review is essential for our discussion of the corresponding vector wave functions. These functions are defined by

(18) $$\bar{M}_{\substack{e\\o}m\lambda}(h) = \nabla \times [\psi_{\substack{e\\o}m\lambda}(h)\hat{z}]$$

(19) $$\bar{N}_{\substack{e\\o}m\lambda}(h) = \frac{1}{\kappa}\nabla \times \bar{M}_{\substack{e\\o}m\lambda}(h)$$

where $\psi_{\substack{e\\o}m\lambda}(h)$ is defined by (2). They are, of course, solutions of the vector wave equations in elliptical-cylinder coordinate systems. The complete expressions of these functions are given by

(20) $$\bar{M}_{\substack{e\\o}m\lambda}(h) = \frac{1}{\beta}\left[R_{\substack{e\\o}m\lambda}\frac{\partial S_{\substack{e\\o}m\lambda}}{\partial v}\hat{u} - S_{\substack{e\\o}m\lambda}\frac{\partial R_{\substack{e\\o}m\lambda}}{\partial u}\hat{v}\right]e^{ihz}$$

(21) $$\bar{N}_{\substack{e\\o}m\lambda}(h) = \frac{1}{\kappa\beta}\left[ihS_{\substack{e\\o}m\lambda}\frac{\partial R_{\substack{e\\o}m\lambda}}{\partial u}\hat{u} + ihR_{\substack{e\\o}m\lambda}\frac{\partial S_{\substack{e\\o}m\lambda}}{\partial v}\hat{v} + \beta\lambda^2 R_{\substack{e\\o}m\lambda}S_{\substack{e\\o}m\lambda}\hat{z}\right]e^{ihz}.$$

The orthogonal properties of these functions are stated by the following equations:

(22) $$\iiint \bar{M}_{\substack{e\\o}m\lambda}(h)\cdot \bar{N}_{\substack{e\\o}m'\lambda'}(-h')\,dV = 0$$

(23) $$\iiint \bar{M}_{\substack{e\\o}m\lambda}(h)\cdot \bar{M}_{\substack{e\\o}m'\lambda'}(-h')\,dV = \begin{cases} 0, & m \neq m' \\ \pi^2\lambda I_{\substack{e\\o}m\lambda}\delta(h-h')\delta(\lambda-\lambda'), & m=m' \end{cases}$$

(24) $$\iiint \bar{N}_{\substack{e\\o}m\lambda}(h)\cdot \bar{N}_{\substack{e\\o}m'\lambda'}(-h')\,dV = \begin{cases} 0, & m \neq m' \\ \pi^2\lambda I_{\substack{e\\o}m\lambda}\delta(h-h')\delta(\lambda-\lambda'), & m=m' \end{cases}$$

where $I_{\substack{e\\o}m\lambda}$ are defined by (10) and (11).

These relations are analogous to (23-1) through (23-3) except that the normalization constants are different. The fine structure, as characterized by the two delta functions in the h and λ domain, is exactly the same. The orthogonal relations can easily be proved because of (9) through (11). It is sufficient for us to prove (23) for the case $m = m'$. The proof of (24) is given as an exercise. We consider, for example, the integral

$$I = \iiint \bar{M}_{em\lambda}(h) \cdot \bar{M}_{em\lambda'}(-h')\, dV$$

where the volume of integration is extended through the entire space. Using (2) and the differential relation that

$$\begin{aligned} dV &= h_1 h_2 h_3 \; du \; dv \; dz \\ &= \beta^2 \; du \; dv \; dz \end{aligned}$$

we have

$$I = \int_0^\infty \int_0^{2\pi} \int_{-\infty}^\infty \left[R_{em\lambda} R_{em\lambda'} \frac{\partial S_{em\lambda}}{\partial v} \frac{\partial S_{em\lambda'}}{\partial v} + S_{em\lambda} S_{em\lambda'} \frac{\partial R_{em\lambda}}{\partial u} \frac{\partial R_{em\lambda'}}{\partial u} \right] e^{i(h-h')z} \, du \, dv \, dz. \tag{25}$$

To simplify (25) we first make use of the relations that

$$\begin{aligned} \int_0^{2\pi} \frac{\partial S_{em\lambda}}{\partial v} \frac{\partial S_{em\lambda'}}{\partial v} dv - \frac{\partial S_{em\lambda}}{\partial v} S_{em\lambda'} \Big|_0^{2\pi} - \int_0^{2\pi} S_{em\lambda'} \frac{\partial^2 S_{em\lambda}}{\partial v^2} dv \\ = \int_0^{2\pi} (b_{em} - c^2 \lambda^2 \cos^2 v) S_{em\lambda} S_{em\lambda'} \, dv \end{aligned}$$

and

$$\begin{aligned} \int_0^\infty \frac{\partial R_{em\lambda}}{\partial u} \frac{\partial R_{em\lambda'}}{\partial u} du = R_{em\lambda'} \frac{\partial R_{em\lambda}}{\partial u} \Big|_0^\infty - \int_0^\infty R_{em\lambda'} \frac{\partial^2 R_{em\lambda}}{\partial u^2} du \\ = - \int_0^\infty (b_{em} - c^2 \lambda^2 \cosh^2 u) R_{em\lambda} R_{em\lambda'} \, du. \end{aligned}$$

Thus (25) can be written in the form

$$I = \int_0^\infty \int_0^{2\pi} \int_{-\infty}^\infty \lambda^2 \beta^2 R_{em\lambda} R_{em\lambda'} S_{em\lambda} S_{em\lambda'} e^{i(h-h')z} \, du \, dv \, dz$$

where

$$\beta = c\,(\cosh^2 u - \cos^2 v)^{1/2}$$

Using (16), and changing the domain of integration from the elliptical-cylinder system to the circular-cylinder system with the relation that

$$\beta^2 du\, dv = r\, dr\, d\phi$$

we obtain

$$I = \pi^2\delta(h-h')\int_0^\infty \int_0^{2\pi} \lambda^2 \left[\sum_n{}' (i)^{m-n} D_n^m(\lambda) J_n(\lambda r) \cos n\phi\right] \cdot \left[\sum_n{}' (i)^{m-n} D_n^m(\lambda') J_n(\lambda' r) \cos n\phi\right] r\, dr\, d\phi .$$

We have put into evidence that the coefficients D_n^m are functions of λ. As a result of the trigonomic functions and the integral representation of the weighted two-dimensional delta function, we obtain

$$I = (1+\delta_0)\pi^3 \lambda \delta(h-h')\delta(\lambda-\lambda') \sum_n{}' [D_n^m(\lambda)]^2 = \pi^2 \lambda I_{en\lambda} \delta(h-h')\delta(\lambda-\lambda')$$

where $I_{en\lambda}$ is defined by (10). The proofs of the remaining relations in (23) and (24) are similar. Once the orthogonal properties of the vector wave functions are known, the way to the eigenfunction expansion of the free-space dyadic Green's function is straightforward.

30. THE DYADIC GREEN'S FUNCTION OF THE FIRST KIND FOR A PERFECTLY CONDUCTING ELLIPTICAL CYLINDER

Following the procedures described in Sec. 24 for the perfectly conducting circular cylinder, we shall first obtain the eigenfunction expansion for the free-space dyadic Green's function. We let

$$(1)\quad \bar{\bar{I}}\delta(\bar{R}-\bar{R}') = \int_0^\infty d\lambda \int_{-\infty}^\infty dh \sum_m \left[\bar{M}_{{}^e_o m\lambda}(h)\bar{A}_{{}^e_o m\lambda}(h) + \bar{N}_{{}^e_o m\lambda}(h)\bar{B}_{{}^e_o m\lambda}(h)\right] .$$

The unknown coefficients $\bar{A}$ and $\bar{B}$ are determined by taking the anterior product of (1) with respect to $\bar{M}_{{}^e_o m'\lambda'}(-h')$ and $\bar{N}_{{}^e_o m'\lambda'}(-h')$ in turn and integrating the resultant equations through the entire space. As a result of (29-22) through (29-24) we obtain

$$\bar{A}_{{}^e_o m\lambda}(h) = \frac{1}{\pi^2 \lambda I_{{}^e_o m\lambda}} \bar{M}'_{{}^e_o m\lambda}(-h)$$

$$\bar{B}_{{}^e_o m\lambda}(h) = \frac{1}{\pi^2 \lambda I_{{}^e_o m\lambda}} \bar{N}'_{{}^e_o m\lambda}(-h) .$$

Thus we have

$$\bar{\bar{G}}_0(\bar{R}|\bar{R}') = \frac{1}{\pi^2}\int_0^\infty d\lambda \int_{-\infty}^{\infty} dh \sum_m \frac{1}{\lambda(h^2+\lambda^2-k^2)I_{\substack{e\\o}m\lambda}} \cdot [\bar{M}_{\substack{e\\o}m\lambda}(h)\bar{M}'_{\substack{e\\o}m\lambda}(-h) + \bar{N}_{\substack{e\\o}m\lambda}(h)\bar{N}'_{\substack{e\\o}m\lambda}(-h)]\,. \tag{2}$$

For the purpose of constructing the function of the first kind, we eliminate the λ-integration using the same technique as described in Sec. 24 in the derivation of (24-5) which yields

$$\bar{\bar{G}}_0(\bar{R}|\bar{R}') = \frac{i}{2\pi}\int_{-\infty}^{\infty} dh \sum_m \frac{1}{\eta^2 I_{\substack{e\\o}m\eta}} \cdot \begin{cases} \bar{M}^{(1)}_{\substack{e\\o}m\eta}(h)\bar{M}'_{\substack{e\\o}m\eta}(-h) + \bar{N}^{(1)}_{\substack{e\\o}m\eta}(h)\bar{N}'_{\substack{e\\o}m\eta}(-h) & u>u' \\ \bar{M}_{\substack{e\\o}m\eta}(h)\bar{M}'^{(1)}_{\substack{e\\o}m\eta}(-h) + \bar{N}_{\substack{e\\o}m\eta}(h)\bar{N}'^{(1)}_{\substack{e\\o}m\eta}(-h) & u<u' \end{cases} \tag{3}$$

where

$$\eta = \sqrt{k^2-h^2}$$

The vector wave functions with superscript (1) are defined with the radial functions described by (29-14) and (29-15). By applying the method of scattering superposition, we can find the function of the first kind satisfying the vector Dirichlet boundary condition on the surface of a conducting elliptical cylinder defined by $u=u_0$. The major axis and the minor axis of the cylinder are given, respectively, by

$$a = c\cosh u_0$$
$$b = c\sinh u_0\,.$$

Omitting the details, we find

$$\bar{\bar{G}}_1(\bar{R}|\bar{R}') = \frac{i}{2\pi}\int_{-\infty}^{\infty} dh \sum_m \frac{1}{\eta^2 I_{\substack{e\\o}m\eta}} \cdot \begin{cases} \bar{M}^{(1)}_{\substack{e\\o}m\eta}(h)[\bar{M}'_{\substack{e\\o}m\eta}(-h) + \alpha_{\substack{e\\o}m\eta}\bar{M}'^{(1)}_{\substack{e\\o}m\eta}(-h)] \\ [\bar{M}_{\substack{e\\o}m\eta}(h) + \alpha_{\substack{e\\o}m\eta}\bar{M}^{(1)}_{\substack{e\\o}m\eta}(h)]\,\bar{M}'^{(1)}_{\substack{e\\o}m\eta}(-h) \end{cases}$$
$$\begin{matrix} + \bar{N}^{(1)}_{\substack{e\\o}m\eta}(h)[\bar{N}'_{\substack{e\\o}m\eta}(-h) + \beta_{\substack{e\\o}m\eta}\bar{N}'^{(1)}_{\substack{e\\o}m\eta}(-h)] & u>u' \\ [+\bar{N}_{\substack{e\\o}m\eta}(h) + \beta_{\substack{e\\o}m\eta}\bar{N}^{(1)}_{\substack{e\\o}m\eta}(h)]\,\bar{N}'^{(1)}_{\substack{e\\o}m\eta}(-h) & u<u' \end{matrix} \tag{4}$$

where

$$\alpha_{\substack{e\\o}mn} = - \left[\frac{\partial R_{\substack{e\\o}mn}(u)}{\partial u} \Bigg/ \frac{\partial R^{(1)}_{\substack{e\\o}mn}(u)}{\partial u} \right]_{u=u_0}$$

$$\beta_{\substack{e\\o}mn} = - \quad R_{\substack{e\\o}mn}(u_0) \,/\, R^{(1)}_{\substack{e\\o}mn}(u_0) \quad .$$

To find the asymptotic expression for (4), we substitute in (4) the asymptotic values of the radial functions $R^{(1)}_{\substack{e\\o}mn}$ assuming (η cosh u) to be large. The problem is straightforward, and is assigned as an exercise.

Once the function of the first kind is known, the function of the second kind can be found by making use of the symmetry relation between these two functions. As was mentioned at the beginning of the chapter, it is impossible to find an orthogonal expansion for the function of the third kind associated with a dielectric elliptical cylinder because both the angular functions and the radial functions depend on the wave number, which has different values for the two regions interior and exterior to the cylinder. The mixed-boundary condition, therefore, cannot be satisfied if we use expressions analogous to (24-12) and (24-13) for the dielectric cylinder. As far as we know, the dyadic Green's function of the third kind has been found so far only for the flat earth, the circular cylinder, and the sphere. We shall treat the spherical case in Chapter 11.

chapter **10**

Perfectly Conducting Wedge and the Half-Sheet

In his book on Optics (p. 266), Sommerfeld remarked that the generalization (of the edge diffraction problem) to three dimensions is directly possible only for scalar or acoustic problems. The master did not reveal the reason why this cannot be done for vector or electromagnetic problems. It was this remark that inspired the present author in 1954 to resolve the problem using the then relatively new technique of the dyadic Green's function. This material now forms the main body of this chapter. This chapter starts with a derivation of the dyadic Green's function of the first kind for a perfectly conducting wedge of arbitrary wedge angle. The function for a half-sheet is just a limiting case when the wedge angle approaches zero. Radiation from dipoles, both electric and magnetic, and from apertures in the presence of a half-sheet, are treated in detail. The examples illustrate quite adequately the versatility of the dyadic Green's function technique. The numerical results obtained for the half-sheet may also be of interest to engineers designing antennas that are similar to the ones considered here.

31. PERFECTLY CONDUCTING WEDGE

For reasons that will become clear later, we judiciously choose the boundary of a perfectly conducting wedge corresponding to $\phi=0$ and $\phi=2\pi-\phi_0$ as shown in Fig. 31-1 where ϕ_0 denotes the angle of the wedge.

The vector wave functions which will be used in the eigenfunction expansion of the dyadic Green's functions of the first and second kind pertaining to the wedge are defined by

$$\bar{M}_{\substack{e\\o}\nu\lambda}(h)=\nabla\times\left[J_\nu(\lambda r)\begin{matrix}\cos\\ \sin\end{matrix}\nu\phi e^{ihz}\hat{z}\right]$$

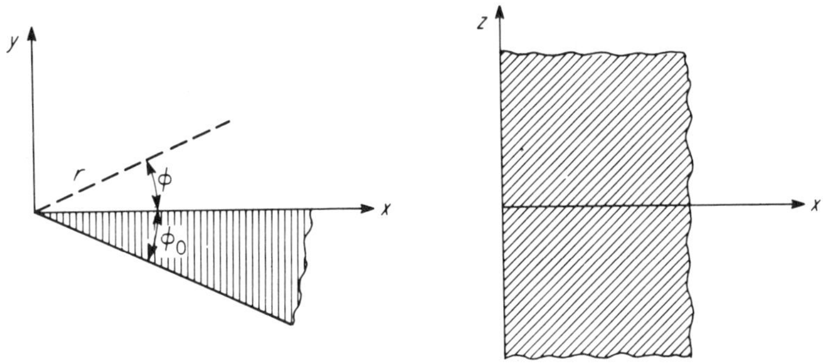

FIG. 31-1. Cross-sectional views of a perfectly conducting wedge.

$$\bar{N}_{\substack{e\\o}\nu\lambda}(h)=\frac{1}{\kappa}\nabla\times\nabla\times\left[J_\nu(\lambda r)\begin{matrix}\cos\\ \sin\end{matrix}\,\nu\phi e^{ihz}\hat{z}\right]$$

where

$$\kappa^2=\lambda^2+h^2$$

$$\nu=n\Big/\left(2-\frac{\phi_0}{\pi}\right),\ n=0,1,2\ldots$$

$$\phi_0=\text{angular span of the conducting wedge.}$$

The complete expressions of these functions are given by

$$\text{(1)}\qquad \bar{M}_{\substack{e\\o}\nu\lambda}(h)=\left[\mp\frac{\nu}{r}J_\nu(\lambda r)\begin{matrix}\sin\\ \cos\end{matrix}\,\nu\phi\hat{r}-\frac{\partial J_\nu(\lambda r)}{\partial r}\begin{matrix}\cos\\ \sin\end{matrix}\,\nu\phi\hat{\phi}\right]e^{ihz}$$

$$\text{(2)}\qquad \bar{N}_{\substack{e\\o}\nu\lambda}(h)=\frac{1}{\kappa}\left[ih\frac{\partial J_\nu(\lambda r)}{\partial r}\begin{matrix}\cos\\ \sin\end{matrix}\,\nu\phi\hat{r}\mp\frac{ih\nu}{r}J_\nu(\lambda r)\begin{matrix}\sin\\ \cos\end{matrix}\,\nu\phi\hat{\phi}\right.$$
$$\left.+\lambda^2 J_\nu(\lambda r)\begin{matrix}\cos\\ \sin\end{matrix}\,\nu\phi\hat{z}\right]e^{ihz}$$

These functions satisfy the boundary conditions that on the surface of the wedge, corresponding to $\phi=0$ and $\phi=2\pi-\phi_0$,

$$\hat{n}\times\bar{M}_e=0$$
$$\hat{n}\times\bar{N}_o=0$$
$$\hat{n}\times\nabla\times\bar{M}_o=0$$
$$\hat{n}\times\nabla\times\bar{N}_e=0.$$

This shows why we have chosen the wedge geometry with respect to the coordinate system as shown in Fig. 31-1 so that the required vector wave functions would have a relatively simple form. Unlike the case for a circular cylinder or a flat earth, the construction of the functions of the first and the second kind for the wedge will be pursued directly using these functions. It would be more complicated in this case to find the free-space dyadic Green's function first and then to apply the method of scattering superposition to determine the functions of the first and second kind. We consider now the differential equation for the function of the first kind

$$\nabla \times \nabla \times \bar{\bar{G}}_1(\bar{R}|\bar{R}') - k^2\bar{\bar{G}}_1(\bar{R}|\bar{R}') = \bar{I}\delta(\bar{R}|\bar{R}'). \tag{3}$$

Since $\bar{M}_{ev\lambda}$ and $\bar{N}_{ov\lambda}$ satisfy the Dirichlet boundary condition on the surface of the wedge they are the proper functions to be used in the eigenfunction expansion of the dyadic delta function. Thus we let

$$\bar{I}\delta(\bar{R}|\bar{R}') = \int_{-\infty}^{\infty} dh \int_0^{\infty} d\lambda \sum_{v} [\bar{M}_{ev\lambda}(h)\bar{A}_{ev\lambda}(h) + \bar{N}_{ov\lambda}(h)\bar{B}_{ov\lambda}(h)] \tag{4}$$

where $\bar{A}_{ev\lambda}(h)$ and $\bar{B}_{ov\lambda}(h)$ are two unknown functions to be determined. The two sets of vector wave functions used in (4) satisfy the orthogonal relations that

$$\iiint \bar{M}_{ev\lambda}(h) \cdot \bar{N}_{ov'\lambda'}(-h')\ dV = 0 \tag{5}$$

$$\iiint \bar{M}_{ev\lambda}(h) \cdot \bar{M}_{ev'\lambda'}(-h')\ dV = \begin{cases} 0, & v \neq v' \\ (1+\delta_0)\pi(2\pi - \phi_0)\lambda\delta(\lambda - \lambda')\delta(h - h'), & v = v' \end{cases} \tag{6}$$

$$\iiint \bar{N}_{ov\lambda}(h) \cdot \bar{N}_{ov'\lambda'}(-h')\ dV = \begin{cases} 0, & v \neq v' \\ (1+\delta_0)\pi(2\pi - \phi_0)\lambda\delta(\lambda - \lambda')\delta(h - h'), & v = v' \end{cases} \tag{7}$$

where the domain of integration is extended through the space exterior to the wedge. The derivation of (5) to (7) follows the same procedure as described in Sec. 23. It will not be repeated here. By taking the anterior product of (4) with $\bar{M}_{ev'\lambda'}(-h')$ and $\bar{N}_{ov'\lambda'}(-h')$ in turn and integrating the resultant equations through the space exterior to the wedge, we obtain, as a result of (5) through (7)

$$\bar{A}_{ev\lambda}(h) = \frac{2 - \delta_0}{2\pi(2\pi - \phi_0)\lambda}\, \bar{M}'_{ev\lambda}(-h)$$

$$\bar{B}_{ev\lambda}(h) = \frac{2 - \delta_0}{2\pi(2\pi - \phi_0)\lambda}\, \bar{N}'_{ov\lambda}(-h).$$

Knowing the eigenfunction expansion of the dyadic delta function, we can determine the corresponding expansion for $\bar{\bar{G}}_1(\bar{R}|\bar{R}')$ in view of (3). The result is given by

$$\bar{\bar{G}}_1(\bar{R}|\bar{R}')=\frac{1}{2\pi(2\pi-\phi_0)}\int_{-\infty}^{\infty} dh \int_0^{\infty} d\lambda \sum_{\nu} \frac{2-\delta_0}{\lambda(\lambda^2+h^2-k^2)} \cdot\left[\bar{M}_{e\nu\lambda}(h)\bar{M}'_{e\nu\lambda}(-h)+\bar{N}_{o\nu\lambda}(h)\bar{N}'_{o\nu\lambda}(-h)\right]. \tag{8}$$

The integration with respect to λ can be eliminated as in the case of the circular cylinder which yields

$$\bar{\bar{G}}_1(\bar{R}|\bar{R}')=\frac{i}{4(2\pi-\phi_0)}\int_{-\infty}^{\infty} dh \sum_{\nu} \frac{2-\delta_0}{\eta^2} \cdot\begin{cases} \bar{M}^{(1)}_{e\nu\eta}(h)\bar{M}'_{e\nu\eta}(-h)+\bar{N}^{(1)}_{o\nu\eta}(h)\bar{N}'_{o\nu\eta}(-h) & r>r' \\ \bar{M}_{e\nu\eta}(h)\bar{M}'^{(1)}_{e\nu\eta}(-h)+\bar{N}_{o\nu\eta}(h)\bar{N}'^{(1)}_{o\nu\eta}(-h), & r<r' \end{cases} \tag{9}$$

where

$$\eta^2=k^2-h^2$$

and the vector wave functions with superscript (1) are defined with respect to the Hankel function of the first kind, that is,

$$\bar{M}^{(1)}_{e\nu\eta}(h)=\nabla\times\left[H^{(1)}_{\nu}(\eta r)\cos\nu\phi\, e^{ihz}\hat{z}\right]$$

$$\bar{N}^{(1)}_{o\nu\eta}(h)=\frac{1}{k}\nabla\times\nabla\times\left[H^{(1)}_{\nu}(\eta r)\sin\nu\phi\, e^{ihz}\hat{z}\right]$$

with

$$\nu=n\Big/\left(2-\frac{\phi_o}{\pi}\right),\quad n=0,1,2,\ldots,$$

$$k^2=\eta^2+h^2.$$

By making use of the symmetrical relation between the functions of the first and the second kind, we can find the expression for $\bar{\bar{G}}_2(\bar{R}|\bar{R}')$ which is given by replacing the even $\bar{M}$ functions and the odd $\bar{N}$ functions in (9), respectively, by the odd $\bar{M}$ functions and the even $\bar{N}$ functions. Like the case of a dielectric elliptical cylinder, there is no way of constructing the dyadic Green's function of the third kind for a dielectric wedge because of the lack of proper orthogonal sets of vector wave functions.

32. THE HALF-SHEET

When the angle of a wedge approaches zero, the wedge degenerates into a half-sheet. The functions of the first and second kind are thus obtained by putting $\phi_0=0$ in (31-9) which yields

(1) $$\bar{\bar{G}}_1(\bar{R}|\bar{R}') = \frac{i}{8\pi}\int_{-\infty}^{\infty} dh \sum_{\nu=0}^{\infty} \frac{2-\delta_0}{\eta^2}$$

$$\cdot \begin{cases} \bar{M}^{(1)}_{e\nu\eta}(h)\bar{M}'_{e\nu\eta}(-h) + \bar{N}^{(1)}_{o\nu\eta}(h)\bar{N}'_{o\nu\eta}(-h), & r>r' \\ \bar{M}_{e\nu\eta}(h)\bar{M}'^{(1)}_{e\nu\eta}(-h) + \bar{N}_{o\nu\eta}(h)\bar{N}'^{(1)}_{o\nu\eta}(-h), & r<r' \end{cases}$$

(2) $$\bar{\bar{G}}_2(\bar{R}|\bar{R}') = \frac{i}{8\pi}\int_{-\infty}^{\infty} dh \sum_{\nu=0}^{\infty} \frac{2-\delta_0}{\eta^2}$$

$$\cdot \begin{cases} \bar{M}^{(1)}_{o\nu\eta}(h)\bar{M}'_{o\nu\eta}(-h) + \bar{N}^{(1)}_{e\nu\eta}(h)\bar{N}'_{e\nu\eta}(-h), & r>r' \\ \bar{M}_{o\nu\eta}(h)\bar{M}'^{(1)}_{o\nu\eta}(-h) + \bar{N}_{e\nu\eta}(h)\bar{N}'^{(1)}_{e\nu\eta}(-h), & r<r' \end{cases}$$

where

$$\nu = \frac{n}{2}, \quad n = 0, 1, 2, \ldots$$

$$\eta^2 = k^2 - h^2.$$

The functions $\bar{M}_{\substack{e\\o}\nu\eta}$ and $\bar{N}_{\substack{e\\o}\nu\eta}$ are defined by (31-1) and (31-2) with λ replaced by η and the functions $\bar{M}^{(1)}_{\substack{e\\o}\nu\eta}$ and $\bar{N}^{(1)}_{\substack{e\\o}\nu\eta}$ are obtained by replacing $J_\nu(\eta r)$ by $H^{(1)}_\nu(\eta r)$. It is seen that only Bessel functions and Hankel functions of half order and integer order are involved in these expressions. Once the expressions for $\bar{\bar{G}}_1$ and $\bar{\bar{G}}_2$ are known, the discussion of the radiation from current elements with known distributions or apertures with known field distributions reduces to that of evaluating some definite complex integrals. In general, because of the Fourier integrals involved in the expressions for the dyadic Green's functions, we cannot evaluate them in a closed form. However, for certain restricted ranges of the parameters, corresponding to the far-zone range, most of the integrals can be evaluated in a closed form in terms of tabulated functions. For convenience, the problems to be considered will be divided into five categories:

A. Radiation from electric dipoles in the presence of a half-sheet

B. Radiation from magnetic dipoles in the presence of a half-sheet.

C. Radiation from slots cut in a half-sheet.

D. Radiation resistance of linear antennas placed near a half-sheet and radiation conductance of notch antennas.

E. Diffraction of a plane wave by a half-sheet.

33. RADIATION FROM ELECTRIC DIPOLES IN THE PRESENCE OF A HALF-SHEET

1. Longitudinal Electric Dipole

Let us consider an infinitesimal electric dipole with a current moment c oriented along the direction of the longitudinal axis of the half-sheet, and

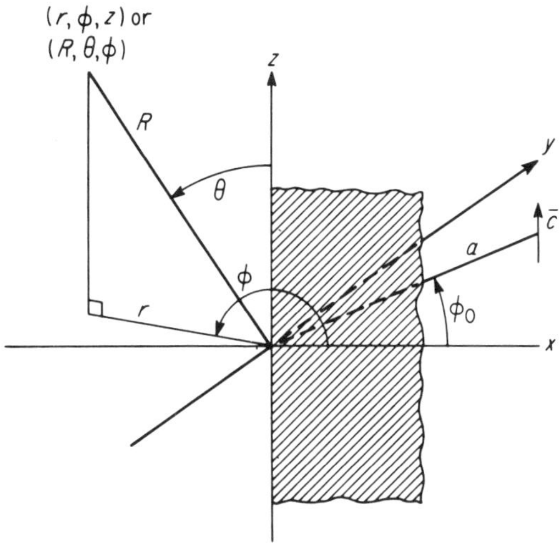

FIG. 33-1. A longitudinal electric dipole located at $(a, \phi_0, 0)$ in the presence of a half-sheet.

located at $(a,\phi_0,0)$ as shown in Fig. 33-1. The electric current density can be written as

(1) $$\bar{J}(\bar{R}') = c\,\frac{\delta(r'-a)\delta(\phi'-\phi_0)\delta(z'-0)}{r'}\,\hat{z}.$$

The delta functions have a weighting factor $1/r'$ so that

$$\iiint \bar{J}(\bar{R}')\,dV' = \iiint \bar{J}(\bar{R}')r'\,dr'\,d\phi'\,dz' = c\hat{z}.$$

Substituting (32-1) and (1) into

(2) $$\bar{E}(\bar{R}) = i\omega\mu_0 \iiint \bar{\bar{G}}_1(\bar{R}|\bar{R}')\cdot\bar{J}(\bar{R}')\,dV'$$

we obtain, for $r > a$,

$$\bar{E}(\bar{R}) = \frac{-\omega\mu_0 c}{4\pi k}\int_{-\infty}^{\infty} dh \sum_{v=0}^{\infty} \bar{N}^{(1)}_{ov\eta}(h) J_v(\eta a)\sin v\phi_0 .$$

The TM modes are not excited in this case because $\bar{M}'_{ev\eta}$ is transverse to the longitudinal axis. For numerical calculation, we shall investigate only the z-component of the electric field which is given by

$$(3)\quad E_z(\bar{R}) = \frac{-\omega\mu_0 c}{4\pi k^2}\int_{-\infty}^{\infty}\eta^2 e^{ihz}\sum_{n=1,2\ldots}\sin\left(\frac{n\phi_o}{2}\right)\sin\left(\frac{n\phi}{2}\right)\cdot J_{n/2}(\eta a)H^{(1)}_{n/2}(\eta r)\,dh$$

where we have already converted ν into $n/2$. The Fourier integral contained in (3) cannot be evaluated in a closed form for arbitrary values of r and z. However, when the point of observation is far away from the origin, the integration can be done by means of the saddle-point method of integration. The procedure is the same as the one described in Sec. 25. Without repeating the details, we obtain

$$(4)\qquad E_z(\bar{R}) = \frac{i\omega\mu_0 c e^{ikR}}{2\pi R}\sin^2\theta\sum_{n=1}^{\infty}(-i)^{n/2}\sin\left(\frac{n\phi_0}{2}\right)\sin\left(\frac{n\phi}{2}\right)\cdot J_{n/2}(ka\sin\theta)$$

where (R,θ,ϕ) denotes the point of observation in the spherical coordinate system as shown in Fig. 33-1. The series contained in (4) can be transformed into a function involving some Fresnel integrals by applying an expansion theorem due to Hargreaves (1918). The analysis is found in Morse and Feshbach (1953, p. 1386). To transform (4) into the desirable form, we let

$$(5)\qquad S(\rho,\phi) = \frac{1}{2}\sum_{n=0}^{\infty}(2-\delta_0)(-i)^{n/2}\cos\frac{n\phi}{2}\,J_{n/2}(\rho)\,.$$

The integral representation of (5), according to Hargreaves' theorem, is given by

$$(6)\qquad S(\rho,\phi) = (\pi)^{-1/2}e^{-i(\rho\cos\phi+\pi/4)}\int_{-\infty}^{\sqrt{2\rho}\cos\phi/2} e^{is^2}\,ds.$$

The integral contained in (6) can be expressed in terms of the Fresnel integral functions defined by

$$(7)\qquad C(x) = \int_0^x \frac{\cos t}{\sqrt{2\pi t}}\,dt, \qquad S(x) = \int_0^x \frac{\sin t}{\sqrt{2\pi t}}\,dt\,.$$

A short table of these functions is found in Watson (1922, p. 744). By a change of the variable of integration (6) can be written in the form

(8) $$S(\rho,\phi)=2^{-1/2}e^{-i(\rho\cos\phi+\pi/4)}\cdot\left\{\frac{1}{2}(1+i)\pm[C(\rho+\rho\cos\phi)+iS(\rho+\rho\cos\phi)]\right\}$$

where the $\pm$ sign in (8) is chosen according to the appropriate range of ϕ:

$$(+)\to\pi\geqslant\phi\geqslant 0$$
$$(-)\to 2\pi\geqslant\phi\geqslant\pi\,.$$

Regarding $S(\rho,\phi)$ as a basic function in this work, we can transform (4) into an expression containing two of these functions with different arguments owing to the simple trigonometric identity that

$$2\sin\left(\frac{n\phi_0}{2}\right)\sin\left(\frac{n\phi}{2}\right)=\cos\frac{n(\phi-\phi_0)}{2}-\cos\frac{n(\phi+\phi_0)}{2}$$

The final expression for $E_z(\bar{R})$ is given by

(9) $$E_z(\bar{R})=\frac{i\omega\mu_0 ce^{ikR}}{4\pi R}\sin^2\theta\cdot[S(ka\sin\theta,\phi-\phi_0)-S(ka\sin\theta,\phi+\phi_0)].$$

In the principal plane, corresponding to $\theta=\pi/2$, the pattern of the electric field is described by

(10) $$F(\phi)=S(ka,\phi-\phi_0)-S(ka,\phi+\phi_0)\,.$$

By using (8) instead of (5) we can determine very accurately the numerical values of the pattern function. It should be mentioned that the pattern function (10) which we obtained here is equivalent to the one obtained by Harrington (1953) for a line source in the presence of a half-sheet. In fact, the complex integral involved in Harrington's work based on the method of Weiner-Hopf integral equation can readily be reduced to the Fresnel integral.

2. Horizontal Electrical Dipole

The orientation and the location of the dipole for this case is shown in Fig. 33-2 which displays only a cross-sectional view in the plane $z=0$.

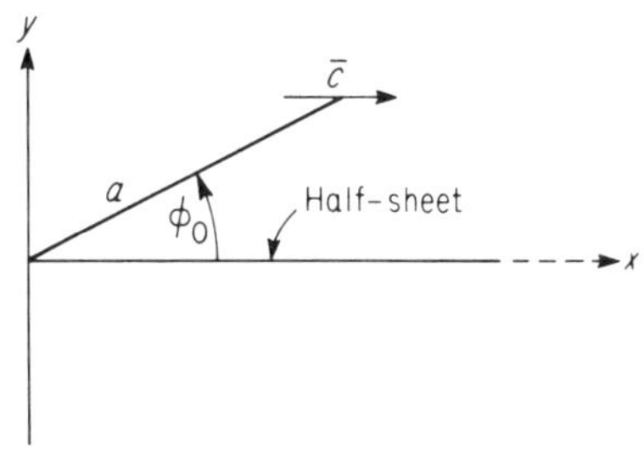

FIG. 33-2. A horizontal electric dipole located at $(a,\ \phi_0,\ 0)$.

In the principal plane, corresponding to $\theta=\pi/2$, the far-zone electric field has only a ϕ-component. Its expression is given by

$$E_\phi\left(R,\frac{\pi}{2},\phi\right)=\frac{\omega\mu_0 c e^{ikR}}{4\pi kaR}$$

$$\cdot\left[-\cos\phi_0\frac{\partial}{\partial\phi}\sum_{n=0}^{\infty}(2-\delta_0)(-i)^{n/2}J_{n/2}(\rho)\sin\frac{n\phi_0}{2}\sin\frac{n\phi}{2}\right.$$

$$\left.+\rho\sin\phi_0\frac{\partial}{\partial\rho}\sum_{n=0}^{\infty}(2-\delta_0)(-i)^{n/2}J_{n/2}(\rho)\cos\frac{n\phi_0}{2}\cos\frac{n\phi}{2}\right]_{\rho=ka}$$

The two series contained in the brackets can be expressed in terms of the S-function defined by (5). Thus we have

$$\frac{\partial}{\partial\phi}\sum_{n=0}^{\infty}(2-\delta_0)(-i)^{n/2}J_{n/2}(\rho)\sin\frac{n\phi_0}{2}\sin\frac{n\phi}{2}=\frac{\partial}{\partial\phi}[S(\rho,\phi-\phi_0)-S(\rho,\phi+\phi_0)]$$

and

$$\frac{\partial}{\partial\rho}\sum_{n=0}^{\infty}(2-\delta_0)(-i)^{n/2}J_{n/2}(\rho)\cos\frac{n\phi_0}{2}\cos\frac{n\phi}{2}=\frac{\partial}{\partial\rho}[S(\rho,\phi-\phi_0)+S(\rho,\phi+\phi_0)]$$

By taking the partial derivatives of the S-functions as represented by (6) or (8) and simplifying the result, we obtain

$$E_\phi\left(R,\frac{\pi}{2},\phi\right)=\frac{-i\omega\mu_0 c e^{ikR}}{4\pi R}\cdot\left\{\sin\phi\,[S(\rho,\phi-\phi_0)-S(\rho,\phi+\phi_0)]\right.$$

$$\left.+\left(\frac{2}{\pi\rho}\right)^{1/2}e^{i(\rho+\pi/4)}\sin\frac{\phi_0}{2}\sin\frac{\phi}{2}\right\}_{\rho=ka}. \tag{11}$$

The pattern function in this case is therefore described by

$$F_2(\phi)=\sin\phi\,[S(ka,\phi-\phi_0)-S(ka,\phi+\phi_0)]$$

$$+\left(\frac{2}{\pi ka}\right)^{1/2}e^{i(ka+\pi/4)}\sin\frac{\phi_0}{2}\sin\frac{\phi}{2}. \tag{12}$$

3. Vertical Electric Dipole

The orientation and the location of the dipole for this case are shown in Fig. 33-3. The far-zone electric field in the principal plane again has only a ϕ-component. Its expression is given by

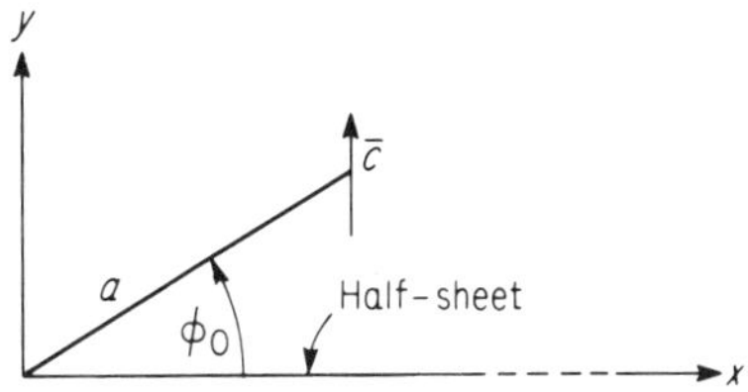

FIG. 33-3. A vertical electric dipole located at $(a, \phi_0, 0)$.

(13) $$E_\phi\left(R, \frac{\pi}{2}, \phi\right) = \frac{i\omega\mu_0 c e^{ikR}}{4\pi R} \cdot \left\{\cos\phi\,[S(ka, \phi - \phi_0) + S(ka, \phi + \phi_0)] + \left(\frac{2}{\pi ka}\right)^{1/2} e^{i(ka + \pi/4)} \cos\frac{\phi_0}{2}\cos\frac{\phi}{2}\right\}.$$

The pattern function in the principal plane is therefore represented by

(14) $$F_3(\phi) = \cos\phi\,[S(ka, \phi - \phi_0) + S(ka, \phi + \phi_0)] + \left(\frac{2}{\pi ka}\right)^{1/2} e^{i(ka + \pi/4)} \cos\frac{\phi_0}{2}\cos\frac{\phi}{2}.$$

Equations (10), (12), and (14), after being normalized with respect to the maximum value of each individual pattern, are plotted in Figs. 33-4, 33-5, and 33-6, for several different values of ka and ϕ_0. The collection provides us an adequate view as to how a half-sheet affects the radiation pattern of a short electric dipole oriented in three distinct directions with respect to the sheet.

It should be mentioned that the problem of the diffraction of a dipole field by a perfectly conducting half-plane has also been investigated thoroughly by Senior (1953). The relationship between the present formulation and the one based on the method of potentials is discussed by Bowman, Senior, and Uslenghi (1969).

34. RADIATION FROM MAGNETIC DIPOLES IN THE PRESENCE OF A HALF-SHEET

In principle, (33-2) can be used to calculate $\bar{E}$ for any current distribution. In practice for currents in the form of small loops it is more convenient to use an alternative formula. As was discussed in Sec. 7, by introducing an equivalent magnetization vector $\bar{M}$ the wave equation for $\bar{H}$ can be written in the form

$$\nabla \times \nabla \times \bar{H} - k^2 \bar{H} = k^2 \bar{M}.$$

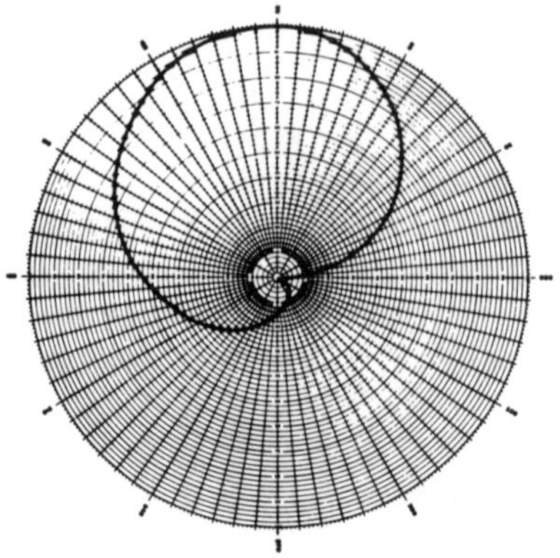

$\phi_0 = 5°, ka = \frac{\pi}{5}$

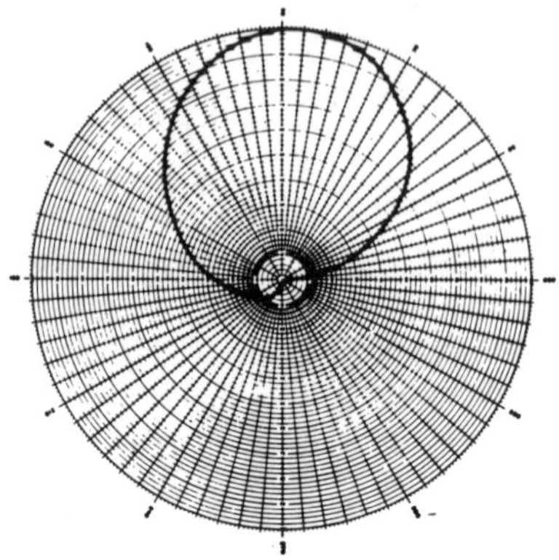

$\phi_0 = 5°, ka = \frac{\pi}{\sqrt{2}}$

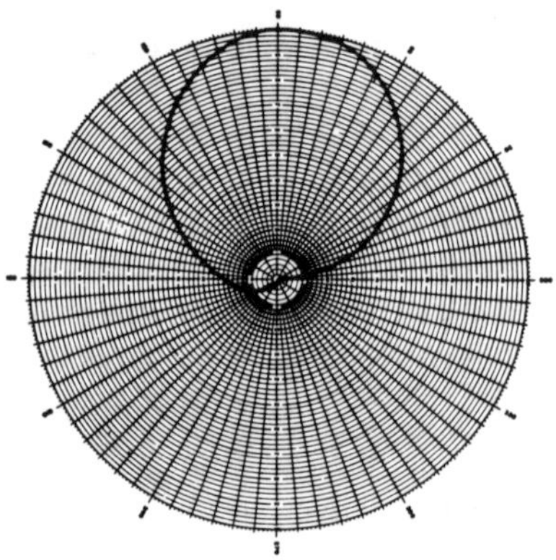

$\phi_0 = 5°, ka = \pi$

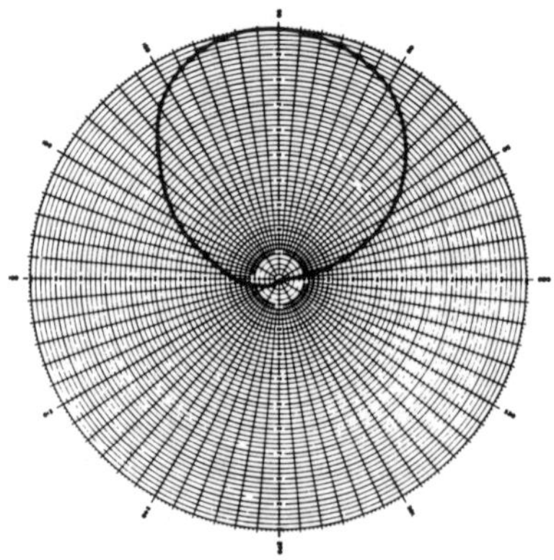

$\phi_0 = 5°, ka = 2\pi$

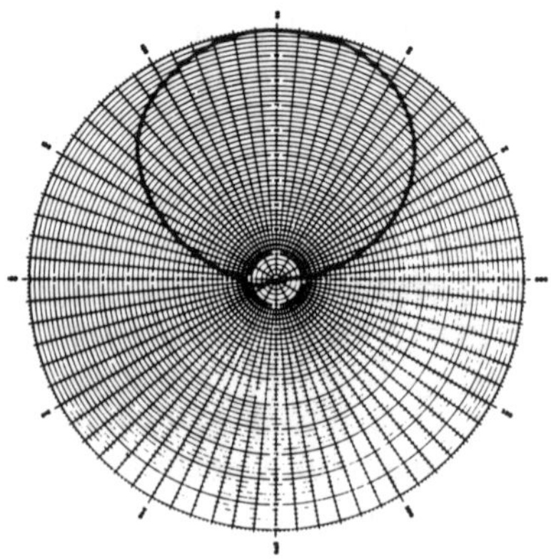

$\phi_0 = 5°, ka = 4\pi$

FIG. 33-4a. Radiation pattern of a longitudinal dipole placed in front of a half-sheet.

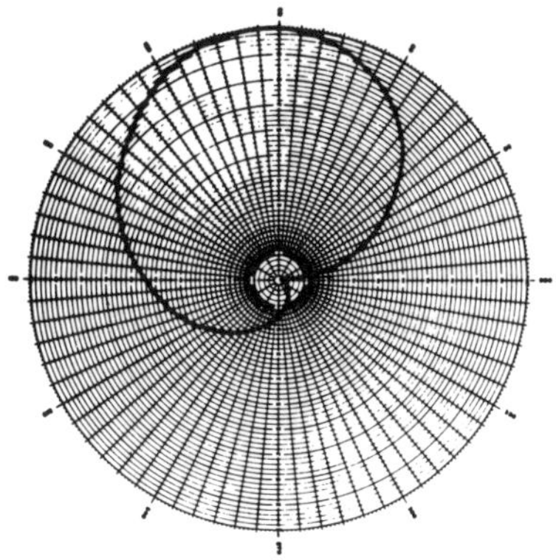

$\phi_0 = 10°,\ ka = \frac{\pi}{5}$

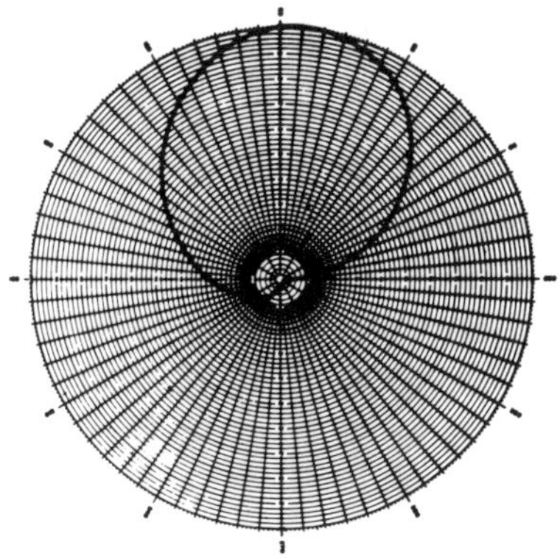

$\phi_0 = 10°,\ ka = \frac{\pi}{\sqrt{2}}$

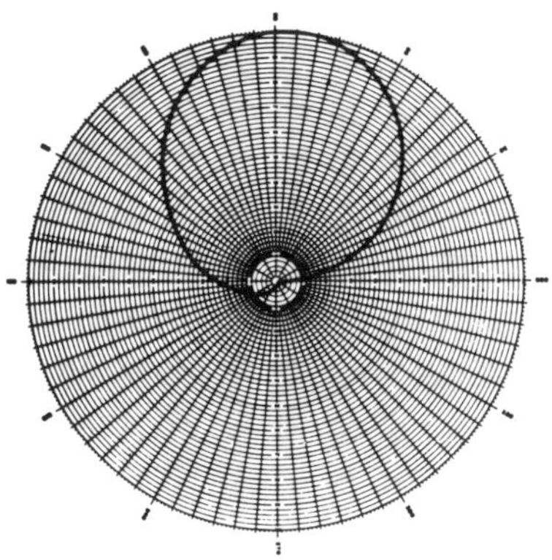

$\phi_0 = 10°,\ ka = \pi$

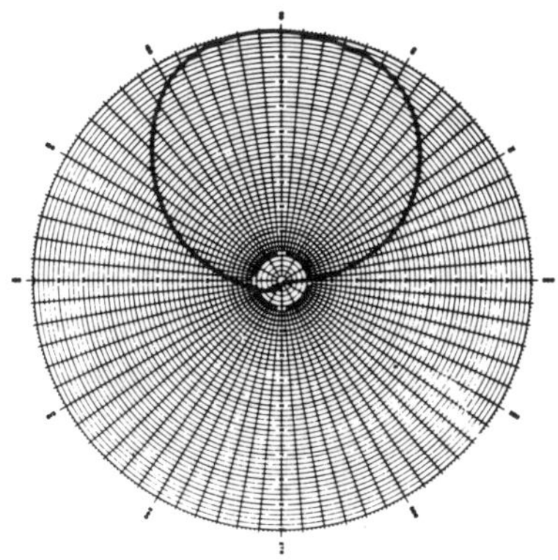

$\phi_0 = 10°,\ ka = 2\pi$

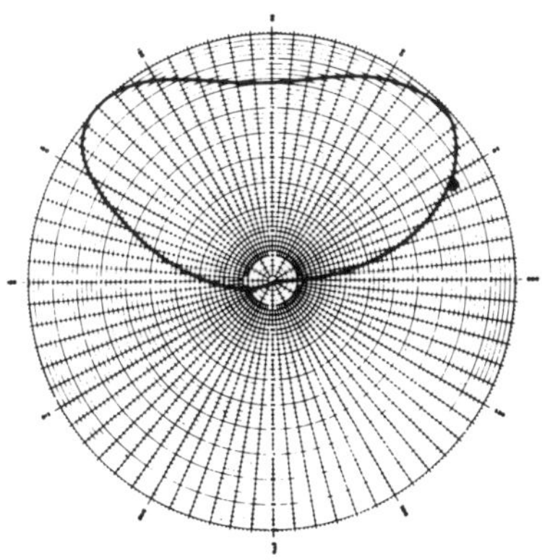

$\phi_0 = 10°,\ ka = 4\pi$

FIG. 33-4b

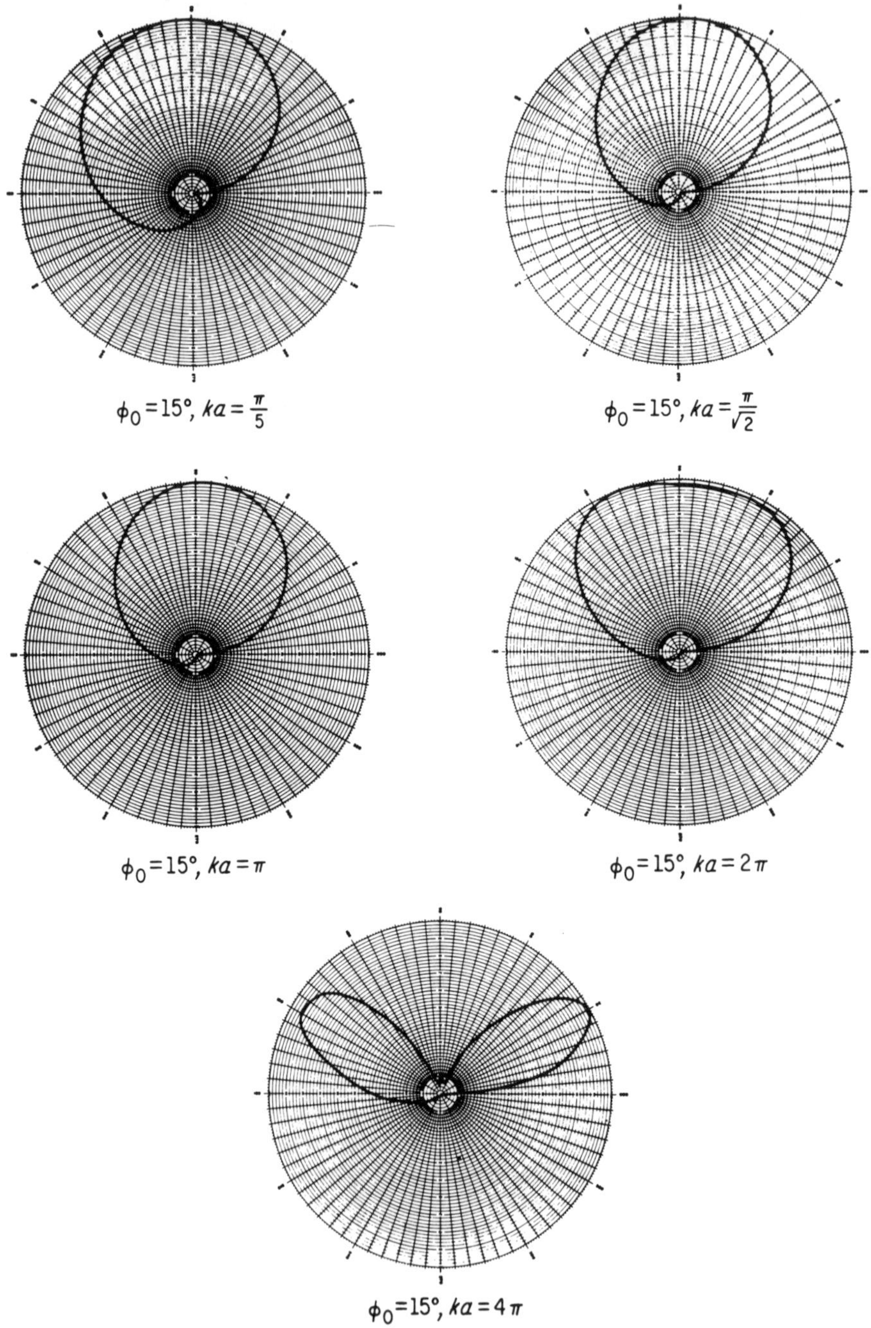
$\phi_0 = 15°, ka = \frac{\pi}{5}$
$\phi_0 = 15°, ka = \frac{\pi}{\sqrt{2}}$
$\phi_0 = 15°, ka = \pi$
$\phi_0 = 15°, ka = 2\pi$
$\phi_0 = 15°, ka = 4\pi$

FIG. 33-4c

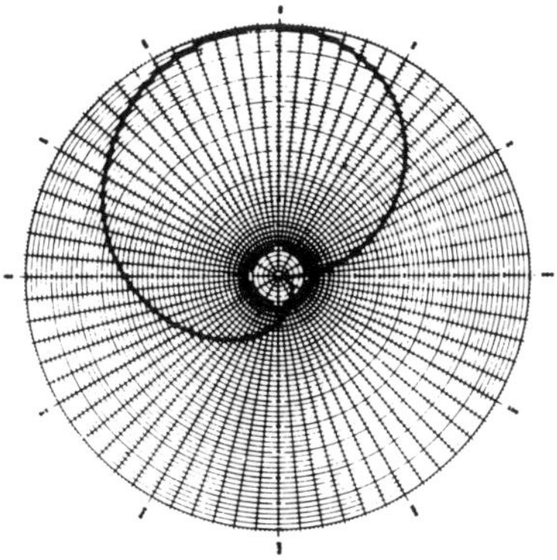

$\phi_0 = 45°,\ ka = \frac{\pi}{5}$

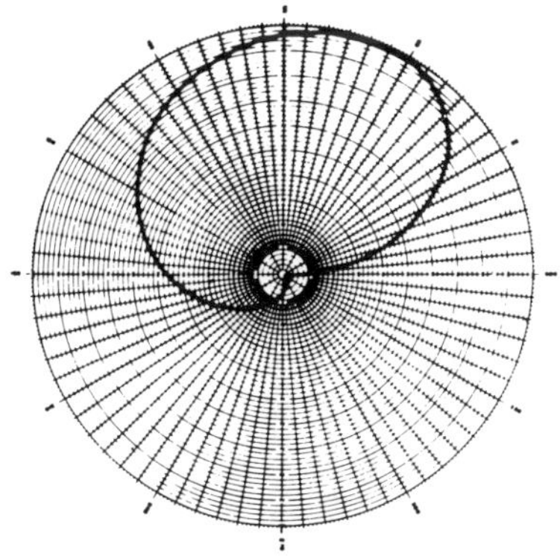

$\phi_0 = 45°,\ ka = \frac{\pi}{\sqrt{2}}$

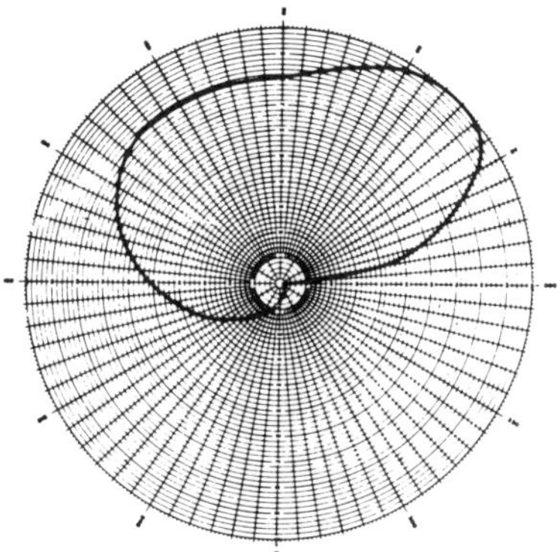

$\phi_0 = 45°,\ ka = \pi$

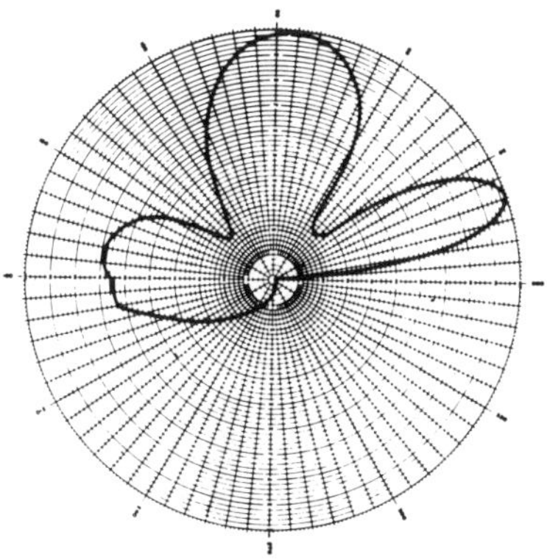

$\phi_0 = 45°,\ ka = 2\pi$

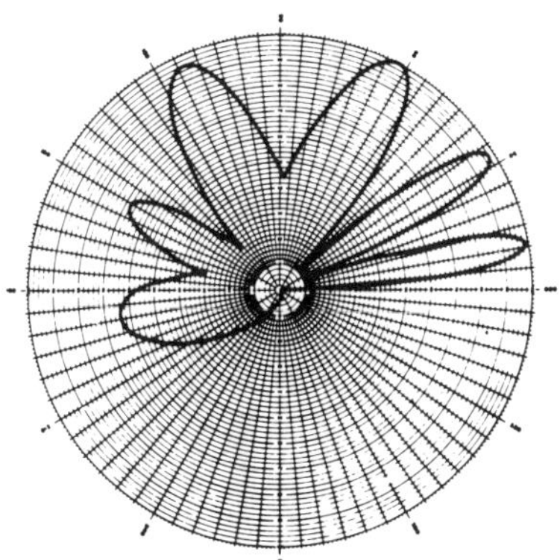

$\phi_0 = 45°,\ ka = 4\pi$

FIG. 33-4d

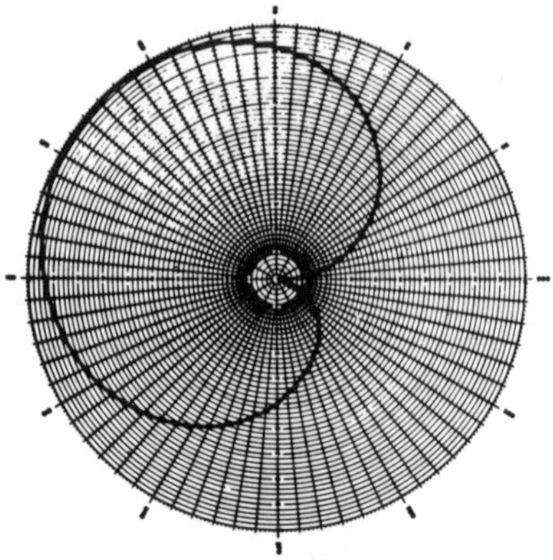

$\phi_0 = 135°,\ ka = \frac{\pi}{5}$

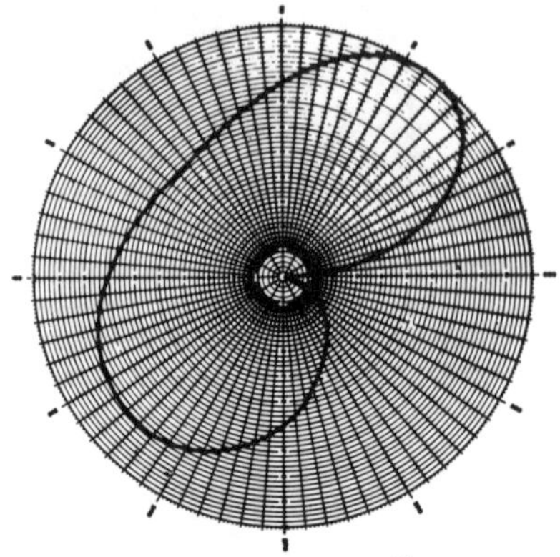

$\phi_0 = 135°,\ ka = \frac{\pi}{\sqrt{2}}$

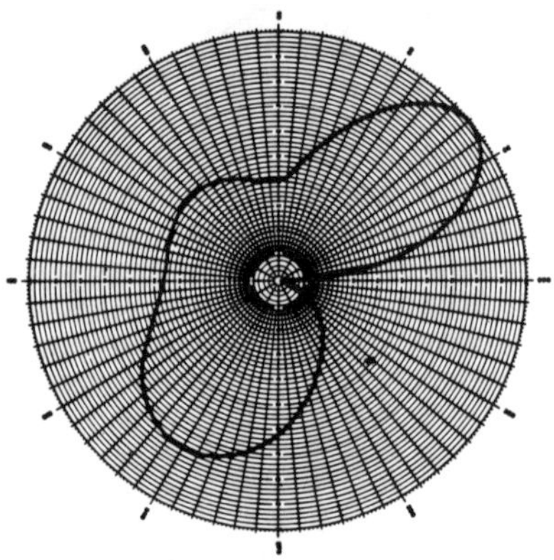

$\phi_0 = 135°,\ ka = \pi$

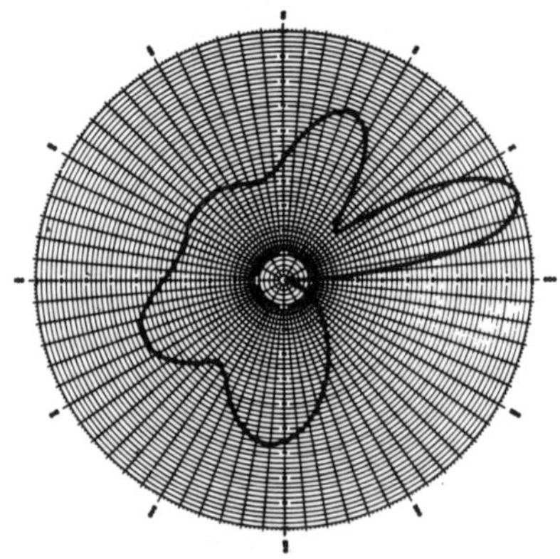

$\phi_0 = 135°,\ ka = 2\pi$

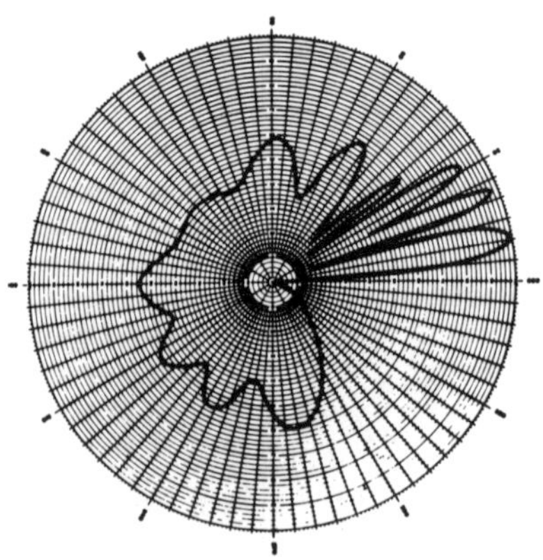

$\phi_0 = 135°,\ ka = 4\pi$

FIG. 33-4e

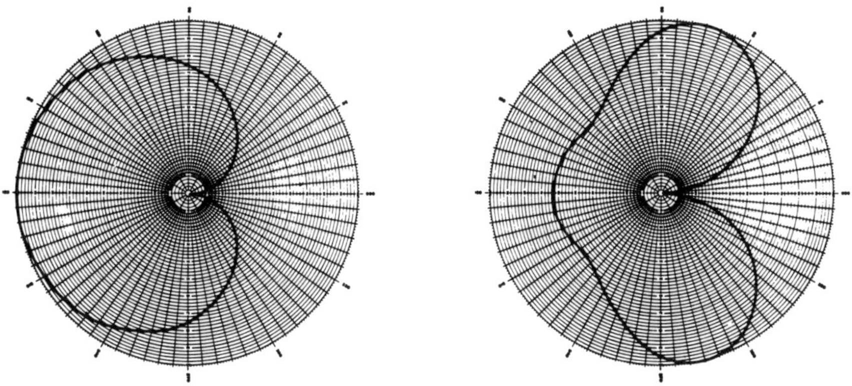

$\phi_0 = 180°, ka = \frac{\pi}{5}$ FIG. 33-4f $\phi_0 = 180°, ka = \pi$

The application of the dyadic Green's function technique to this equation yields

$$\bar{H}(\bar{R}) = k^2 \iiint \bar{\bar{G}}_2(\bar{R}|\bar{R}') \cdot \bar{M}(\bar{R}')\, dV'. \tag{1}$$

This equation can be obtained directly from (17-10) if we let $\nabla' \times \bar{J}(\bar{R}') = k^2(\bar{M}(\bar{R}')$. When the point of observation is not situated within the source region, the corresponding expression for $\bar{E}(\bar{R})$ is given by

$$\bar{E}(\bar{R}) = \frac{i}{\omega\varepsilon_o} \nabla \times \bar{H}(\bar{R}) = i\omega\mu_o \iiint \nabla \times \bar{\bar{G}}_2(\bar{R}|\bar{R}') \cdot \bar{M}(\bar{R}')\, dV'. \tag{2}$$

This is the desired formula which will be used to evaluate the electric field due to a small current loop in the presence of a perfectly conducting diffracting body. For a half-sheet, on account of (32-2), we have

$$\nabla \times \bar{\bar{G}}_2(\bar{R}|\bar{R}') = \frac{ik}{8\pi} \int_{-\infty}^{\infty} dh \sum_{\nu=0}^{\infty} \frac{2-\delta_0}{\eta^2} \cdot \begin{cases} \bar{N}^{(1)}_{o\nu\eta}(h) \bar{M}'_{o\nu\eta}(-h) + \bar{M}^{(1)}_{e\nu\eta}(h) \bar{N}'_{e\nu\eta}(-h) & r > r' \\ \bar{N}_{o\nu\eta}(h) \bar{M}'^{(1)}_{o\nu\eta}(-h) + \bar{M}_{e\nu\eta}(h) \bar{N}'^{(1)}_{e\nu\eta}(-h & r < r'. \end{cases} \tag{3}$$

The orientation and location of the magnetic dipoles to be considered are the same as those for electric dipoles shown in Figs. 33-1, 33-2, and 33-3 except that we replace $\bar{c}$ by $\bar{m}$ where $\bar{m}$ denotes the magnetic dipole moment. For a small current loop carrying current I and of area A, m is numerically equal to IA. Thus, for a longitudinal magnetic dipole located at $(a, \phi_0, 0)$ we let

$$\bar{M}(\bar{R}') = m \frac{\delta(r'-a)\delta(\phi'-\phi_o)\delta(z-0)}{r'} \hat{z}.$$

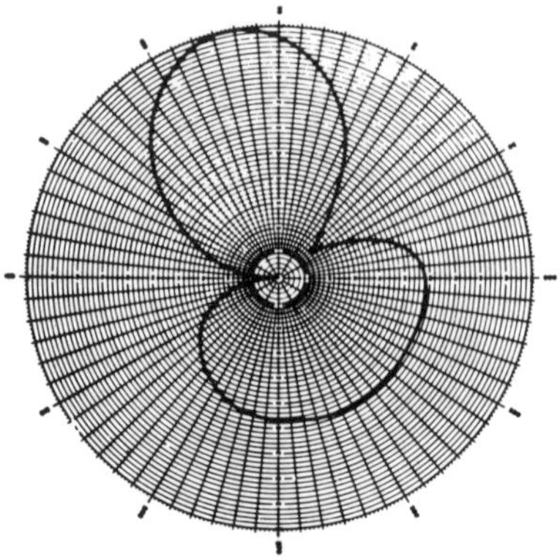

$\phi_0 = 5°, ka = \frac{\pi}{5}$

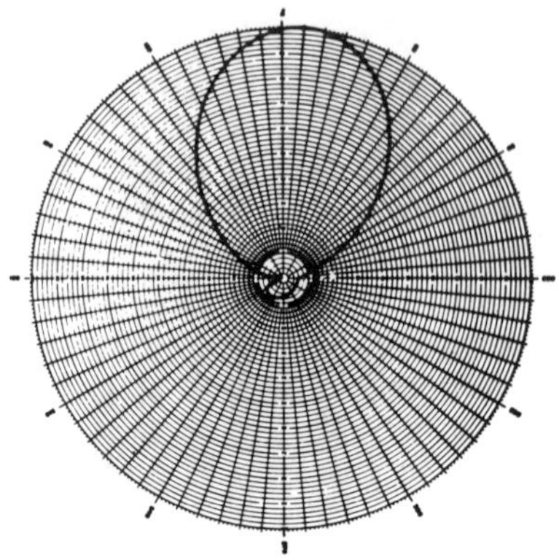

$\phi_0 = 5°, ka = \frac{\pi}{\sqrt{2}}$

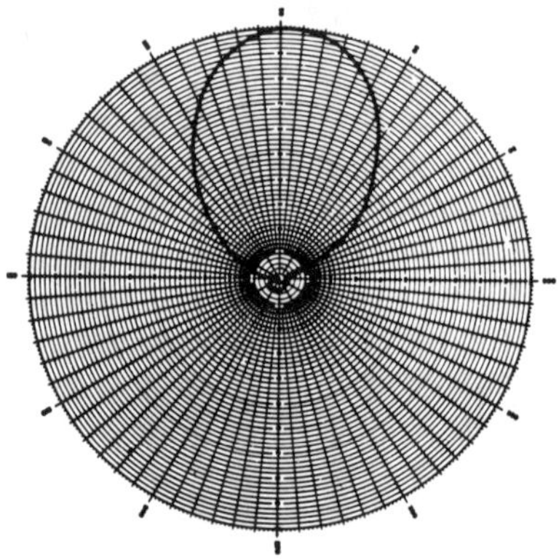

$\phi_0 = 5°, ka = \pi$

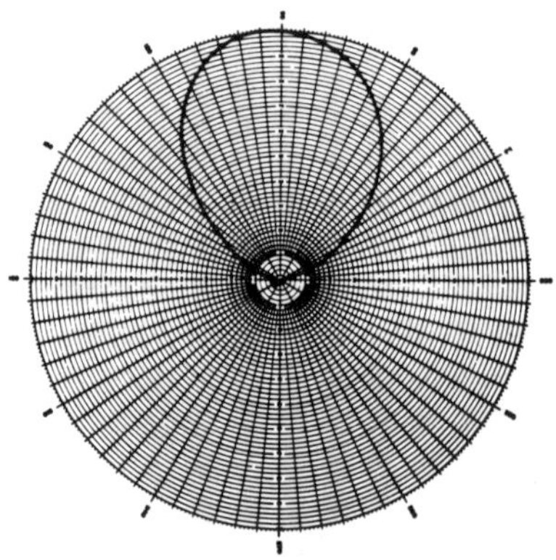

$\phi_0 = 5°, ka = 2\pi$

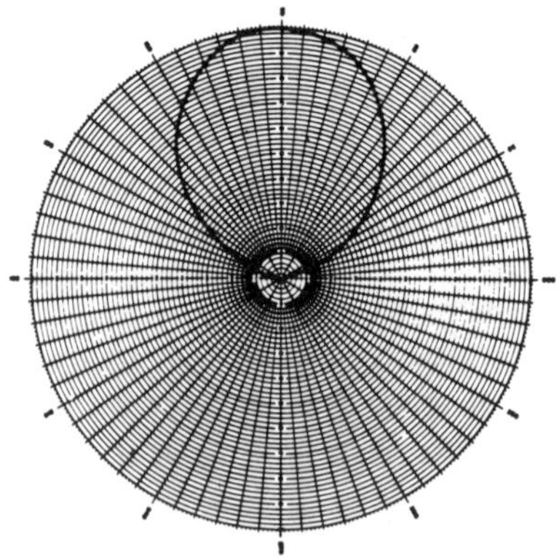

$\phi_0 = 5°, ka = 4\pi$

FIG. 33-5a. Radiation pattern of a horizontal dipole placed in front of a half-sheet,

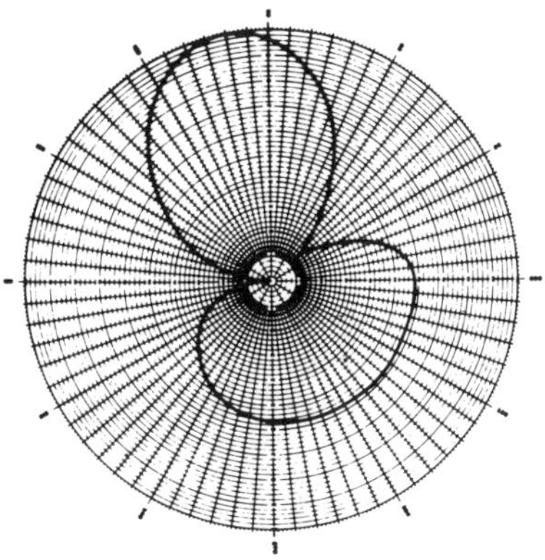

$\phi_0 = 10°$, $ka = \frac{\pi}{5}$

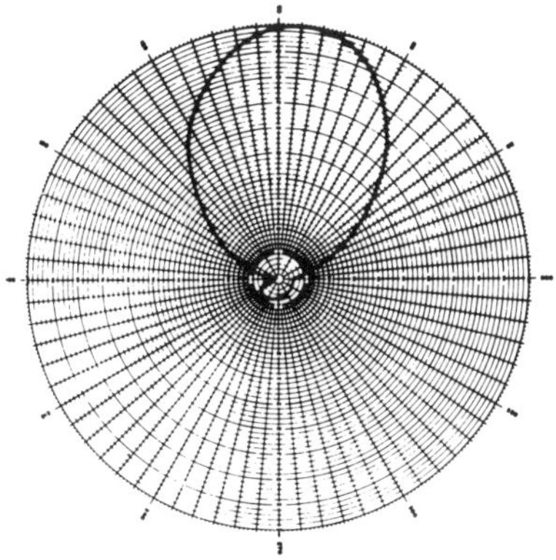

$\phi_0 = 10°$, $ka = \frac{\pi}{\sqrt{2}}$

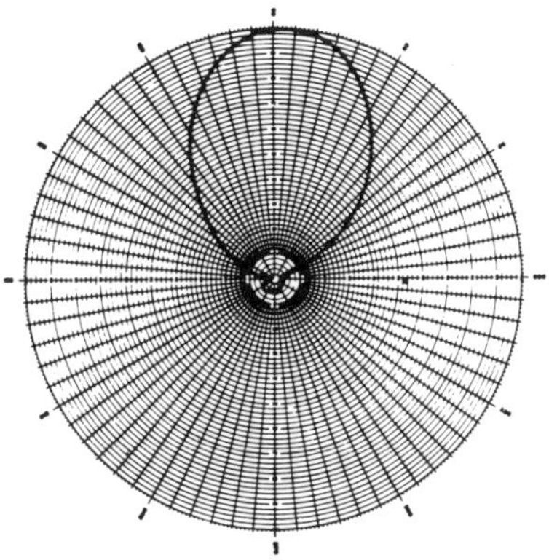

$\phi_0 = 10°$, $ka = \pi$

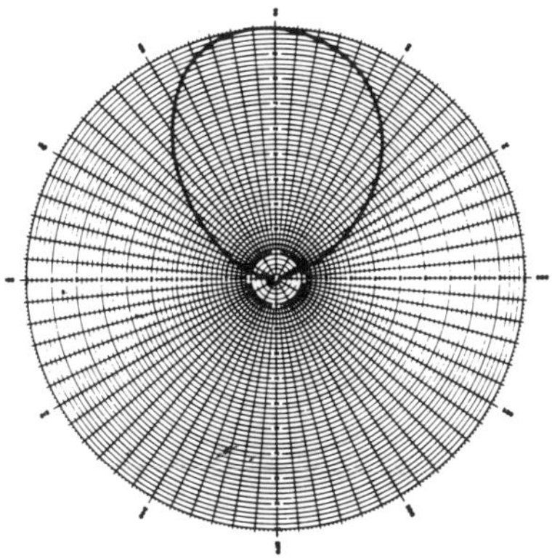

$\phi_0 = 10°$, $ka = 2\pi$

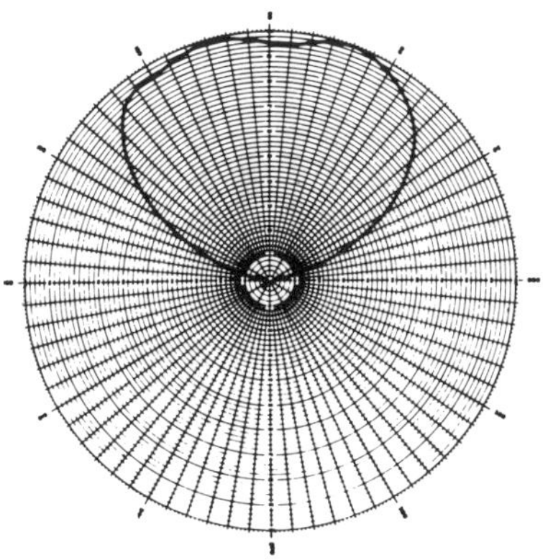

$\phi_0 = 10°$, $ka = 4\pi$

FIG. 33-5b

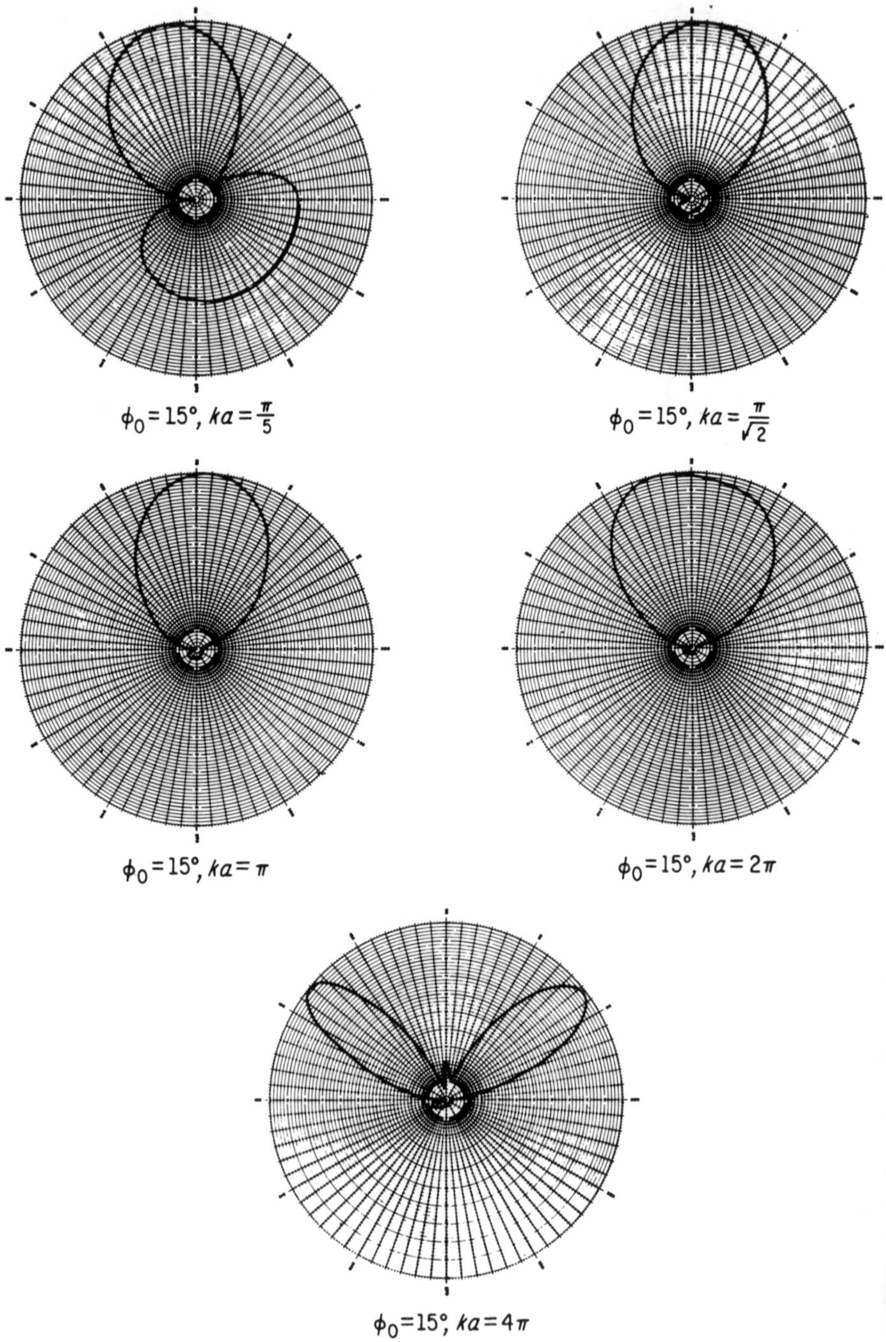
$\phi_0 = 15°, ka = \frac{\pi}{5}$
$\phi_0 = 15°, ka = \frac{\pi}{\sqrt{2}}$
$\phi_0 = 15°, ka = \pi$
$\phi_0 = 15°, ka = 2\pi$
$\phi_0 = 15°, ka = 4\pi$

FIG. 33-5c

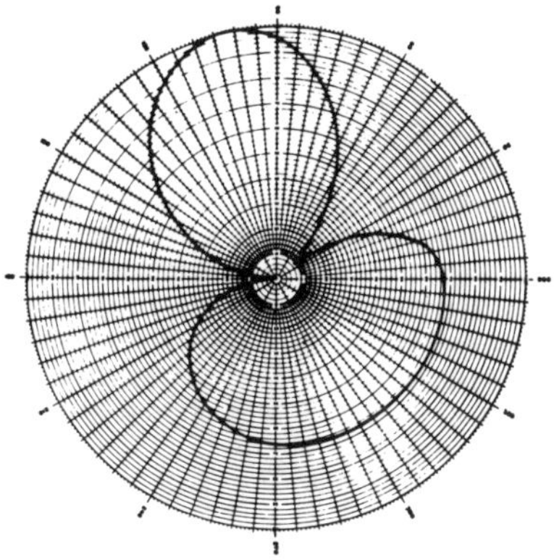

$\phi_0 = 45°, ka = \frac{\pi}{5}$

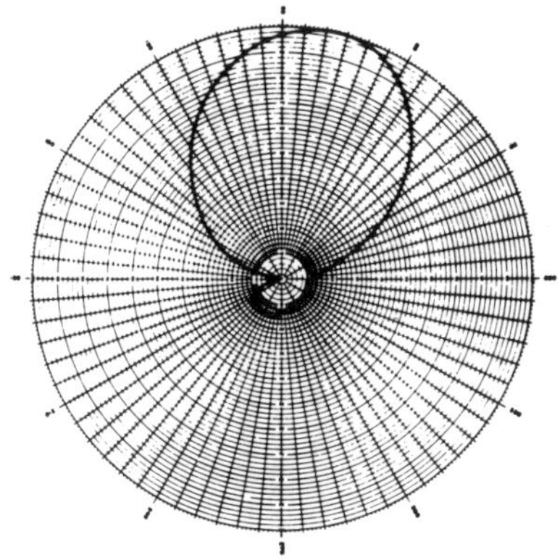

$\phi_0 = 45°, ka = \frac{\pi}{\sqrt{2}}$

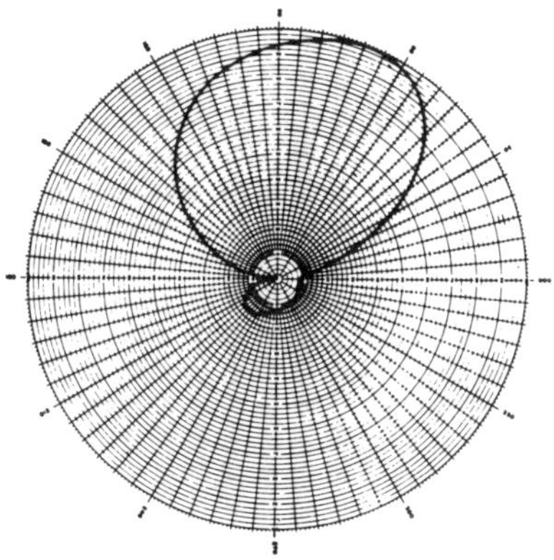

$\phi_0 = 45°, ka = \pi$

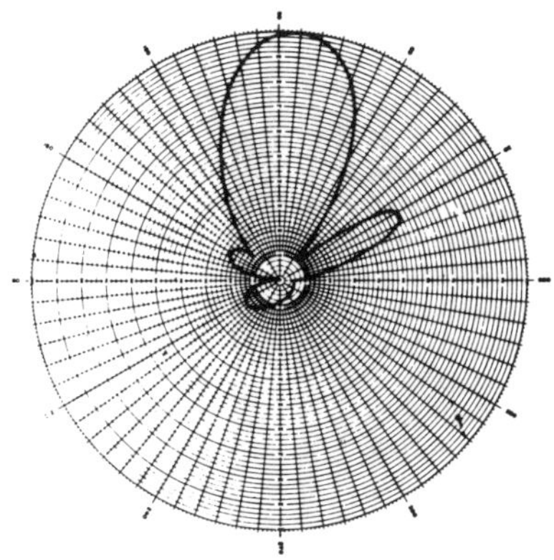

$\phi_0 = 45°, ka = 2\pi$

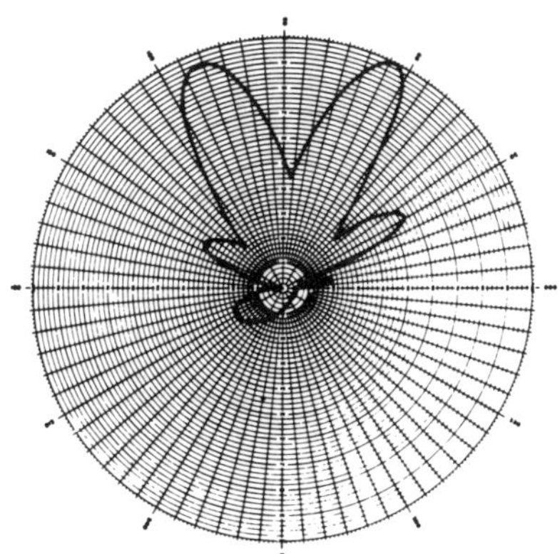

$\phi_0 = 45°, ka = 4\pi$

FIG. 33-5d

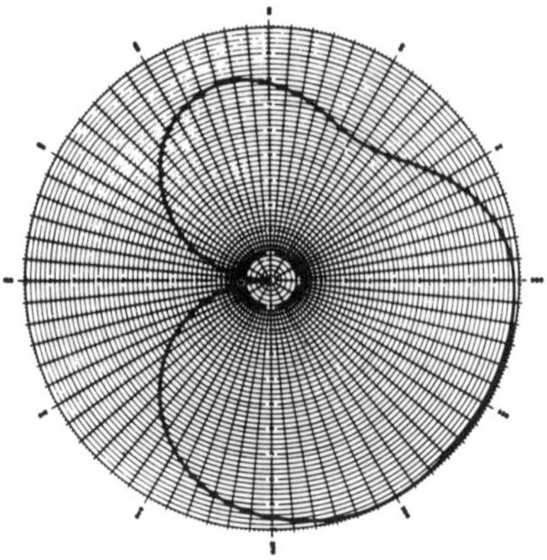

$\phi_0 = 135°,\ ka = \frac{\pi}{5}$

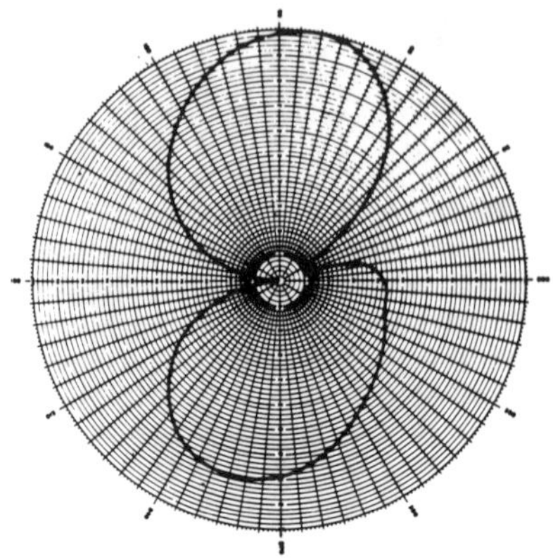

$\phi_0 = 135°,\ ka = \frac{\pi}{\sqrt{2}}$

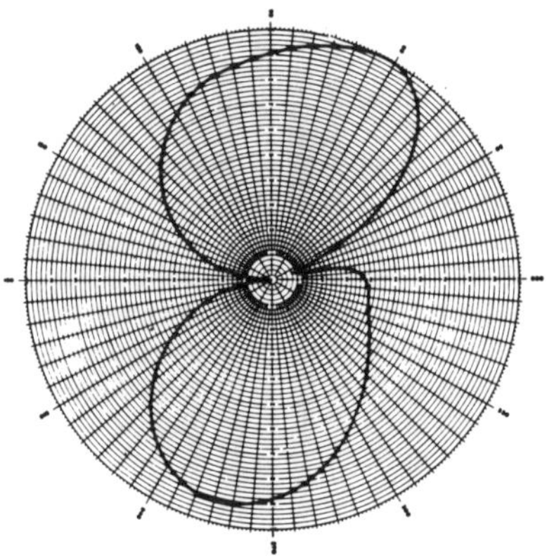

$\phi_0 = 135°,\ ka = \pi$

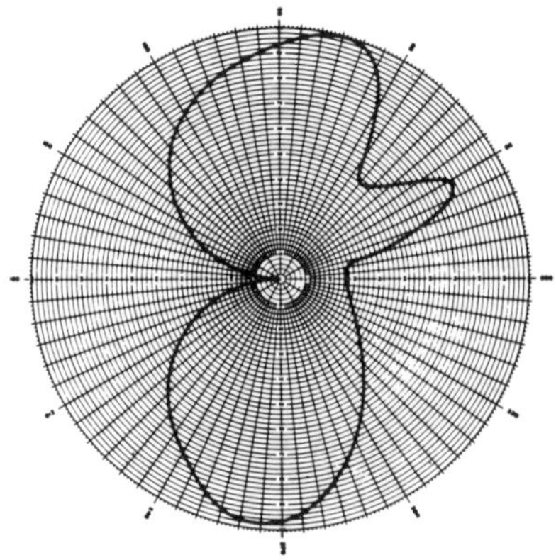

$\phi_0 = 135°,\ ka = 2\pi$

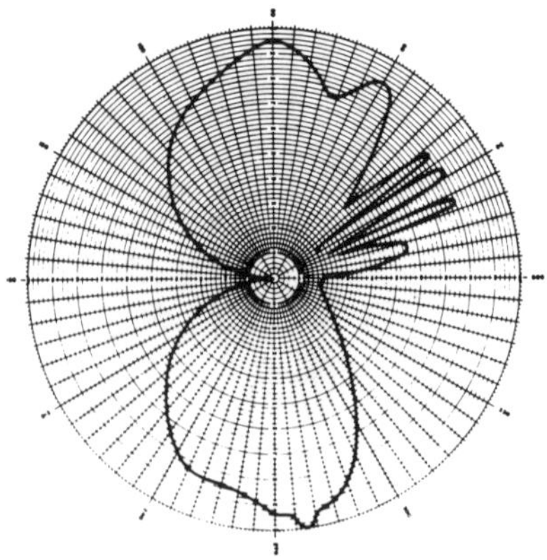

$\phi_0 = 135°,\ ka = 4\pi$

FIG. 33-5e

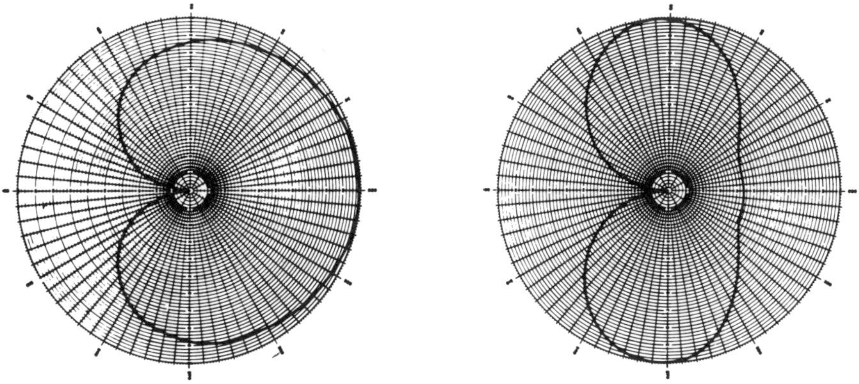

$\phi_0 = 180°,\ ka = \frac{\pi}{5}$ $\qquad$ $\phi_0 = 180°,\ ka = \pi$

FIG. 33-5f

Since the analysis is almost the same as that for the electric dipoles except that we use (2) and (3) as the basic equations, the details will be omitted. We summarize below the expressions for the far-zone electric field in the principal plane which will be used later in our discussion of radiating apertures.

1. Longitudinal Magnetic Dipole

$$E_\phi\left(R,\frac{\pi}{2},\phi\right) = \frac{\omega\mu_0 k m e^{ikR}}{4\pi R}\left\{S(ka,\phi-\phi_0)+S(ka,\phi+\phi_0)\right\}. \tag{4}$$

2. Horizontal Magnetic Dipole

$$E_z\left(R,\frac{\pi}{2},\phi\right) = \frac{\omega\mu_0 k m e^{ikR}}{4\pi R}\left\{\sin\phi\left[S(ka,\phi-\phi_0)+S(ka,\phi+\phi_0)\right] + \left(\frac{2}{\pi ka}\right)^{\frac{1}{2}} e^{i(ka+\pi/4)}\cos\frac{\phi_0}{2}\cos\frac{\phi}{2}\right\}. \tag{5}$$

3. Vertical Magnetic Dipole

$$E_z\left(R,\frac{\pi}{2},\phi\right) = \frac{-\omega\mu_0 k m e^{ikR}}{4\pi R}\left\{\cos\phi\left[S(ka,\phi-\phi_0)-S(ka,\phi+\phi_0)\right] - \left(\frac{2}{\pi ka}\right) e^{i(ka+\pi/4)}\sin\frac{\phi_0}{2}\sin\frac{\phi}{2}\right\}. \tag{6}$$

In the following section we will show that when a longitudinal magnetic dipole or a horizontal magnetic dipole is placed at the surface of a half-sheet, corresponding to $\phi_0 = 0$ or $\phi_0 = 2\pi$, the pattern function is the same as that of a small slot cut in the sheet.

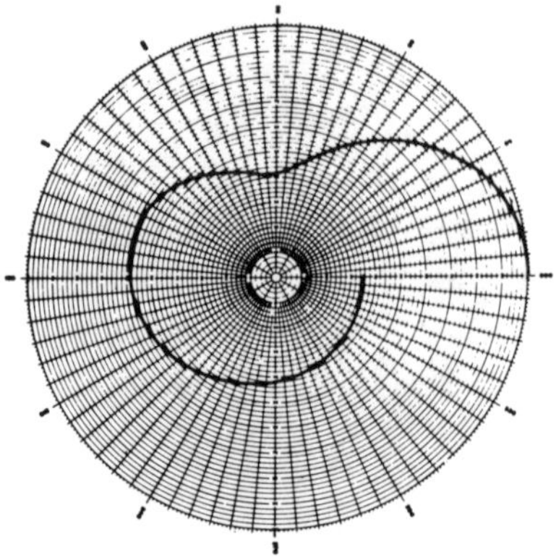

$\phi_0 = 0°, ka = \frac{\pi}{5}$

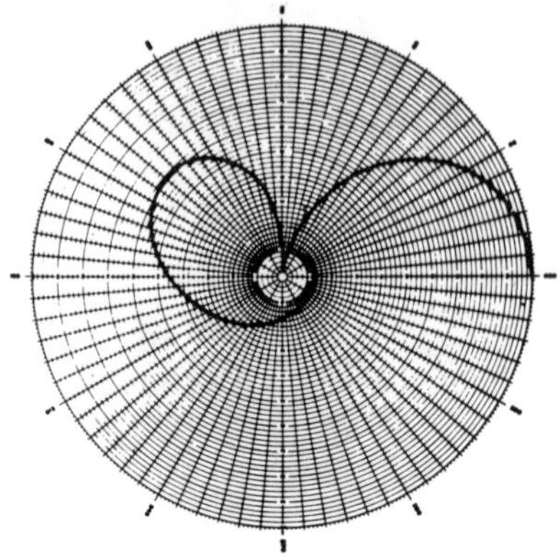

$\phi_0 = 0°, ka = \frac{\pi}{\sqrt{2}}$

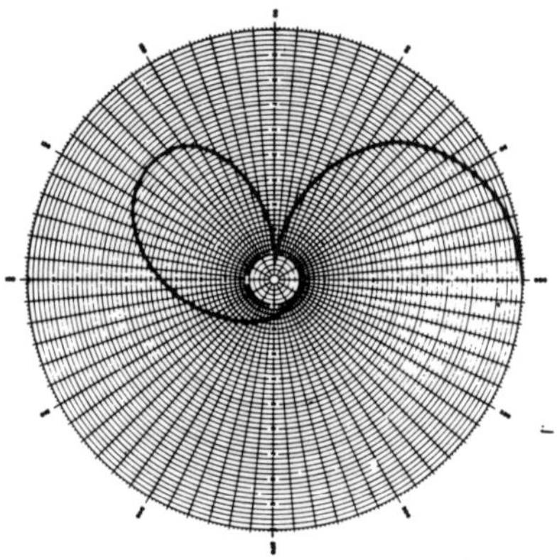

$\phi_0 = 0°, ka = \pi$

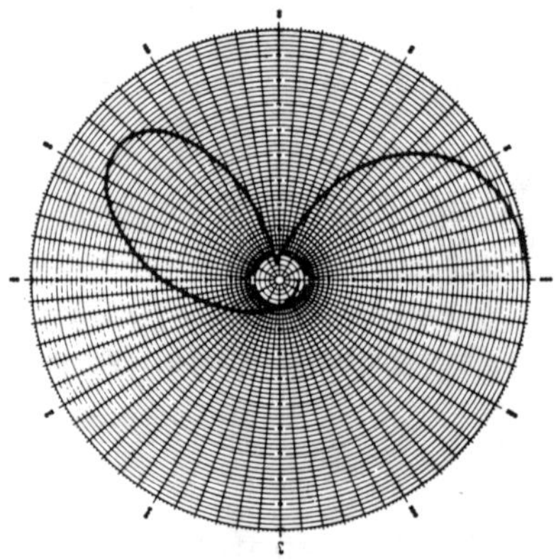

$\phi_0 = 0°, ka = 2\pi$

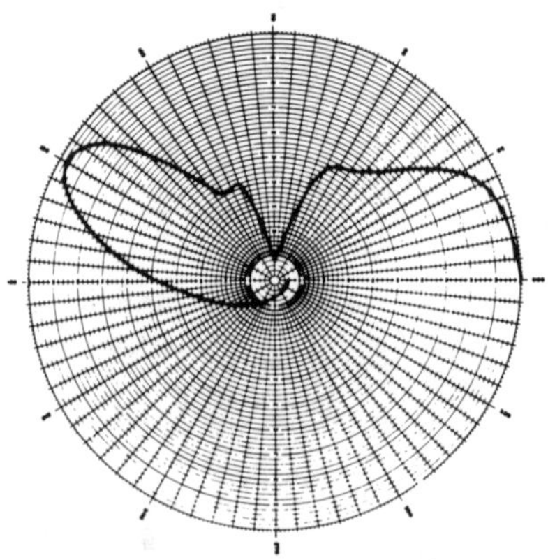

$\phi_0 = 0°, ka = 4\pi$

FIG. 33-6a. Radiation pattern of a vertical dipole placed in front of a half-sheet.

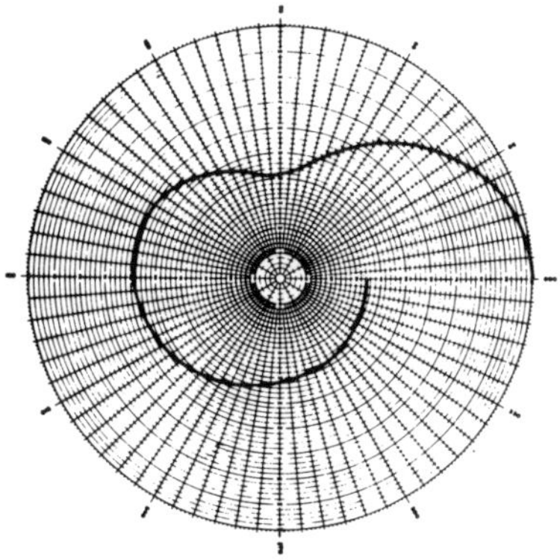

$\phi_0 = 5°, ka = \frac{\pi}{5}$

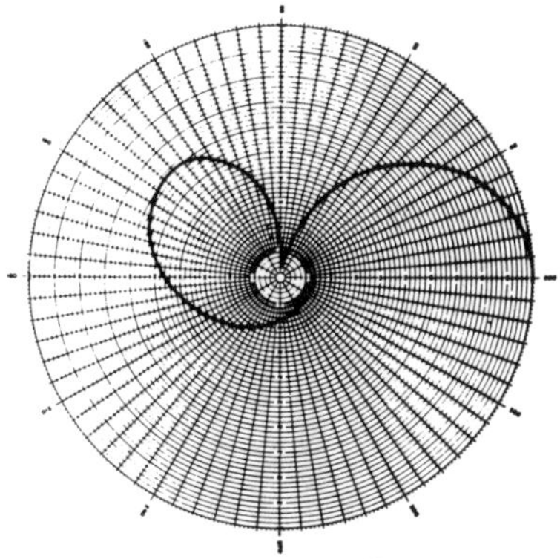

$\phi_0 = 5°, ka = \frac{\pi}{\sqrt{2}}$

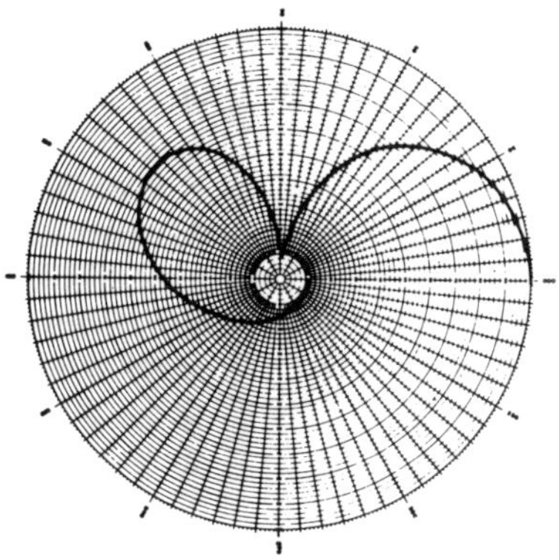

$\phi_0 = 5°, ka = \pi$

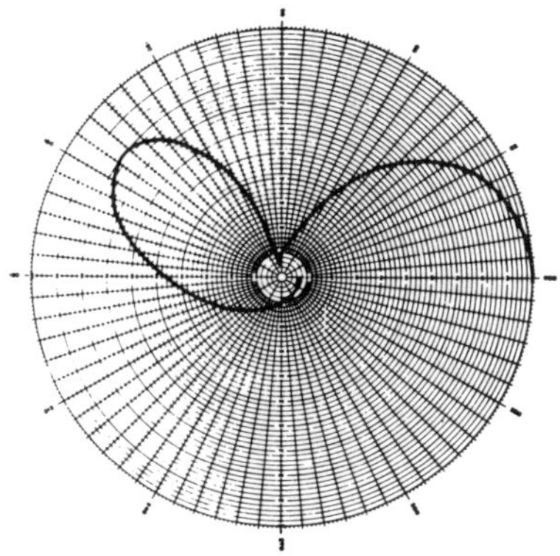

$\phi_0 = 5°, ka = 2\pi$

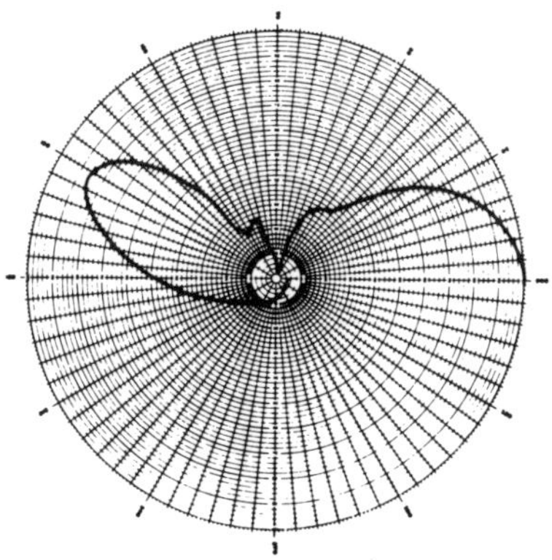

$\phi_0 = 5°, ka = 4\pi$

FIG. 33-6b

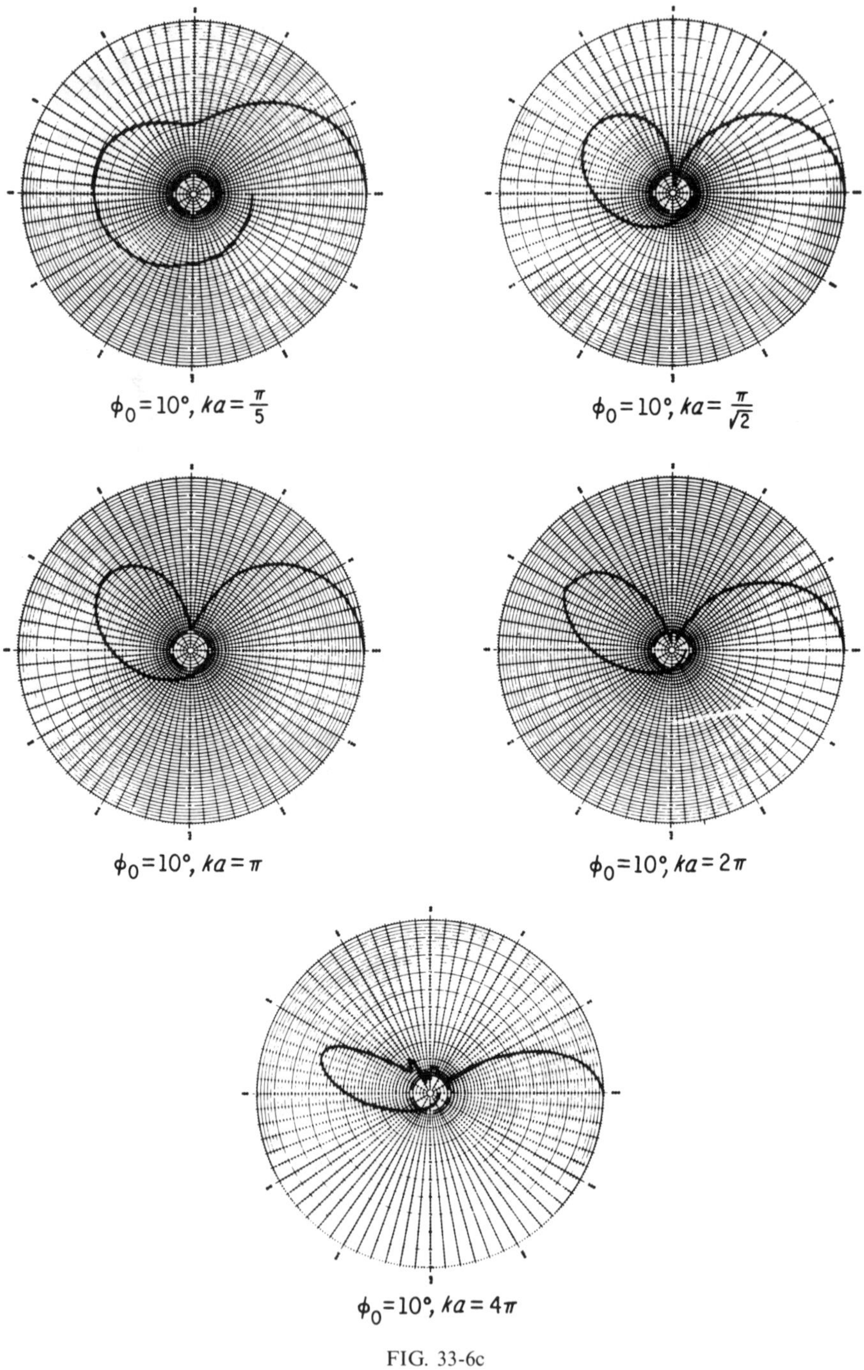
$\phi_0 = 10°$, $ka = \frac{\pi}{5}$
$\phi_0 = 10°$, $ka = \frac{\pi}{\sqrt{2}}$
$\phi_0 = 10°$, $ka = \pi$
$\phi_0 = 10°$, $ka = 2\pi$
$\phi_0 = 10°$, $ka = 4\pi$

FIG. 33-6c

35. RADIATION FROM SLOTS CUT IN A HALF-SHEET

To find the electric field due to radiating apertures cut on a conducting body we can use the general formula presented in Sec. 17. For clarity, we will discuss some of the materials using the specific type of Green's functions which are now at our disposal. For aperture problems we identify the Green's function in (17-2) to be the function of the first kind. When there is no volume distribution of current, the volume integral in that equation disappears. By interchanging the roles of $\bar{R}$ and $\bar{R}'$ and imposing the radiation condition at infinity, we obtain

$$\text{(1)} \qquad \bar{E}(\bar{R}) = -\iint_{S_d} \left\{ [\hat{n}_i \times \bar{E}(\bar{R}')] \cdot \nabla' \times \bar{\bar{G}}_1(\bar{R}'|\bar{R}) + [\hat{n}_i \times \nabla' \times \bar{E}(\bar{R}')] \cdot \bar{\bar{G}}_1(\bar{R}'|\bar{R}) \right\} dS'$$

where we have specifically used $\hat{n}_i$ to denote the inward normal to the surface of the diffracting body. By letting $\hat{n} = -\hat{n}_i$, and rearranging the terms (1) can be written in the form

$$\bar{E}(\bar{R}) = \iint_{S_d} \left\{ \tilde{\nabla}' \times \bar{\bar{G}}_1(\bar{R}'|\bar{R}) \cdot [\hat{n} \times \bar{E}(\bar{R}')] - [\hat{n} \times \tilde{\bar{\bar{G}}}_1(\bar{R}'|\bar{R}] \cdot \nabla' \times \bar{E}(\bar{R}') \right\} dS'.$$

In view of the symmetrical properties of the dyadic Green's functions listed in Table 17-1, we can convert it to

$$\bar{E}(\bar{R}) = \iint_{S_d} \left\{ \nabla \times \bar{\bar{G}}_2(\bar{R}|\bar{R}') \cdot [\hat{n} \times \bar{E}(\bar{R}')] - [\hat{n} \times \bar{\bar{G}}_1(\bar{R}|\bar{R}')] \cdot \nabla' \times \bar{E}(\bar{R}') \right\} dS'.$$

Since, by definition, $\hat{n} \times \bar{\bar{G}}_1 = 0$ on the surface of the body under consideration, we finally obtain

$$\text{(2)} \qquad \bar{E}(\bar{R}) = \iint_{S_d} \nabla \times \bar{\bar{G}}_2(\bar{R}|\bar{R}') \cdot [\hat{n} \times \bar{E}(\bar{R}')] \, dS'.$$

This is the basic formula which will be used in determining the electric field due to a radiating aperture cut on a conducting body. In that case, $\hat{n} \times \bar{E}(\bar{R}')$ vanishes elsewhere except across the aperture. In fact, by treating S_d as a general surface, (2) represents an alternative statement of the Huygen's principle using the function of the second kind instead of the free-space dyadic Green's function as the piloting function. By doing so, only the tangential component of the electric field is needed; acting as the virtual source of radiation. By comparing (2) with (34-2) it is obvious that the aperture electric field can also be interpreted as an equivalent layer of surface magnetization. Thus if we let the magnetization vector be distributed on a surface so that (34-2) becomes

$$\bar{E}(\bar{R}) = i\omega\mu_0 \iint \nabla \times \bar{\bar{G}}_2(\bar{R}|\bar{R}') \cdot \bar{M}_s(\bar{R}')\, dS' \tag{3}$$

then (2) and (3) are equivalent, if we assume

$$\hat{n} \times \bar{E}(\bar{R}') = i\omega\mu_0 \bar{M}_s(\bar{R}'). \tag{4}$$

Equation (4) can actually be obtained from (7-11) if we consider a volume integration of that equation in the neighborhood of an aperture and apply the same technique that was used in the derivation of (6-16) relating the discontinuity of the H-field with a surface current-density function.

In applying (2) to radiating apertures on a half-sheet, we shall distinguish two types of excitations, designated as "one-sided" and "two-sided." In practice, the one-sided excitation corresponds to an aperture radiation from a waveguide terminated on a half-sheet. If the opening on the sheet is excited by a two-wire transmission line, we have a two-sided excitation. The two cases are illustrated in Fig. 35-1.

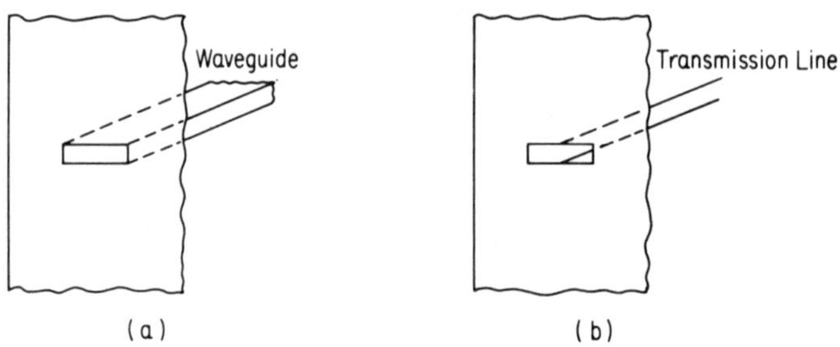

FIG. 35-1. (a) One-sided excitation; (b) two-sided excitation.

For numerical calculation, we shall only consider apertures in the form of narrow slots. This simplifies our analysis considerably. The types of slots to be studied are shown in Fig. 35-2, where the polarization of the aperture electric field is indicated by the arrows.

1. Longitudinal Slot

The field distribution for an infinitesimally narrow longitudinal slot can be written in the form

$$\bar{E}(\bar{R}') = f(z')\delta(r' - a)\hat{r}'$$

where $f(z')$ is a given function of z', depending on the actual field distribution

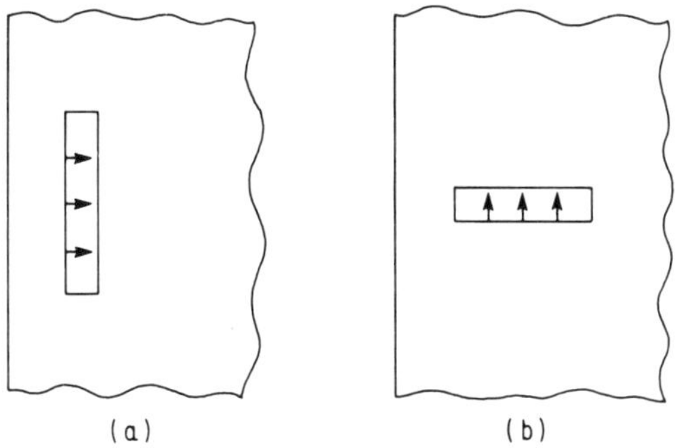

FIG. 35-2. (a) Longitudinal slot; (b) horizontal slot.

along the slot. In the case of a so-called half-wave resonant slot, we let

$$f(z') = v_0 \cos kz', \quad \frac{\lambda}{4} \geqslant z' \geqslant -\frac{\lambda}{4}.$$

Since we are going to discuss only the far-zone field in the principal plane, the exact knowledge of $f(z')$ is not needed. We merely demand that it is an even function of z'. For a one-sided slot, we assume that the opening is facing the $+y$ direction. In this case, the radiation pattern is the same as that of a longitudinal magnetic dipole placed at $r = a$ and $\phi_0 = 0$. According to (34-4), the field pattern is given by

(5) $$F_1(\phi) = S(ka, \phi).$$

For a two-sided excitation, the pattern function would be

(6) $$F_2(\phi) = S(ka, \phi) - S(ka, 2\pi - \phi).$$

2. Horizontal Slot

In general it is much more difficult to analyze the field pattern due to a horizontal slot. For an infinitesimally narrow slot we let

(7) $$\bar{E}(\bar{R}') = f(r')\delta(z' - 0)\,\hat{z},$$

then the far-zone electric field in the principal plane for a one-sided slot is represented by the expression

(8) $$E_z\left(R, \frac{\pi}{2}, \phi\right) = \frac{e^{ikR}}{4\pi R} \sum_{n=1}^{\infty} n(-i)^{n/2} \sin\frac{n\phi}{2} \int f(r') J_{n/2}(kr') \frac{dr'}{r'}.$$

This expression is obtained by substituting (7) into (2) with the Fourier integral simplified by the method of saddle-point integration. The radial integral contained in (8) depends strongly on the specific form of the function

$f(r')$. In general, it cannot be evaluated in a closed form. If we assume that the slot is infinitesimally short then $f(r')$ can be replaced by $\delta(r'-a)$ where a denotes the location of the small hole from the edge of the half-sheet. In that case we have

(9)

$$E_z\left(R,\frac{\pi}{2},\phi\right) = \frac{e^{ikR}}{4\pi aR}\sum_{n=1}^{\infty} n(-i)^{n/2}\sin\frac{n\phi}{2}J_{n/2}(ka) = -\frac{e^{ikR}}{4\pi aR}\frac{\partial}{\partial\phi}S(ka,\phi)\,.$$

The pattern function in this case is the same as that of a horizontal magnetic dipole, pointed in the x-direction, placed at the surface of the half-sheet, that is, by putting $\phi_0=0$ in (34-5). A convenient form of the pattern function to be used for numerical calculation is

(10) $$F_1(\phi)=\frac{1}{ika}\frac{\partial}{\partial\phi}S(ka,\phi)=\sin\phi\,S(ka,\phi)+\frac{e^{i(ka+\pi/4)}}{(2\pi ka)^{1/2}}\sin\frac{\phi}{2}\,.$$

For a two-sided infinitesimal horizontal slot the pattern function can most conveniently be written in the form

(11) $$F_2(\phi) = |\sin\phi|\,e^{-ika\cos\phi}\left[C(x)+iS(x)\right]_{x=ka(1+\cos\phi)} + \frac{ie^{ika}}{(\pi ka)^{1/2}}\sin\frac{\phi}{2}\,.$$

Numerical calculations based on (5), (6), (10), and (11) for several different values of ka are plotted in Figs. 35-3 to 35-6.

It is of some practical interest to examine in detail the radiation pattern of a slot when it is located far away from the edge of a half-sheet. A typical pattern for a longitudinal slot placed at a distance corresponding to $ka=30$ is shown in Fig. 35-7. The locations of the maxima and minima of such a pattern can be ascertained in a relatively simple manner. In the case of a one-sided longitudinal slot, the extreme values are determined by the equation

(12) $$\frac{\partial|S(ka,\phi)|}{\partial\phi} = 0\,.$$

Using the explicit expressions for $S(ka,\phi)$ represented by (33-8), we find that (12) yields

(13) $$\frac{0.5+C(x)}{0.5+S(x)} = -\tan x, \qquad x=ka(1+\cos\phi).$$

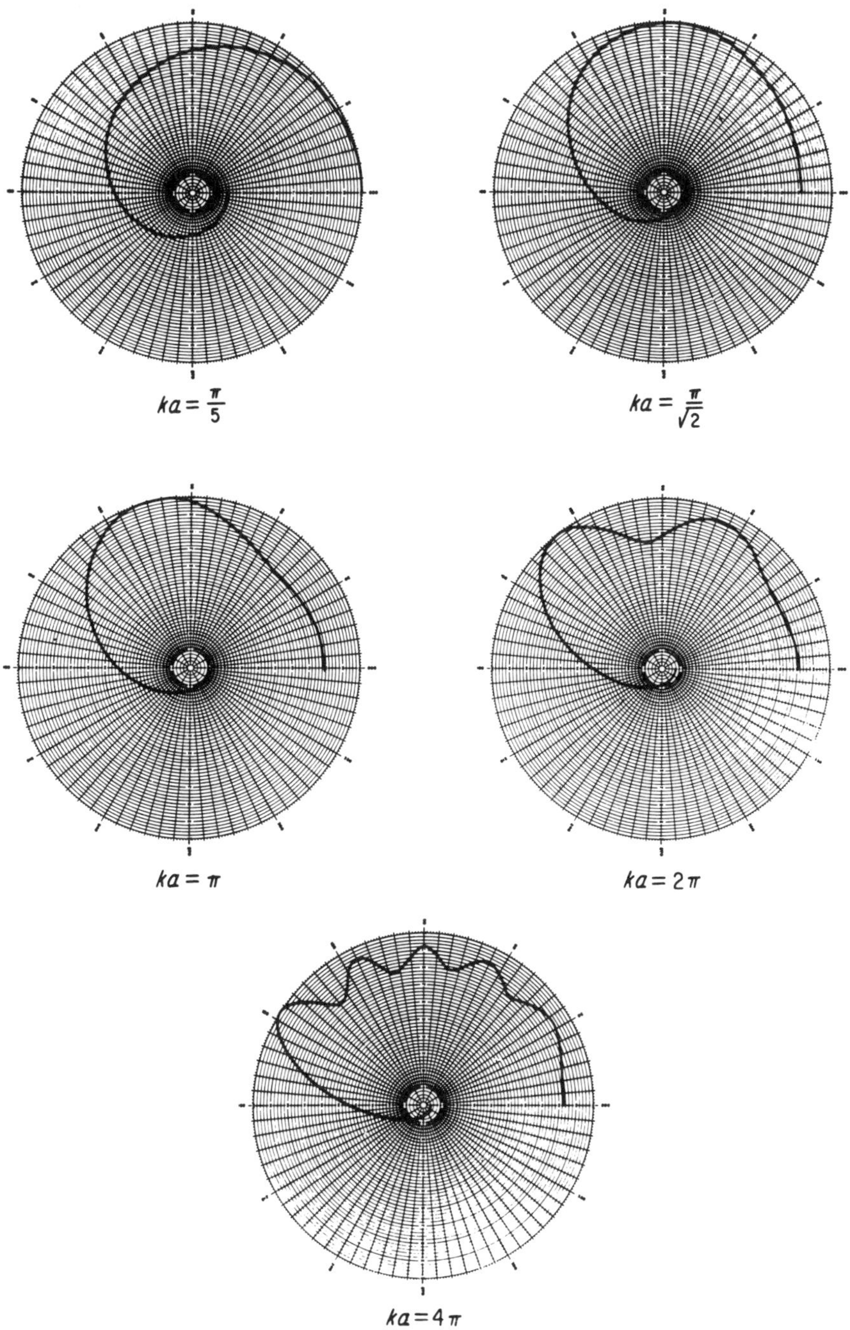

FIG. 35-3. Radiation pattern of a one-sided longitudinal slot.

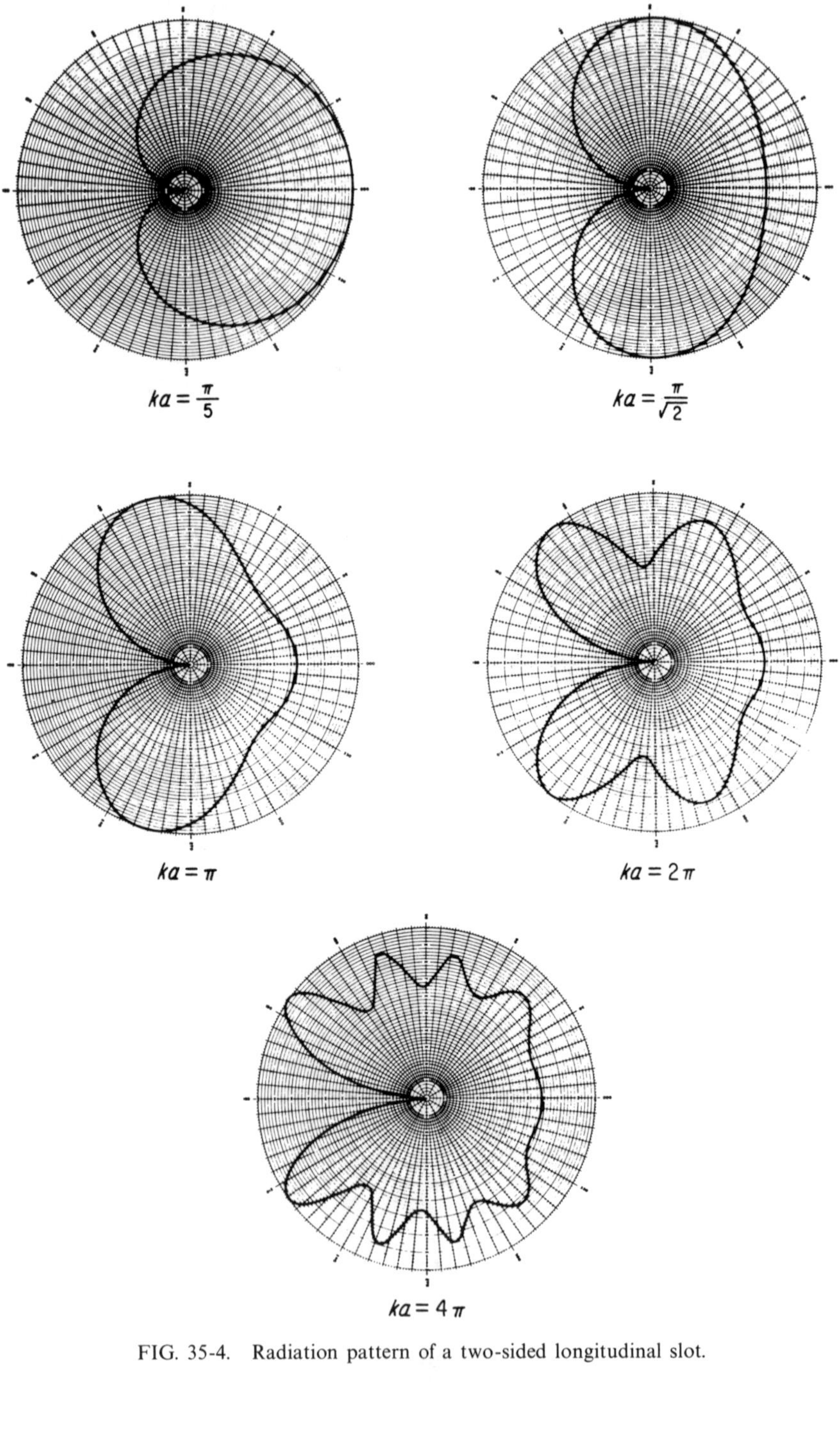

FIG. 35-4. Radiation pattern of a two-sided longitudinal slot.

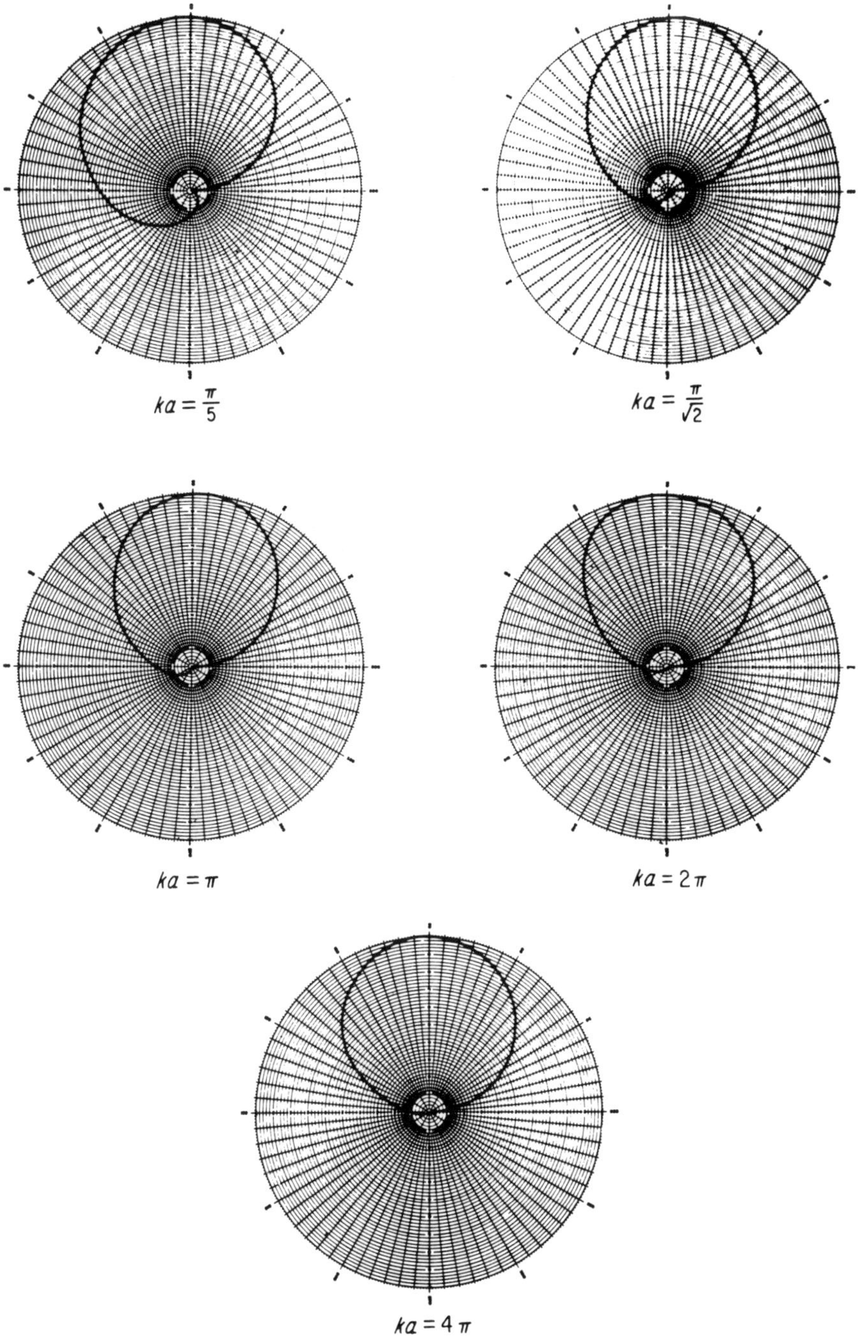

FIG. 35-5. Radiation pattern of a small one-sided horizontal aperture.

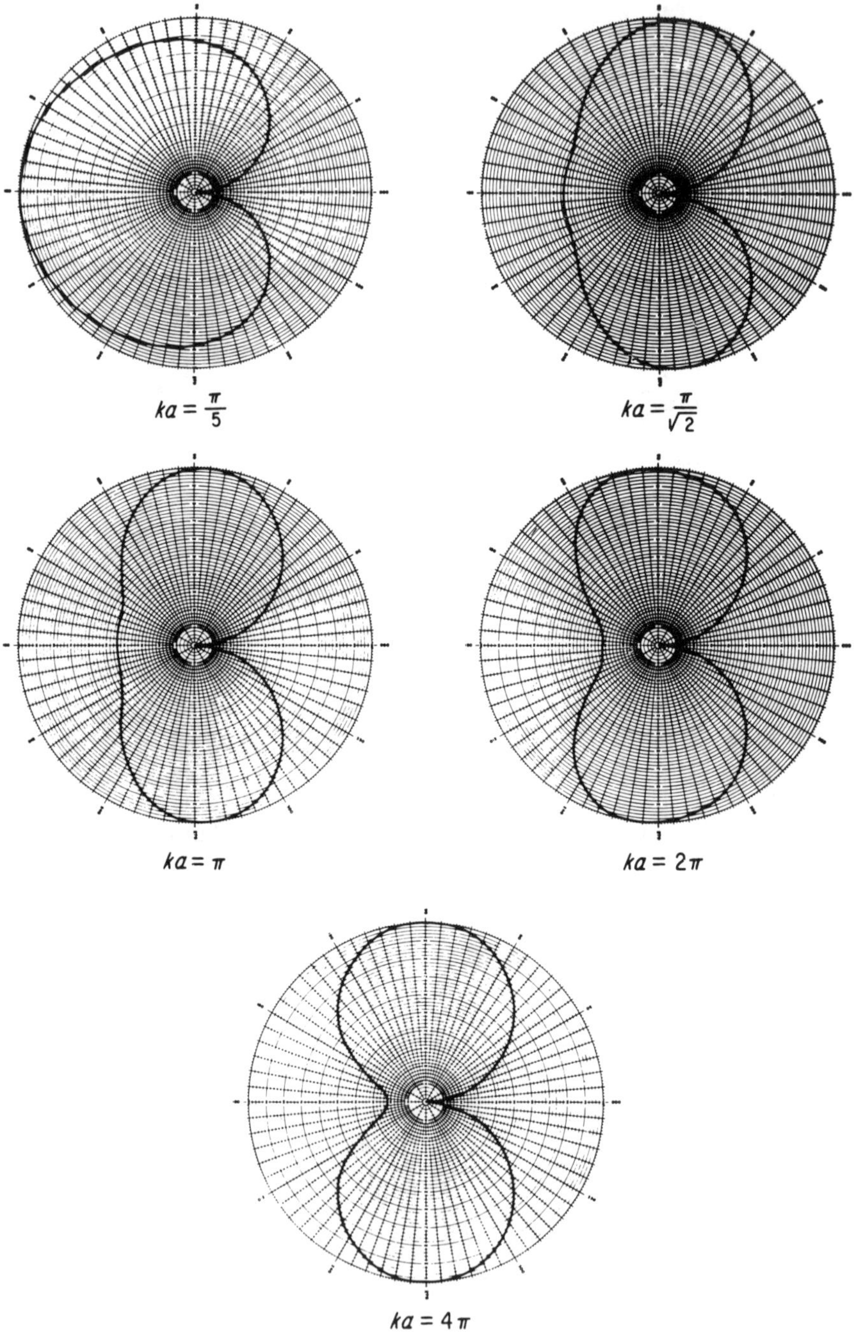

FIG. 35-6. Radiation pattern of a small two-sided horizontal aperture.

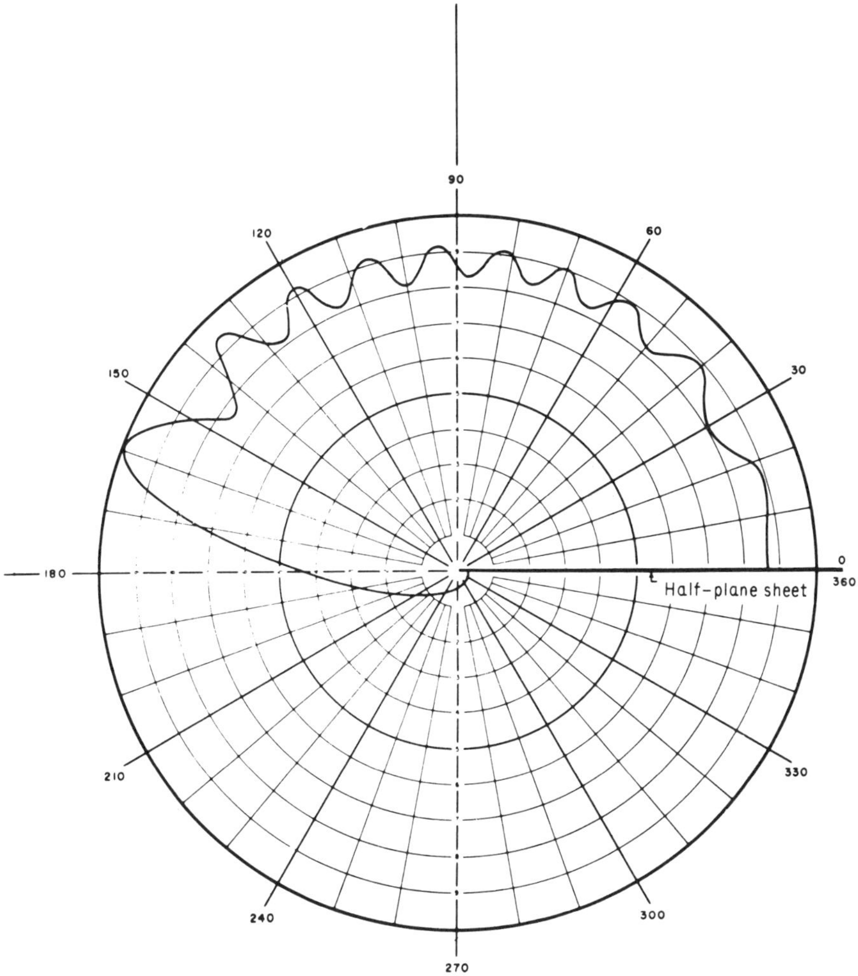

FIG. 35-7. Radiation pattern of a longitudinal slot placed at $ka = 30$.

The graphical solution for the roots of (13) is shown in Fig. 35-8. Denoting these roots by χ_m, it is seen from the display that the approximate solutions for these roots are given by

$$\chi_m = (4m - 1)\frac{\pi}{4}, \quad m = 1, 2, \ldots$$

or

$$\cos\phi_m = (4m - 1)\frac{\pi}{4ka} - 1, \quad m = 1, 2, \ldots \tag{14}$$

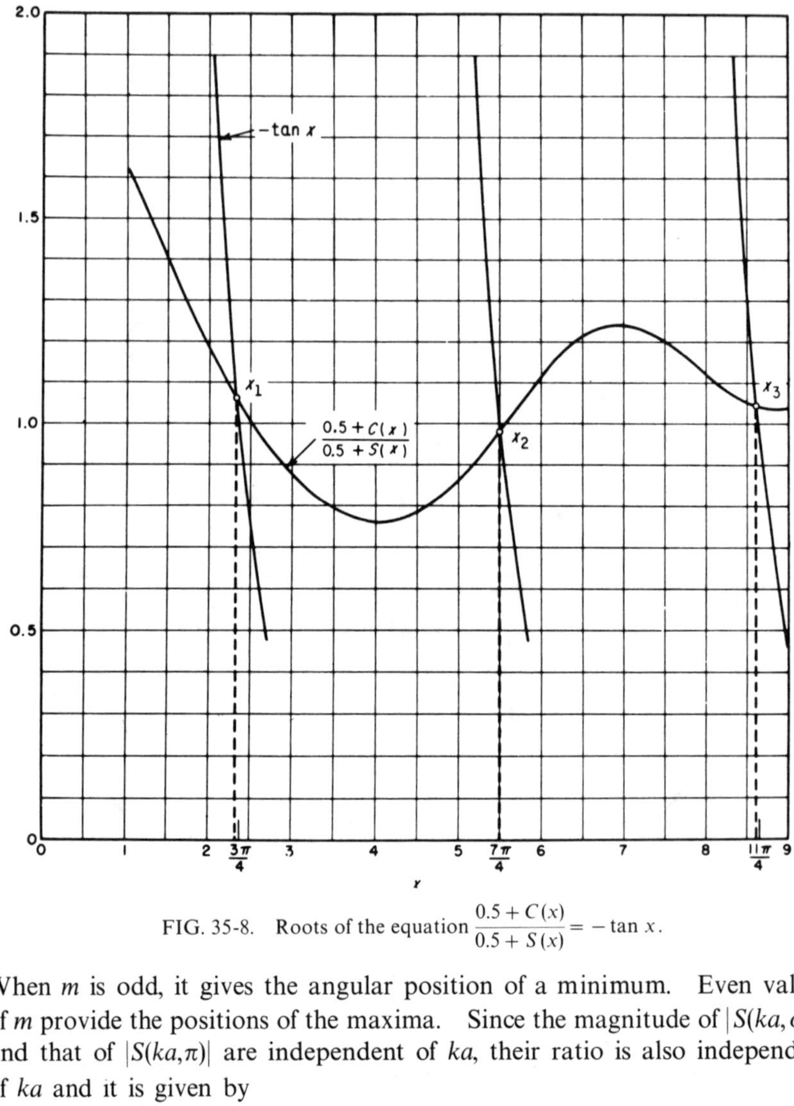

FIG. 35-8. Roots of the equation $\dfrac{0.5 + C(x)}{0.5 + S(x)} = -\tan x$.

When m is odd, it gives the angular position of a minimum. Even values of m provide the positions of the maxima. Since the magnitude of $|S(ka, \phi_1)|$ and that of $|S(ka, \pi)|$ are independent of ka, their ratio is also independent of ka and it is given by

$$\frac{|S(ka, \pi)|}{|S(ka, \phi_1)|} = 0.428 \tag{15}$$

This number can be used as a measure of the rate of decay of the field intensity from its peak value to the value observed at the grazing direction. Equation (14) shows that when ka is large, ϕ_1 approaches π, but there is always a sharp decay of the field as the point of observation changes from the lit region to the shadow region. Such an effect must be taken into consideration in the design of a flush-mounted antenna over a flat surface of finite dimension.

36. RADIATION RESISTANCE OF LINEAR ANTENNAS PLACED NEAR A HALF-SHEET AND RADIATION CONDUCTANCE OF NOTCH ANTENNAS

In this paragraph we will discuss the radiation resistance of a quarter-wave monopole attached to the edge of a half-sheet, and the radiation resistance of a half-wave dipole placed in front of the sheet with its axis parallel to the edge of the sheet as shown in Fig. 36-1, (a) and (b). The

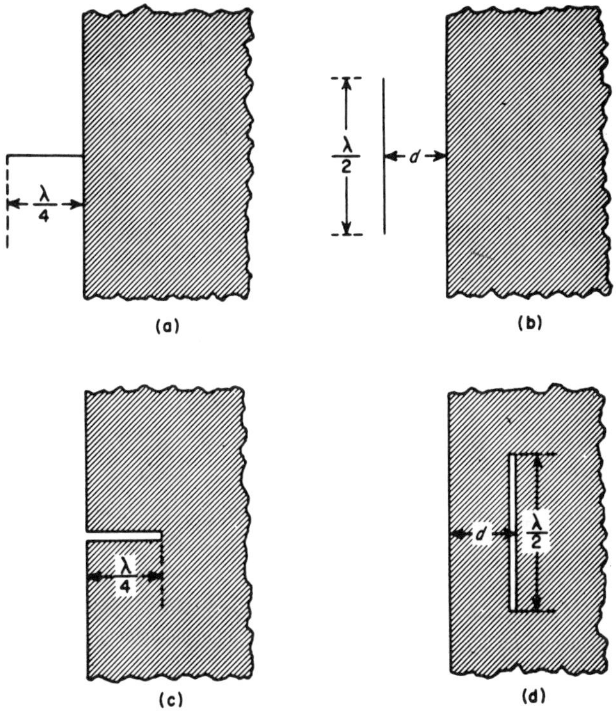

FIG. 36-1. Dipoles and slots with a half-sheet.

complementary problems of a quarter-wave slot normal to the edge of the sheet, and a half-wave slot cut in the sheet are shown in (c) and (d) of the same figure. The radiation resistance of the dipole and the radiation conductance of the complementary slot satisfies the Babinet-Booker relationship

$$\frac{R_d}{G_s}=\frac{\eta_0^2}{4},$$

where $\eta_0=120\pi\,\Omega$. The analysis that follows for the radiation resistances of the monopole and the dipole therefore are equally useful for evaluating the radiation conductances of the complementary slots. It is, of course, understood that these slots are two-sided.

To evaluate the radiation resistance of the two linear antennas under consideration, we shall apply the Poynting theorem to calculate the total radiated power of these antennas in the presence of the half sheet. For a base-driven quarter-wave monopole it will be assumed that the current distribution is sinusoidal. The current-density function to be used is represented by

$$\bar{J}(\bar{R}')=I_0\cos kr'\,\frac{\delta(z'-0)\delta(\phi'-\pi)}{r'}\,\hat{r}',\qquad \frac{\lambda}{4}\geqslant r'\geqslant 0\,.$$

Substituting it into (33-2) and simplifying the Fourier integrals by the method of saddle-point integration, we find that the far-zone electric field, when expressed in spherical coordinate system, is represented by

$$(1)\qquad E_\theta=-\frac{\eta_0 I_0 e^{ikR}}{2\pi R}\sum_{n=1,3\ldots}(i)^{n/2-1}\sin\left(\frac{n\phi}{2}\right)\cos\theta\,K_{n/2}(\theta)$$

$$(2)\qquad E_\phi=-\frac{\eta_0 I_0 e^{ikR}}{2\pi R}\sum_{n=1,3\ldots}\left(\frac{n}{2}\right)(i)^{n/2-1}\cos\left(\frac{n\phi}{2}\right)\frac{1}{\sin\theta}\,I_{n/2}(\theta)$$

where

$$(3)\qquad K_{n/2}(\theta)=k\int_0^{\lambda/4}J'_{n/2}(kr\sin\theta)\cos kr\,dr$$

$$(4)\qquad I_{n/2}(\theta)=\int_0^{\lambda/4}J_{n/2}(kr\sin\theta)\cos kr\,dr\,.$$

The prime in the integral function $K_{n/2}(\theta)$ denotes the derivative of the function with respect to the entire argument. The functions defined in (3) and (4) can be evaluated in terms of the Fresnel integrals. This is due to the fact that the half-order Bessel functions can be expressed in finite terms of algebraic and trigonometric functions (Stratton, p. 405). The first two low-order functions are

$$J_{1/2}(x)=\left(\frac{2}{\pi x}\right)^{1/2}\sin x$$

$$J_{3/2}(x) = \left(\frac{2}{\pi x}\right)^{1/2} \left(-\cos x + \frac{\sin x}{x}\right).$$

It is obvious that $I_{1/2}(\theta)$ can be expressed in terms of the Fresnel cosine-integral as defined by (33-7). To evaluate $I_{3/2}(\theta)$, we can use the method of integration by parts to reduce the algebraic function to a lower order. The complete expressions for $K_{1/2}$, $K_{3/2}$, $I_{1/2}$, and $I_{3/2}$ are tabulated in Appendix C. Using (1) and (2), we can determine the radiation resistance of the quarter-wave monopole defined in terms of the base current I_0. Its expression is

$$(5) \qquad R_m = \frac{2}{\eta_0 I_0^2} \int_0^{2\pi} \int_0^{\pi/2} (E_\theta^2 + E_\phi^2) R^2 \sin\theta \, d\theta \, d\phi$$

$$= \frac{\eta_0}{2\pi} \sum_{n=1,3\ldots} \left[\left(\frac{n}{2}\right)^2 \int_0^{\pi/2} I_{n/2}^2(\theta) \frac{d\theta}{\sin\theta} \right.$$

$$\left. + \int_0^{\pi/2} K_{n/2}^2(\theta) \cos^2\theta \sin\theta \, d\theta \right].$$

The integrals contained in (5) cannot be simplified further, and must be evaluated by numerical methods. The integrands

$$\frac{1}{\sin\theta}\left[\frac{n}{2} I_{n/2}(\theta)\right]^2 \quad \text{and} \quad \left[K_{n/2}(\theta)\cos\theta\right]^2 \sin\theta$$

are plotted in Fig. 36-2 for $n=1$ and $n=3$. The graph shows that contribution to R_m by terms with $n=5$ and higher are negligible. The sum of the four leading terms gives a value of R_m equal to 86.3 ohms. It is interesting to observe that the value is greater than the free-space radiation resistance of a half-wave dipole which is equal to 73.2 ohms. For a quarter-wave monopole driven against a ground plane, the radiation resistance, is, of course, only 36.6 ohms. The input resistance of a monopole driven against a large but finite sheet of metal has been measured experimentally using a model operated at 700 MHz. The measured value at resonance is equal to 92 ohms; that occurs when the length of the monopole is approximately equal to a quarter-wavelength. The proper trend is thus confirmed by the experiment.

For a half-wave dipole placed near a half-sheet as shown in Fig. 36-1b, we let

$$\bar{J}(\bar{R}') = I_0 \cos kz' \, \frac{\delta(r'-d)\delta(\phi-\pi)}{r'} \, \hat{z}, \qquad \frac{\lambda}{4} \geqslant z' \geqslant -\frac{\lambda}{4}.$$

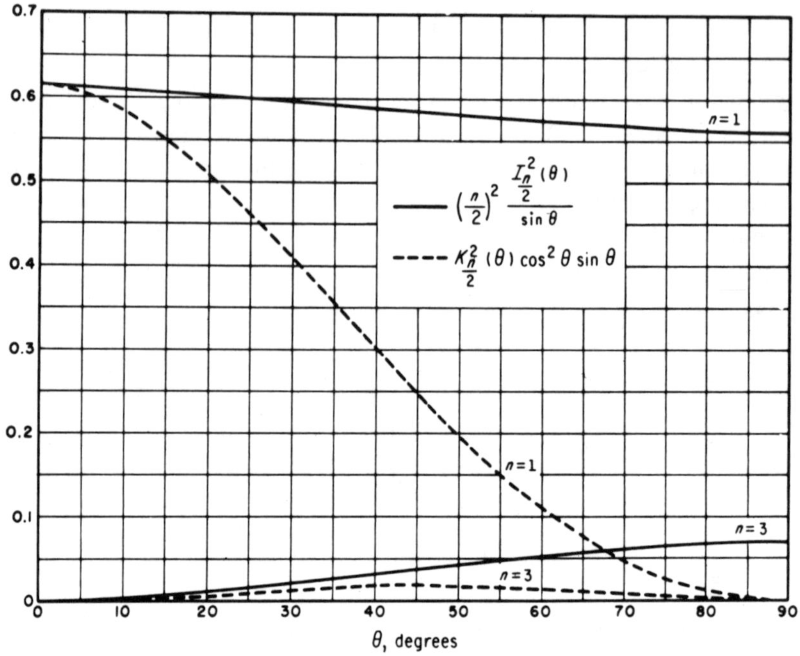

FIG. 36-2. The functions $\left(\frac{n}{2}\right)^2 I^2_{n/2}(\theta)/\sin\theta$ and $K^2_{n/2}\,\theta\cos^2\theta\sin\theta$.

The far-zone electric field in this case has only a θ-component. Following the same treatment, we obtain

$$(6)\qquad E_{\theta} = -\frac{\eta_0 I_0 e^{ikR}}{\pi R}\,\frac{\cos\left(\frac{\pi}{2}\cos\theta\right)}{\sin\theta}\sum_{n=1,3\ldots}(i)^{n/2}J_{n/2}(kd\sin\theta)\sin\frac{n\phi}{2}.$$

The radiation resistance of the dipole can be evaluated according to the formula

$$(7)\qquad R_d = \frac{2}{\eta_0 I_0^2}\int_0^{2\pi}\int_0^{\pi/2} E_\theta^2 R^2 \sin\theta\, d\theta\, d\phi$$

$$= \frac{2\pi}{\eta_0}\sum_{n=1,3\ldots}\int_0^{\pi/2}\frac{\cos^2\left(\frac{\pi}{2}\cos\theta\right)}{\sin\theta}J^2_{n/2}(kd\sin\theta)\,d\theta.$$

The integrals contained in (7) have been calculated numerically with the aid of a table of half-order Bessel functions. The final result for R_d as a function of kd is plotted in Fig. 36-3. For comparison, the radiation

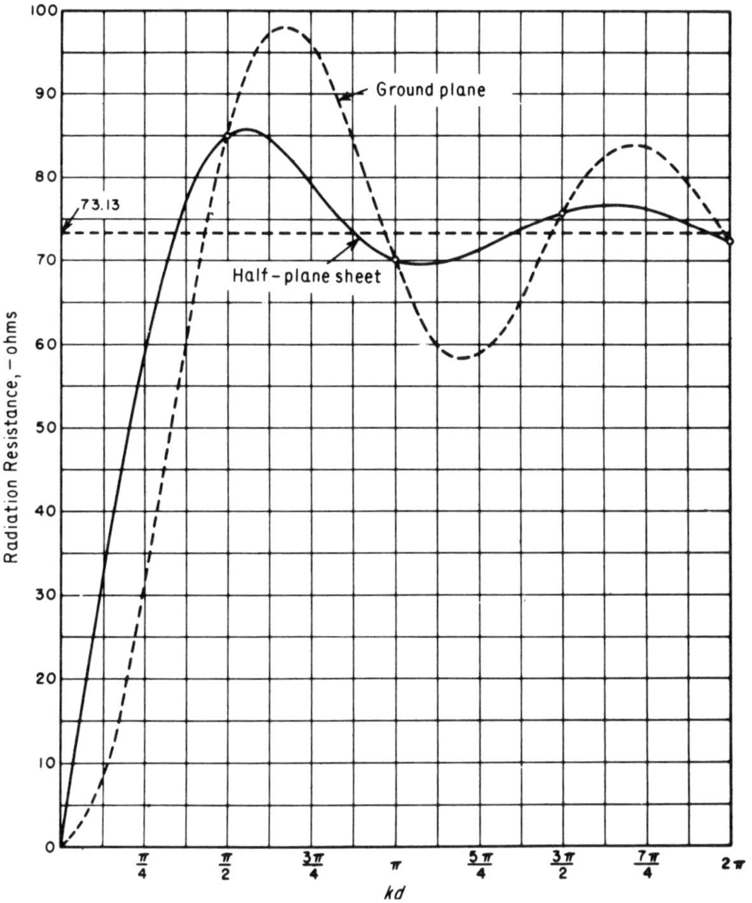

FIG. 36-3. Radiation resistance of a half-wave dipole placed in front of a half-sheet and in front of a ground plane.

resistance of a horizontal dipole placed above a ground plane is also plotted in the same figure. The two curves seem to intersect at those points where d is an integral multiple of quarter-wavelengths. It is believed that the identity probably can be proved analytically but we have not yet been able to supply such a proof. Physically it is difficult to explain why the two configurations produce the same radiated power for these cases, since the field distributions of the two structures are entirely different. Needless to say, the analysis discussed here can be applied equally well to a conducting wedge of arbitrary wedge angle. When the wedge angle is greater than 180°, the structure forms a corner reflector. Problems involving a horizontal dipole placed inside a corner reflector can, therefore, be investigated in the same manner, particularly when the angle of reflection does not satisfy the relation that $\alpha = \pi/n$, with $n = 1, 2, \ldots$. When $\alpha = \pi/n$, the problem can be analyzed more conveniently by the method of images.

37. DIFFRACTION OF A PLANE WAVE BY A HALF-SHEET

The conventional method of treating the diffraction of a plane electromagnetic wave by a simple geometrical body is to expand the plane wave in terms of the proper vector wave functions pertaining to the body and then find the appropriate vector wave functions for the diffracted field such that both the radiation condition and the boundary conditions at the surface of the body are satisfied. An alternative method, which is usually more clumsy, is to determine the far-zone field of an electric dipole placed in the neighborhood of the diffracting body; then by means of the reciprocity theorem the solution for the diffraction of a plane electromagnetic wave by the same body can be found. In the second method we must have at our disposal the far-zone field of a dipole with an arbitrary orientation with respect to the diffracting body. The dyadic Green's function technique actually comprises both of these two formulations. Without applying the reciprocity theorem, let us examine the diffraction problem for a half-sheet as an independent problem within the framework of our general formulation.

We consider an electric dipole which is perpendicular to the radial vector $\bar{R}_0$ as shown in Fig. 37-1. As $\bar{R}_0$ recedes to infinity from the origin, the primary field of the dipole would degenerate into a plane wave.

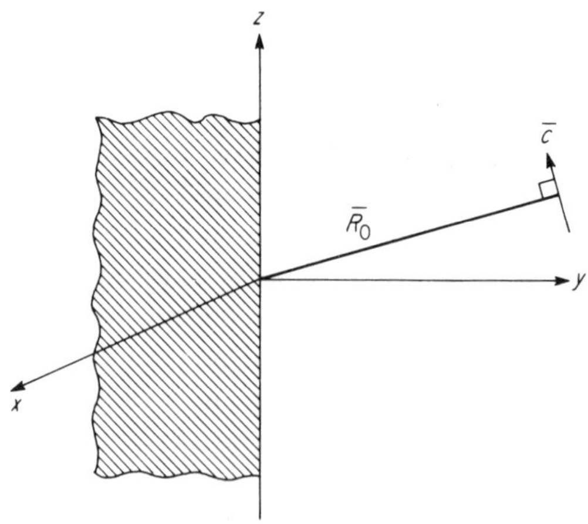

FIG. 37-1. An electric dipole oriented in a direction perpendicular to a radial vector $\bar{R}_0$.

The problem shown in Fig. 37-1 is similar to those treated in Sec. 33 except that we demand a certain orientation of the dipole with respect to its position vector. To identify the plane wave field with the field of a dipole measured at a large distance, we go back to (8-15). According to that formula the far-zone of a field of a dipole with current moment $\bar{c}$ as observed in a broadside direction can be written as

(1) $$\bar{E} = \frac{i\omega\mu_0\bar{c}e^{ikR_d}}{4\pi R_d}$$

where R_d is the perpendicular distance measured from the dipole. If we let

$$R_d = R_0 + d$$

and assume $d \ll R_0$, then

(2) $$\bar{E} = \left(\frac{i\omega\mu_0 c e^{ikR_0}}{4\pi R_0}\right)\hat{c}e^{ikd} = \left(\frac{i\omega\mu_0 c e^{ikR_0}}{4\pi R_0}\right)\hat{c}e^{i\bar{k}\cdot\bar{R}},$$

where the relations between R_d, R_0, d, and $\bar{R}$ are shown graphically in Fig. 37-2. When R_0 is very large, we designate the quantity inside the brackets of (2) as the amplitude of an incident plane wave propagating in the direction $\hat{k}$, that is

(3) $$\bar{E} = E_0\hat{c}e^{i\bar{k}\cdot\bar{R}}$$

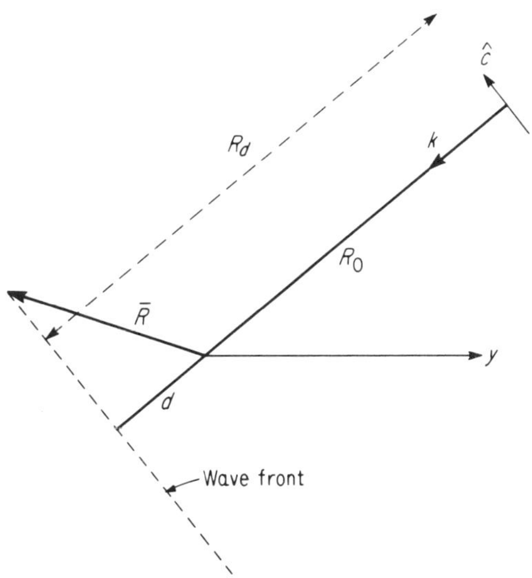

FIG. 37-2. Relations between various position vectors.

where

(4) $$E_0 = \frac{i\omega\mu_0 c e^{ikR_0}}{4\pi R_0}.$$

We now consider the asymptotic solution of the electric field resulting from

$$\bar{E}(\bar{R}) = i\omega\mu_0 \iiint \bar{\bar{G}}_1(\bar{R}|\bar{R}') \cdot \bar{J}(\bar{R}')\, dV'$$

with $\bar{\bar{G}}_1(\bar{R}|\bar{R}')$ given by (32-1) under the condition that $r < r'$. The analysis is very similar to the one covered in Sec. 33 except that the roles of $\bar{R}$ and $\bar{R}'$ are now interchanged. For illustration, let us treat the case that

$$\bar{J}(\bar{R}') = c\,\frac{\delta(r' - R_0)\delta(\phi - \phi_0)\delta(z' - 0)}{r'}\,\hat{z}$$

then

(5) $$\bar{E}(\bar{R}) = -\frac{\omega\mu_0 c}{8\pi k}\int_{-\infty}^{\infty} \sum_{\nu=0}^{\infty} \lambda^2 \bar{N}_{o\nu\eta}(h) H_\nu^{(1)}(\lambda R_0) \sin \nu\phi_0 \, dh$$

where

$$\lambda^2 = k^2 - h^2$$

$$\nu = \frac{n}{2}, \quad n = 1, 2, \ldots.$$

For large values of kR_0, the asymptotic expression for (5) is given by

(6) $$\bar{E}(\bar{R}) = \frac{i\omega\mu_0 c e^{ikR_0}}{2\pi R_0} \sum_{n=1}^{\infty} (-i)^{n/2} \sin\frac{n\phi_0}{2} \sin\frac{n\phi}{2} J_{n/2}(kr)\,\hat{z}.$$

Of course, if we invoke the reciprocity theorem, (6) can be obtained from (33-4) by putting $\theta = \pi/2$ in that equation, and substituting for R and a, respectively, R_0 and r. In view of (4) and (33-5), (6) can be converted to

(7) $$E_z(\bar{R}) = 2E_0 \sum_{n=1}^{\infty} (-i)^{n/2} \sin\frac{n\phi_0}{2} \sin\frac{n\phi}{2} J_{n/2}(kr)$$

$$= E_0[S(kr, \phi - \phi_0) - S(kr, \phi + \phi_0)].$$

This represents the total electric field resulting from the incidence of a plane wave

$$\bar{E}_i = E_0 e^{-ikr\cos(\phi - \phi_0)}\,\hat{z}$$

on the half-sheet. Because our method of approach is different from Sommerfeld's original formulation (Sommerfeld, *Optics*, p. 249) our final expression also appears different in form although they are equivalent.

In addition to Sommerfeld's original and masterful treatment of this subject, the problem under consideration was also covered very thoroughly by Baker and Copson (1950), particularly from the point of view of geometrical theory of diffraction. For completeness, we shall repeat some of these discussions but based on our formulation. For large values of x, the Fresnel integrals defined by (33-7) have the asymptotic form

$$C(x) + iS(x) = \frac{1}{2}(1+i) - \frac{ie^{ix}}{\sqrt{2\pi x}}, \qquad x \gg 1 .$$

The asymptotic formula for the function $S(\rho, \phi)$ defined by (33-8) is therefore given by

$$S(\rho,\phi) = \begin{cases} e^{-i\rho\cos\phi} - \dfrac{e^{i(\rho+\pi/4)}}{2\sqrt{\pi\rho(1+\cos\phi)}}, & \pi > \phi \geqslant 0 \\[2ex] \dfrac{e^{i(\rho+\pi/4)}}{2\sqrt{\pi\rho(1+\cos\phi)}}, & 2\pi \geqslant \phi > \pi . \end{cases} \tag{8}$$

It is understood that (8) is valid only if $\rho(1+\cos\phi) \gg 1$. Thus it cannot be applied to the region where ϕ is near π. To discuss the asymptotic expression for (7), we shall divide the region of observation into three distinct zones as shown in Fig. 37-3.

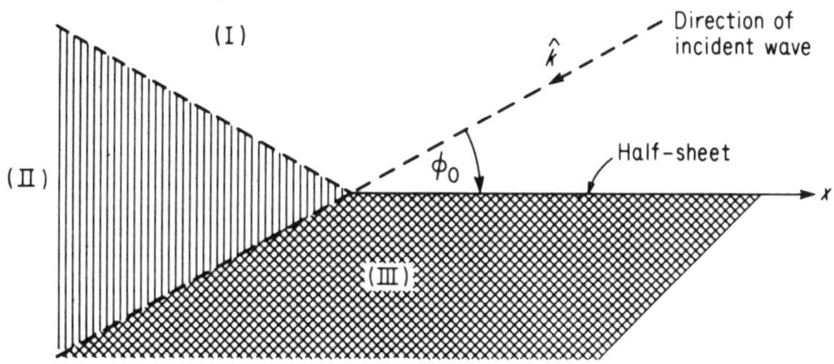

FIG. 37-3. Three distinct zones.

(I) Reflection Zone; $\phi < \pi - \phi_0$.

In this zone

$$S(kr, \phi - \phi_0) = e^{-ikr\cos(\phi-\phi_0)} - \frac{e^{i(kr+\pi/4)}}{2\sqrt{2\pi kr}\cos\left(\dfrac{\phi-\phi_0}{2}\right)}$$

$$S(kr, \phi + \phi_0) = e^{-ikr\cos(\phi + \phi_0)} - \frac{e^{i(kr + \pi/4)}}{2\sqrt{2\pi kr}\cos\left(\frac{\phi + \phi_0}{2}\right)}$$

provided that $kr[1 + \cos(\phi + \phi_0)] \gg 1$. This inequality implies that ϕ cannot be too close to $\pi - \phi_0$. Substituting these two expressions into (7), we obtain

$$E_z(\bar{R}) = E_0\left[e^{-ikr\cos(\phi - \phi_0)} - e^{-ikr\cos(\phi + \phi_0)} + \frac{e^{i(kr + \pi/4)}}{2\sqrt{2\pi kr}} \cdot \left(\frac{1}{\cos\frac{\phi + \phi_0}{2}} - \frac{1}{\cos\frac{\phi - \phi_0}{2}}\right)\right] \cdot \tag{9}$$

The first term in the above expression is simply the incident wave. The second is the same as the reflected wave from a perfectly conducting full-sheet. The remaining term is the diffracted field attributed to the edge of the half-sheet. It has the form of a cylindrical wave emerging from the edge.

(II) Interference Zone: $\pi - \phi_0 < \phi < \pi + \phi_0$.

In this zone

$$S(kr, \phi - \phi_0) = e^{-ikr\cos(\phi - \phi_0)} - \frac{e^{i(kr + \pi/4)}}{2\sqrt{2\pi kr}\cos\frac{\phi - \phi_0}{2}}$$

$$S(kr, \phi + \phi_0) = \frac{e^{i(kr + \pi/4)}}{2\sqrt{2\pi kr}\left|\cos\frac{\phi + \phi_0}{2}\right|} = -\frac{e^{i(kr + \pi/4)}}{2\sqrt{2\pi kr}\cos\frac{\phi + \phi_0}{2}}$$

provided that $kr[1 + \cos(\phi \pm \phi_0)] \gg 1$. Thus, ϕ cannot be too close to $\pi - \phi_0$ or $\pi + \phi_0$. Substituting these expressions into (7), we obtain

$$E_z(\bar{R}) = E_o\left[e^{-ikr\cos(\phi - \phi_0)} + \frac{e^{i(kr + \pi/4)}}{2\sqrt{2\pi kr}}\left(\frac{1}{\cos\frac{\phi + \phi_0}{2}} - \frac{1}{\cos\frac{\phi - \phi_0}{2}}\right)\right]. \tag{10}$$

We have only the direct wave and the diffracted cylindrical wave in this zone.

(III) Shadow Zone; $\pi + \phi_0 < \phi$.

In this zone

$$S(kr, \phi - \phi_0) = \frac{e^{i(kr + \pi/4)}}{2\sqrt{2\pi kr}\left|\cos\frac{\phi - \phi_0}{2}\right|} = -\frac{e^{i(kr + \pi/4)}}{2\sqrt{2\pi kr}\cos\frac{\phi - \phi_0}{2}}$$

$$S(kr, \phi + \phi_0) = \frac{e^{i(kr + \pi/4)}}{2\sqrt{2\pi kr}\left|\cos\frac{\phi + \phi_0}{2}\right|} = -\frac{e^{i(kr + \pi/4)}}{2\sqrt{2\pi kr}\cos\frac{\phi + \phi_0}{2}}$$

provided that ϕ is not near $\pi + \phi_0$. In this case, we have

(11) $$E_z(\bar{R}) = \frac{e^{i(kr + \pi/4)}}{2\sqrt{2\pi kr}}\left(\frac{1}{\cos\frac{\phi + \phi_0}{2}} - \frac{1}{\cos\frac{\phi - \phi_0}{2}}\right)$$

Equations (9) to (11) are identical to the ones previously given by Sommerfeld and Baker and Copson.

As we have emphasized, these asymptotic expressions are valid approximations only if ϕ is not near $\pi - \phi_0$ or $\pi + \phi_0$. To be more specific, we can define two parabolic contours such that

$$kr\left[1 + \cos(\phi + \phi_0)\right] = K$$

$$kr\left[1 + \cos(\phi - \phi_0)\right] = K$$

where K is a positive constant equal to or greater than 10. Then (9) to (11) are good approximations when the point of observation (r, ϕ) lies outside of these contours. Figure 37-4 shows two typical contours corresponding to $K = 4\pi$ so that

$$\frac{r}{\lambda}\left[1 + \cos(\phi \pm \phi_0)\right] = 2.$$

The three distinct zones are now more clearly defined in this figure. When the point of observation lies inside the parabolic regions, we must use the exact expression as described by (7) to evaluate the field. Near the edge of the sheet, it is sufficient to keep the leading terms of the series expansion of the Fresnel integral given by

(12) $$C(x) + iS(x) = \left(\frac{2x}{\pi}\right)^{1/2} \sum_{n=0}^{\infty} \frac{(ix)^n}{n!(2n+1)}$$

for the numerical calculation of the field intensity. The so-called "edge condition" that characterizes the behavior of an electromagnetic field in the neighborhood of the sharp edge of a conductor can be investigated by this approach.

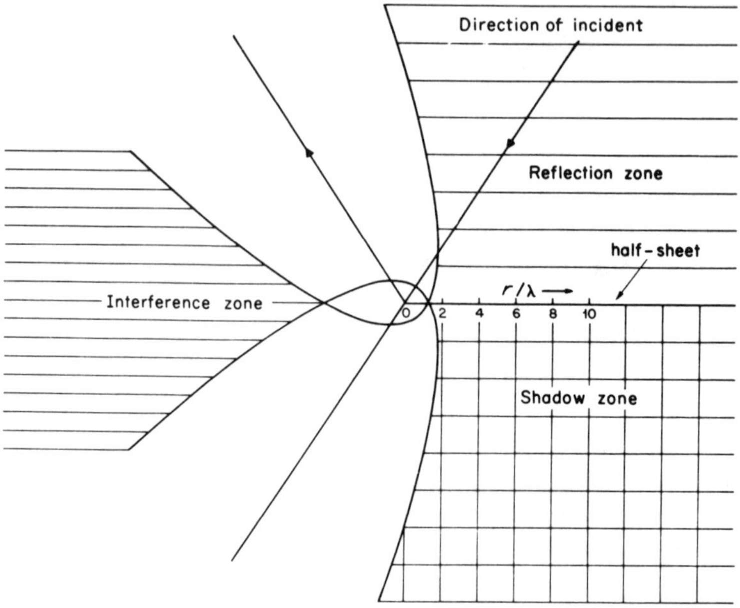

FIG. 37-4. Contours defining the valid regions ot the asymptotic solutions.

38. CIRCULAR CYLINDER AND HALF-SHEET

The dyadic Green's functions which we have derived for the half-sheet can be generalized to include the effect of a cylinder, conducting or dielectric, mounted at the edge of the sheet as shown in Fig. 38-1.

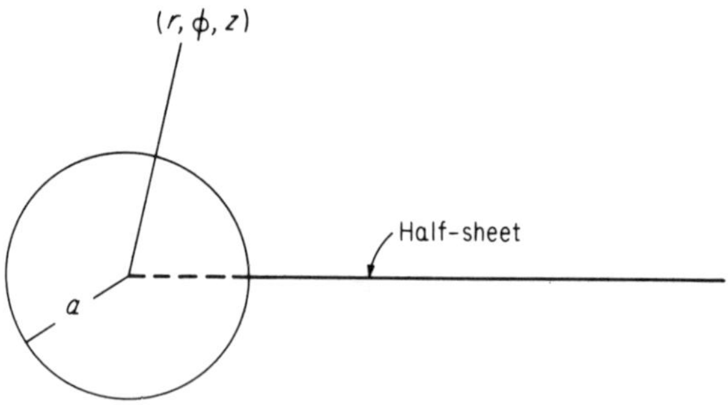

FIG. 38-1. A composite body made of a circular cylinder and a half-sheet.

For illustration, let us assume the cylinder to be perfectly conducting, then by means of the method of scattering superposition we can find the function of the first or the second kind for the composite body without any difficulty. The final expression for the function of the first kind is given by

$$\overline{\overline{G}}_1(\bar{R}|\bar{R}') = \frac{i}{8\pi}\int_{-\infty}^{\infty} dh \sum_{v=0}^{\infty} \frac{2-\delta_0}{\eta^2} \tag{1}$$

$$\cdot \begin{cases} \bar{M}^{(1)}_{ev\eta}(h)\left[\bar{M}'_{ev\eta}(-h) + \alpha \bar{M}'^{(1)}_{ev\eta}(-h)\right] \\ \left[\bar{M}_{ev\eta}(h) + \alpha \bar{M}^{(1)}_{ev\eta}(h)\right]\bar{M}'^{(1)}_{ev\eta}(-h) \end{cases}$$

$$\left.\begin{aligned} &+ \bar{N}^{(1)}_{ov\eta}(h)\left[\bar{N}'_{ov\eta}(-h) + \beta \bar{N}'^{(1)}_{ov\eta}(-h)\right] \\ &+ \left[\bar{N}_{ov\eta}(h) + \beta \bar{N}^{(1)}_{ov\eta}(h)\right]\bar{N}'^{(1)}_{ov\eta}(-h) \end{aligned}\right\} \quad \begin{aligned} r &> r' \\ r &< r' \end{aligned}$$

where

$$\alpha = -\frac{\partial J_v(x)}{\partial x} \bigg/ \frac{\partial H_v^{(1)}(x)}{\partial x}, \qquad x = \eta a$$

$$\beta = -J_v(\eta a) \;/\; H_v^{(1)}(\eta a)$$

$$v = \frac{n}{2}, \quad n = 0, 1, 2, \ldots$$

$$\eta^2 = k^2 - h^2.$$

It is obvious that the same technique can be applied to composite bodies made of a conducting wedge and a cylinder and a cylindrical waveguide partitioned by a conducting wedge. In the next chapter we shall discuss the same for a body made of a sphere and a conducting cone.

chapter **11**

Spheres and Perfectly Conducting Cones

The construction of the dyadic Green's functions pertaining to spheres, perfectly conducting or dielectric, and perfectly conducting cones follows the same procedure used for the cylinder, except that the spherical and the conical vector wave functions are used. While the rectangular and the cylindrical vector wave functions are closely related to the Hertzian potentials, the spherical and the conical ones are akin to the Debye potentials. Once the orthogonal relations of these functions are known, the remaining task is straightforward.

In order to appreciate the elegance of the Ohm-Rayleigh method which we have used so far in finding various dyadic Green's functions, we shall present in this chapter an alternative, but rather tedious, algebraic method of deriving the eigenfunction expansion of the free-space dyadic Green's method. This method avoids the necessity of discussing the orthogonal properties of the vector wave functions, but its execution depends heavily upon our recognition of the intricate recurrence relations between various types of vector wave functions which are all generated from the same scalar wave functions but with different piloting vectors.

39. EIGENFUNCTION EXPANSION OF THE FREE-SPACE DYADIC GREEN'S FUNCTION USING THE SPHERICAL VECTOR WAVE FUNCTIONS

The eigenfunctions which are solutions of the scalar wave equation $\nabla^2\psi + \kappa^2\psi = 0$ in spherical coordinate systems and finite at origin can be written in the form

(1) $$\psi_{{}^{e}_{o}mn}(\kappa) = j_n(\kappa R) P_n^m(\cos\theta) \begin{matrix}\cos\\ \sin\end{matrix}\, m\phi$$

where $j_n(x)$ denotes the spherical Bessel function of order n which satisfies the differential equation

$$\frac{d^2}{dx^2}[xj_n(x)] + \left[1 - \frac{n(n+1)}{x^2}\right] xj_n(x) = 0 .$$

The spherical Bessel function is related to the half-order cylindrical Bessel function by

(2) $$j_n(x) = \left(\frac{\pi}{2x}\right)^{1/2} J_{n+1/2}(x) .$$

Later on, we need the spherical Hankel function of the first kind, denoted by $h_n^{(1)}(x)$, which is related to the cylindrical Hankel function of the first kind in the same way. That is,

$$h_n^{(1)}(x) = \left(\frac{\pi}{2x}\right)^{1/2} H_{n+1/2}^{(1)}(x) .$$

The function P_n^m (cos θ) in (1) denotes the associated Legendre functions of order (n,m). It satisfies the differential equation

(3) $$\frac{1}{\sin\theta}\frac{d}{d\theta}\left[\sin\theta\,\frac{dP_n^m(\cos\theta)}{d\theta}\right] + \left[n(n+1) - \frac{m^2}{\sin^2\theta}\right] P_n^m(\cos\theta) = 0 .$$

A very thorough description of the general properties of the spherical Bessel functions and the associated Legendre functions is found in Stratton (p. 399–411) and will not be repeated here. For problems involving a sphere, n and m are integers. Later on when we apply these functions to conical structures, n, in general, will be fractional. In that case we shall use a different notation. Until then, it is understood that n as well as m represent integers. As was first shown by Stratton (p. 415), two sets of spherical vector wave functions can be defined which are solutions of the vector wave equation $\nabla \times \nabla \times \bar{F} - \kappa^2 \bar{F} = 0$. They are

(4) $$\bar{M}_{\substack{e\\o}mn}(\kappa) = \nabla \times [\psi_{\substack{e\\o}mn}\bar{R}]$$

(5) $$\bar{N}_{\substack{e\\o}mn}(\kappa) = \frac{1}{\kappa}\nabla \times \nabla \times [\psi_{\substack{e\\o}mn}\bar{R}] .$$

These functions, like the other vector wave functions introduced before, satisfy the symmetrical relations

$$\bar{N}_{\substack{e\\o}mn}(\kappa) = \frac{1}{\kappa}\nabla \times \bar{M}_{\substack{e\\o}mn}(\kappa)$$

$$\bar{M}_{\substack{e \\ o}mn}(\kappa) = \frac{1}{\kappa}\nabla \times \bar{N}_{\substack{e \\ o}mn}(\kappa).$$

The complete expressions of these functions are given by

(6)
$$\bar{M}_{\substack{e \\ o}mn}(\kappa) = \mp \frac{m}{\sin\theta} j_n(\kappa R) P_n^m(\cos\theta) \begin{matrix}\sin \\ \cos\end{matrix} m\phi\, \hat{\theta} - j_n(\kappa R) \frac{\partial P_n^m(\cos\theta)}{\partial\theta} \begin{matrix}\cos \\ \sin\end{matrix} m\phi\, \hat{\phi}$$

(7)
$$\bar{N}_{\substack{e \\ o}mn}(\kappa) = \frac{n(n+1)}{\kappa R} j_n(\kappa R) P_n^m(\cos\theta) \begin{matrix}\cos \\ \sin\end{matrix} m\phi\, \hat{R} + \frac{1}{\kappa R}\frac{\partial}{\partial R}[R j_n(\kappa R)] \left[\frac{\partial P_n^m(\cos\theta)}{\partial\theta} \begin{matrix}\cos \\ \sin\end{matrix} m\phi\, \hat{\theta} \mp \frac{m}{\sin\theta} P_n^m(\cos\theta) \begin{matrix}\sin \\ \cos\end{matrix} m\phi\, \hat{\phi} \right].$$

It should be pointed out that the θ and ϕ components of these functions have the radial function, either

$$j_n(\kappa R) \qquad \text{or} \qquad \frac{1}{\kappa R}\frac{\partial}{\partial R}[R j_n(\kappa R)],$$

as their constituent. This characteristic has an important bearing on the boundary conditions which must be satisfied by an electromagnetic field at the surface of a sphere. The two sets of spherical vector wave functions thus defined have the following orthogonal relations:

(8)
$$\iiint \bar{M}_{\substack{e \\ o}mn}(\kappa) \cdot \bar{N}_{\substack{e \\ o}m'n'}(\kappa')\, dV = 0$$

(9)
$$\iiint \bar{M}_{\substack{e \\ o}mn}(\kappa) \cdot \bar{M}_{\substack{e \\ o}m'n'}(\kappa')\, dV = \begin{cases} 0, & m \neq m', \quad n \neq n' \\ \dfrac{(1+\delta_0)\pi^2 n(n+1)(n+m)!}{\kappa^2(2n+1)(n-m)!}\,\delta(\kappa-\kappa'), & m = m', \quad n = n' \end{cases}$$

$$\delta_0 = \begin{cases} 1, & m = 0 \\ 0, & m \neq 0 \end{cases}$$

(10)
$$\iiint \bar{N}_{\substack{e \\ o}mn}(\kappa) \cdot \bar{N}_{\substack{e \\ o}m'n'}(\kappa')\, dV$$

$$
= \begin{cases} 0, & m \neq m', \quad n \neq n' \\ \dfrac{(1+\delta_0)\pi^2 n(n+1)(n+m)!}{\kappa^2(2n+1)(n-m)!}\,\delta(\kappa-\kappa'), & m=m', \quad n=n'. \end{cases}
$$

The domain of integration is understood to be through the entire space. The proof of these relations for $m \neq m'$ and $n \neq n'$ is relatively simple because of the orthogonal properties of the trigonometric functions and those of the associated Legendre functions. It is sufficient for us to demonstrate the cases where $m = m'$ and $n = n'$. We consider, for example, the integral

$$
\begin{aligned} I &= \iiint \bar{N}_{emn}(\kappa) \cdot \bar{N}_{emn}(\kappa')\, dV \\ &= \int_0^\infty \int_0^\pi \int_0^{2\pi} \bar{N}_{emn}(\kappa) \cdot \bar{N}_{emn}(\kappa') R^2 \sin\theta \, dR\, d\theta\, d\phi. \end{aligned} \tag{11}
$$

Substituting the explicit expression for $\bar{N}_{emn}$ as given by (7) into (11), and performing the ϕ-integration we obtain

$$
\begin{aligned} I = \frac{(1+\delta_0)\pi}{\kappa\kappa'} \int_0^\infty \int_0^\pi \Bigg\{ & n^2(n+1)^2 j_n(\kappa R) j_n(\kappa' R)\, [P_n^m(\cos\theta)]^2 \\ & + \frac{\partial}{\partial R}[R j_n(\kappa R)] \frac{\partial}{\partial R}[R j_n(\kappa' R)] \left[\left(\frac{\partial P_n^m}{\partial\theta}\right)^2 + \left(\frac{m P_n^m}{\sin\theta}\right)^2 \right] \Bigg\} \sin\theta\, dR\, d\theta. \end{aligned}
$$

It is known (Stratton, pp. 403, 417) that

$$
\int_0^\pi [P_n^m(\cos\theta)]^2 \sin\theta\, d\theta = \frac{2}{2n+1}\frac{(n+m)!}{(n-m)!}
$$

$$
\int_0^\pi \left[\left(\frac{\partial P_n^m}{\partial\theta}\right)^2 + \left(\frac{m P_n^m}{\sin\theta}\right)^2 \right] \sin\theta\, d\theta = \frac{2n(n+1)}{2n+1}\frac{(n+m)!}{(n-m)!}.
$$

Thus, after the integration with respect to θ, we have

$$
\begin{aligned} I = \frac{2(1+\delta_0)\pi n(n+1)}{\kappa\kappa'(2n+1)} \frac{(n+m)!}{(n-m)!} \int_0^\infty \Bigg\{ & n(n+1) j_n(\kappa R) j_n(\kappa' R) \\ & + \frac{\partial}{\partial R}[R j_n(\kappa R)] \frac{\partial}{\partial R}[R j_n(\kappa' R)] \Bigg\}\, dR. \end{aligned} \tag{12}
$$

From the recurrence formulas for the Bessel functions, we can deduce the following formulas for the spherical Bessel functions:

(13) $$\frac{d}{dx}[xj_n(x)] = \frac{x}{2n+1}[(n+1)j_{n-1}(x) - nj_{n+1}(x)]$$

(14) $$j_n(x) = \frac{x}{2n+1}[j_{n-1}(x) + j_{n+1}(x)] .$$

Using these relations, the integrand in (12) can be changed to

$$\frac{R^2}{2n+1}[(n+1)j_{n-1}(\kappa R)j_{n-1}(\kappa' R) + nj_{n+1}(\kappa R)j_{n+1}(\kappa' R)].$$

In view of the integral representation of the weighted delta function in the three-dimensional case as stated by (3-26), we have

$$\int_0^\infty R^2 j_{n-1}(\kappa R)j_{n-1}(\kappa' R)\, dR = \frac{\pi}{2}\,\delta(\kappa - \kappa')$$

and

$$\int_0^\infty R^2 j_{n+1}(\kappa R)j_{n+1}(\kappa' R)\, dR = \frac{\pi}{2}\,\delta(\kappa - \kappa')$$

hence (12) reduces to

$$I = (1+\delta_0)\pi^2 \frac{n(n+1)}{(2n+1)} \frac{(n+m)!}{(n-m)!} \frac{\delta(\kappa - \kappa')}{\kappa^2}$$

which is the normalization factor found in (10). The proof for (9) follows the same procedure with some minor variations in details.

Knowing the proper vector wave functions to be used and their orthogonal properties, we can find the eigenfunction expansion of the free-space dyadic Green's function by means of the Ohm-Rayleigh method in the same manner as described in Chapter 7. The only difference is that we now have two discrete sets of eigenvalues, m and n, and one set of continuous eigenvalues, κ. Thus we let

(15) $$\bar{\bar{I}}\delta(\bar{R} - \bar{R}') = \int_0^\infty d\kappa \sum_{n=1}^{\infty} \sum_{m=0}^{n} [\bar{M}_{\substack{e\\o}mn}(\kappa)\bar{A}_{\substack{e\\o}mn}(\kappa) + \bar{N}_{\substack{e\\o}mn}(\kappa)\bar{B}_{\substack{e\\o}mn}(\kappa)].$$

n starts with unity because $n=0$ and $m=0$ corresponds to a null set. By taking the anterior scalar product of (5) with $\bar{M}_{\substack{e\\o}m'n'}(\kappa')$ and $\bar{N}_{\substack{e\\o}m'n'}(\kappa')$ in turn and integrating the resultant equations in the entire space, we find, as a result of the orthogonal properties of the spherical vector wave functions that

$$\bar{A}_{\substack{e\\o}mn}(\kappa) = \frac{2-\delta_0}{2\pi^2}\,\frac{(2n+1)}{n(n+1)}\,\frac{(n-m)!}{(n+m)!}\,\kappa^2 \bar{M}'_{\substack{e\\o}mn}(\kappa)$$

$$\bar{B}_{\substack{e\\o}mn}(\kappa) = \frac{2-\delta_0}{2\pi^2}\,\frac{(2n+1)}{n(n+1)}\,\frac{(n-m)!}{(n+m)!}\,\kappa^2 \bar{N}'_{\substack{e\\o}mn}(\kappa)$$

where the primed functions are defined with respect to the primed coordinates (R', θ', ϕ'), the site of the position vector $\bar{R}'$. The free-space dyadic Green's function which satisfies the equation

$$\nabla \times \nabla \times \bar{\bar{G}}_0(\bar{R}|\bar{R}') - k^2 \bar{\bar{G}}_0(\bar{R}|\bar{R}') = \bar{\bar{I}}\delta(\bar{R} - \bar{R}')$$

therefore, can be written in the form

$$\bar{\bar{G}}_0(\bar{R}|\bar{R}') = \frac{1}{2\pi^2}\int_0^\infty \frac{\kappa^2 d\kappa}{\kappa^2 - k^2} \sum_{n=1}^\infty \sum_{m=0}^n (2-\delta_0)\frac{2n+1}{n(n+1)}\frac{(n-m)!}{(n+m)!} \cdot \left[\bar{M}_{\substack{e\\o}mn}(\kappa)\bar{M}'_{\substack{e\\o}mn}(\kappa) + \bar{N}_{\substack{e\\o}mn}(\kappa)\bar{N}'_{\substack{e\\o}mn}(\kappa)\right]. \tag{16}$$

We now invoke the integral relation (4-16) that

$$\int_0^\infty \frac{\kappa^2 d\kappa}{\kappa^2 - k^2} j_n(\kappa R) j_n(\kappa R') = \frac{i\pi k}{2}\begin{cases} h_n^{(1)}(kR) j_n(kR'), & R > R' \\ j_n(kR) h_n^{(1)}(kR'), & R < R' \end{cases} \tag{17}$$

By writing the dyads in (16) such as $\bar{M}_{emn}(\kappa)\ \bar{M}'_{emn}(\kappa)$, in an operational form

$$\bar{M}_{emn}(\kappa)\ \bar{M}'_{emn}(\kappa) = \bar{\bar{T}}_{emn}[j_n(\kappa R) j_n(\kappa R')],$$

we have

$$\int_0^\infty \frac{\kappa^2 d\kappa}{\kappa^2 - k^2} \bar{\bar{T}}_{emn}[j_n(\kappa R) j_n(\kappa R')] = \frac{i\pi k}{2}\begin{cases} \bar{M}^{(1)}_{emn}(k)\bar{M}'_{emn}(k), & R > R' \\ \bar{M}_{emn}(k)\bar{M}'^{(1)}_{emn}(k), & R < R' \end{cases}$$

where $\bar{M}^{(1)}_{emn}(k)$ means that the vector wave function is defined with respect to the spherical Hankel function of the first kind, i.e.,

$$\bar{M}^{(1)}_{emn}(k) = \nabla \times \left[h_n^{(1)}(kR) P_n^m(\cos\theta)\cos m\phi \bar{R}\right].$$

By repeating the same for the dyads involving the $\bar{N}$-functions in (16), we obtain finally an expression for $\bar{\bar{G}}_0$ in a closed form, which reads

$$(18)\quad \bar{\bar{G}}_0(\bar{R}|\bar{R}') = \frac{ik}{4\pi} \sum_{n=1}^{\infty} \sum_{m=0}^{n} (2-\delta_0) \frac{2n+1}{n(n+1)} \frac{(n-m)!}{(n+m)!}$$

$$\cdot \begin{cases} \bar{M}^{(1)}_{\substack{e\\o}mn}(k)\bar{M}'_{\substack{e\\o}mn}(k) + \bar{N}^{(1)}_{\substack{e\\o}mn}(k)\bar{N}'_{\substack{e\\o}mn}(k), & R > R' \\ \bar{M}_{\substack{e\\o}mn}(k)\bar{M}'^{(1)}_{\substack{e\\o}mn}(k) + \bar{N}_{\substack{e\\o}mn}(k)\bar{N}'^{(1)}_{\substack{e\\o}mn}(k), & R < R'. \end{cases}$$

This expression will later be used to construct other kinds of dyadic Green's functions involving a sphere. Before doing so, we shall describe an alternative method that can also be used to find (18) at the expense of a rather tedious analytical manipulation of various vector wave functions.

40. AN ALGEBRAIC METHOD OF FINDING THE EIGENFUNCTION EXPANSION OF THE FREE-SPACE DYADIC GREEN'S FUNCTION

To prepare for the conversion of various vector wave functions that will be involved in this method, we shall first introduce other types of special vector wave functions in addition to the two standard ones defined by (39-4) and (39-5). According to the general discussion of the vector wave functions in the beginning of Sec. 19, functions of the type

$$\bar{M}^{(c)} = \nabla \times (\psi \hat{c})$$

$$\bar{N}^{(c)} = \frac{1}{k} \nabla \times \nabla \times (\psi \hat{c})$$

are solutions of the vector wave equation as long as ψ is a solution of the scalar wave equation $\nabla^2\psi + k^2\psi = 0$ and the piloting vector $\hat{c}$, is a constant vector. If we identify $\hat{c}$ to be $\hat{x}$ or $\hat{y}$ or $\hat{z}$ and ψ to be the eigenfunction of the scalar wave equation in spherical coordinate systems, i.e.,

$$\psi = \psi_{\substack{e\\o}mn}(k) = j_n(kR) P_n^m (\cos\theta) \frac{\cos}{\sin} m\phi$$

then we can define six more spherical vector wave functions as follows:

$$(1)\qquad \bar{M}^{(c)}_{\substack{e\\o}mn}(k) = \frac{1}{k} \nabla \times [\psi_{\substack{e\\o}mn} \hat{c}]$$

$$(2)\qquad \bar{N}^{(c)}_{\substack{e\\o}mn}(k) = \frac{1}{k^2} \nabla \times \nabla \times [\psi_{\substack{e\\o}mn} \hat{c}]$$

with $c = x$, y, z, respectively. For clarity, we shall call these functions the spherical wave functions of the c-type. The additional constant $1/k$ which is included in these functions makes them of the same dimension as that of $\bar{M}$ or $\bar{N}$ defined by (39-4) and (39-5). By converting the unit vectors

$\hat{x}$, $\hat{y}$ and $\hat{z}$ into the unit vectors in the spherical coordinate system with the aid of Table 1-2, we can find the explicit expressions for these functions. We shall now demonstrate some interesting relations between these types of functions and the standard or the radial spherical vector wave functions. These identities are the key relations to be used later in our derivation of the free-space dyadic Green's function based on the algebraic method. We will not prove all these relations in the text, but will give one example in detail. It will show quite clearly our approach to the analysis. We consider, for example, the function

$$\bar{M}^{(x)}_{emn}(k) = \frac{1}{k}\nabla \times (\psi_{emn}\hat{x}) \tag{3}$$

with

$$\psi_{emn} = j_n(kR)P_n^m(\cos\theta)\cos m\phi\,.$$

Substituting into it

$$\hat{x} = \sin\theta\cos\phi\hat{R} + \cos\theta\cos\phi\hat{\theta} - \sin\phi\hat{\phi}$$

we obtain

$$\begin{aligned}\bar{M}^{(x)}_{emn}(k) = \frac{1}{kR}\Bigg\{&\frac{-1}{\sin\theta}\left[\frac{\partial}{\partial\theta}(\sin\theta\sin\phi\,\psi_{emn}) + \frac{\partial}{\partial\phi}(\cos\theta\cos\phi\,\psi_{emn})\right]\hat{R}\\ &+\left[\frac{1}{\sin\theta}\frac{\partial}{\partial\phi}(\sin\theta\cos\phi\,\psi_{emn}) + \frac{\partial}{\partial R}(R\sin\phi\,\psi_{emn})\right]\hat{\theta}\\ &+\left[\frac{\partial}{\partial R}(R\cos\theta\cos\phi\,\psi_{emn}) - \frac{\partial}{\partial\theta}(\sin\theta\cos\phi\,\psi_{emn})\right]\hat{\phi}\Bigg\}.\end{aligned} \tag{4}$$

Let us now study the composition of the radial component of this function first, which is given by

$$\begin{aligned}\hat{R}\cdot\bar{M}^{(x)}_{emn} = \frac{-j_n(\rho)}{\rho\sin\theta}\Bigg[&\sin\phi\cos m\phi\frac{d}{d\theta}(\sin\theta\,P_n^m)\\ &+\cos\theta\,P_n^m\frac{d}{d\phi}(\cos\phi\cos m\phi)\Bigg],\end{aligned} \tag{5}$$

where ρ denotes kR. Using the sum and difference formulas for the trigonometric functions, we can write (5) in the form

$$\hat{R}\cdot\bar{M}^{(x)}_{emn} = \frac{-j_n(\rho)}{2\rho\sin\theta}\Bigg\{\left[\frac{d}{d\theta}(\sin\theta\,P_n^m) - (m+1)\cos\theta\,P_n^m\right]\sin(m+1)\phi \tag{6}$$

$$-\left[\frac{d}{d\theta}(\sin\theta\, P_n^m) + (m-1)\cos\theta\, P_n^m\right]\sin(m-1)\phi\Bigg\}$$

$$= \frac{-j_n(\rho)}{2\rho}\left[\left(\frac{dP_n^m}{d\theta} - m\,\frac{\cos\theta}{\sin\theta}\,P_n^m\right)\sin(m+1)\phi\right.$$

$$\left.-\left(\frac{dP_n^m}{d\theta} + m\,\frac{\cos\theta}{\sin\theta}\,P_n^m\right)\sin(m-1)\phi\right].$$

For further reduction we use the recurrence relations that

$$\frac{dP_n^m}{d\theta} \pm m\,\frac{\cos\theta}{\sin\theta}\,P_n^m = \begin{cases} (n-m+1)(n+m)P_n^{m-1} \\ -P_n^{m-1} \end{cases}$$

which can be obtained by combining the following two formulas found in Stratton (p. 401):

$$\frac{1}{2}\left[(n-m+1)(n+m)P_n^{m-1} \pm P_n^{m+1}\right] = \begin{cases} \dfrac{m\cos\theta}{\sin\theta}\,P_n^m \\[2ex] \dfrac{dP_n^m}{d\theta} \end{cases}$$

Thus (6) can be written in the form

$$\hat{R}\cdot\bar{M}_{emn}^{(x)} = \frac{j_n(\rho)}{2\rho}\left[P_n^{m+1}\sin(m+1)\phi + (n-m+1)(n+m)P_n^{m-1}\sin(m-1)\phi\right].$$

In view of (39-7), this is recognized as the radial component of two odd $\bar{N}$ functions with orders $(n,m+1)$ and $(n,m-1)$ respectively, or

(7) $$\hat{R}\cdot\bar{M}_{emn}^{(x)}(k) = \hat{R}\cdot\frac{1}{2n(n+1)}\left[\bar{N}_{o(m+1)n}(k) + (n+m)(n-m+1)\bar{N}_{o\,(m-1)n}(k)\right].$$

The identification of the composition of the θ-component or the ϕ-component of (4) is much more intricate. We consider the θ-component which is given by

(8) $$\hat{\theta}\cdot\bar{M}_{emn}^{(x)}(k) = \frac{1}{\rho}\left[j_n P_n^m\,\frac{d}{d\phi}(\cos\phi\cos m\phi) + \frac{d}{d\rho}(\rho j_n)P_n^m\sin\phi\cos m\phi\right].$$

A reduction of the trigonometric functions yields

(9) $$\hat{\theta}\cdot\bar{M}_{emn}^{(x)}(k) = \frac{P_n^m}{2\rho}\left\{\left[\frac{d(\rho j_n)}{d\rho} - (m+1)j_n\right]\sin(m+1)\phi\right.$$

$$\left.-\left[\frac{d(\rho j_n)}{d\rho} + (m-1)j_n\right]\sin(m-1)\phi\right\}.$$

This is obviously not equal to the θ-component of the two odd functions contained in (7). To reduce it further we need the following recurrence relations for the associated Legendre functions and the spherical Bessel functions:

(10) $$P_n^m = \frac{1}{(2n+1)\sin\theta}\left[P_{n+1}^{m+1} - P_{n-1}^{m+1}\right]$$

(11) $$P_n^m = \frac{1}{(2n+1)\sin\theta}\left[-(n-m+2)(n-m+1)P_{n+1}^{m-1} + (n+m)(n+m-1)P_{n-1}^{m-1}\right]$$

(12) $$\frac{dP_n^{m-1}}{d\theta} = \frac{n(n+1)}{(2n+1)\sin\theta}\left[\frac{n-m}{n+1}P_{n+1}^{m+1} - \frac{n+m+1}{n}P_{n-1}^{m+1}\right]$$

(13) $$\frac{dP_n^{m-1}}{d\theta} = \frac{n(n+1)}{(2n+1)\sin\theta}\left[\frac{n-m+2}{n+1}P_{n+1}^{m-1} - \frac{n+m-1}{n}P_{n-1}^{m-1}\right]$$

(14) $$\frac{j_n}{\rho} = \frac{(\rho j_n)'}{(n+1)\rho} + \frac{j_{n+1}}{n+1}$$

(15) $$\frac{j_n}{\rho} = \frac{-(\rho j_n)'}{n\rho} + \frac{j_{n-1}}{n},$$

where $(\rho j_n)'$ denotes the derivatives of ρj_n. Equation (10) is one of the relations listed by Stratton (p. 401). Equations (11) to (13) can be derived by a proper combination of the pertinent ones found in the list. Equations (14) and (15) can be derived from (39-13) and (39-14). To reduce (9) to the desired recognizable form, the algebra is long and tedious and is outlined below. Using (10), (11), (14), and (15), we first change (9) to

(16) $$\hat{\theta}\cdot\bar{M}_{emn}^{(x)} = \frac{1}{2(2n+1)}\left\{\frac{P_{n+1}^{m+1}}{\sin\theta}\left[\frac{n-m}{n+1}\frac{(\rho j_n)'}{\rho} - \frac{m+1}{n+1}j_{n+1}\right]\right.$$
$$\left. - \frac{P_{n-1}^{m+1}}{\sin\theta}\left[\frac{n+m+1}{n}\frac{(\rho j_n)'}{\rho} - \frac{m+1}{n}j_{n-1}\right]\right\}\sin(m+1)\phi$$
$$+ \frac{1}{2(2n+1)}\left\{(n-m+2)(n-m+1)\frac{P_{n+1}^{m-1}}{\sin\theta}\left[\frac{n+m}{n+1}\frac{(\rho j_n)'}{\rho}\right.\right.$$
$$\left. + \frac{m-1}{n+1}j_{n+1}\right] - (n+m)(n+m-1)\frac{P_{n-1}^{m-1}}{\sin\theta}\left[\frac{n-m+1}{n}\frac{(\rho j_n)'}{\rho}\right.$$
$$\left.\left. + \frac{m-1}{n}j_{n-1}\right]\right\}\sin(m-1)\phi\,.$$

We rearrange it in the form

$$(17)\quad \hat{\theta}\cdot\bar{M}^{(x)}_{emn}=\frac{1}{2(2n+1)}\left\{-\frac{m+1}{n+1}\frac{P^{m+1}_{n+1}}{\sin\theta}j_{n+1}+\frac{m+1}{n}\frac{P^{m+1}_{n-1}}{\sin\theta}j_{n-1}\right.$$

$$\left.+\frac{(\rho j_n)'}{\rho\sin\theta}\left[\frac{n-m}{n+1}P^{m+1}_{n+1}-\frac{n+m+1}{n}P^{m+1}_{n-1}\right]\right\}\sin(m+1)\phi$$

$$+\frac{1}{2(2n+1)}\left\{\frac{(m-1)(n-m+2)(n-m+1)}{(n+1)}\frac{P^{m-1}_{n+1}}{\sin\theta}j_{n+1}\right.$$

$$-\frac{(m-1)(n+m)(n+m-1)}{n}\frac{P^{m-1}_{n-1}}{\sin\theta}j_{n-1}$$

$$+(n+m)(n-m+1)\frac{(\rho j_n)'}{\rho\sin\theta}\left[\frac{n-m+2}{n+1}\frac{P^{m-1}_{n+1}}{\sin\theta}\right.$$

$$\left.\left.-\frac{n+m-1}{n}\frac{P^{m-1}_{n-1}}{\sin\theta}\right]\right\}\sin(m-1)\phi\,.$$

In view of (12) and (13), it can be written in the form

$$(18)\quad \hat{\theta}\cdot\bar{M}^{(x)}_{emn}=\frac{1}{2}\left\{\frac{m+1}{2n+1}\left[\frac{-1}{n+1}\frac{P^{m+1}_{n+1}}{\sin\theta}j_{n+1}+\frac{1}{n}\frac{P^{m+1}_{n-1}}{\sin\theta}j_{n-1}\right]\right.$$

$$\left.+\frac{1}{n(n+1)}\frac{\partial P^{m+1}_n}{\partial\theta}\frac{(\rho j_n)'}{\rho}\right\}\sin(m+1)\phi$$

$$+\frac{1}{2}\left\{\frac{m-1}{2n+1}\left[\frac{(n-m+2)(n-m+1)}{n+1}\frac{P^{m-1}_{n+1}}{\sin\theta}j_{n+1}\right.\right.$$

$$\left.-\frac{(n+m)(n+m-1)}{n}\frac{P^{m-1}_{n-1}}{\sin\theta}j_{n-1}\right]$$

$$\left.+\frac{(n+m)(n-m+1)}{n(n+1)}\frac{\partial P^{m-1}_n}{\partial\theta}\frac{(\rho j_n)'}{\rho}\right\}\cdot\sin(m-1)\phi\,.$$

Each term in (18) can now be identified with the θ-component of a standard spherical vector wave function. There are six of them. In fact, (18) is described more characteristically by

$$(19)\quad \hat{\theta}\cdot\bar{M}^{(x)}_{emn}=\hat{\theta}\cdot\left\{\frac{1}{2n(n+1)}\left[\bar{N}_{o(m+1)n}+(n+m)(n-m+1)\bar{N}_{o(m-1)n}\right]\right.$$

$$+\frac{1}{2(n+1)(2n+1)}\left[\bar{M}_{e(m+1)(n+1)}-(n-m+1)(n-m+2)\bar{M}_{e(m-1)(n+1)}\right]$$

$$\left.-\frac{1}{2n(2n+1)}\left[\bar{M}_{e(m+1)(n-1)}-(n+m-1)(n+m)\bar{M}_{e(m-1)(n-1)}\right]\right\}$$

It is noted that the two $\bar{N}$-functions in (19) including the coefficients, are the same as the ones appearing in (7). Since the standard $\bar{M}$-functions do not have a radial component, they are, therefore, not involved in (7). If we do the same study for the ϕ-component the result shows that it is of the same form as (19). Thus, we have established a complete relationship between $\bar{M}^{(x)}_{emn}$ and the six radial vector wave functions that are given by (19) with $[\hat{\theta}\,\cdot]$ deleted. The relations between all the vector wave functions of the c-type and the standard ones have been found and they are tabulated in Part C of Appendix B. Knowing these relations, we can proceed to rederive (39-18) based on the algebraic method.

As was shown in Sec. 15 the free-space dyadic Green's function can be written in the form

$$(20)\qquad \bar{\bar{G}}_0(\bar{R}|\bar{R}') = \bar{G}_0^{(x)}(\bar{R}|\bar{R}')\hat{x} + \bar{G}_0^{(y)}(\bar{R}|\bar{R}')\hat{y} + \bar{G}_0^{(z)}(\bar{R}|\bar{R}')\hat{z}$$

where

$$(21)\qquad \bar{G}_0^{(c)}(\bar{R}|\bar{R}') = \left(1 + \frac{1}{k^2}\nabla\nabla\cdot\right)G_0(\bar{R}|\bar{R}')\hat{c}, \qquad c = x, y, z$$

$$(22)\qquad G_0(\bar{R}|\bar{R}') = \frac{e^{ik|\bar{R}-\bar{R}'|}}{4\pi|\bar{R}-\bar{R}'|}.$$

For $\bar{R} \neq \bar{R}'$, (21) is equivalent to

$$(23)\qquad \bar{G}_0^{(c)}(\bar{R}|\bar{R}') = \frac{1}{k^2}\nabla\times\nabla\times\left[G_0(\bar{R}|\bar{R}')\,\hat{c}\right], \qquad c = x, y, z.$$

The three-dimensional free-space Green's function as defined by (22) has a series expansion given by

$$(24)\qquad G_0(\bar{R}|\bar{R}') = \frac{ik}{4\pi}\sum_{n=0}^{\infty}\sum_{m=0}^{n}(2-\delta_0)(2n+1)\frac{(n-m)!}{(n+m)!}$$
$$\cdot P_n^m(\cos\theta)P_n^m(\cos\theta')\left[\cos m\phi\cos m\phi' + \sin m\phi\sin m\phi'\right]$$
$$\cdot\begin{cases} j_n(kR)\,h_n^{(1)}(kR'), & R < R' \\ h_n^{(1)}(kR)\,j_n(kR'), & R > R'. \end{cases}$$

This expression is obtained by combining two expansions given by Stratton (Eq. 46, p. 408 and Eq. 87, p. 414). Substituting (24) into (21), we obtain, for $R < R'$,

$$(25)\qquad \bar{G}_0^{(c)}(\bar{R}|\bar{R}') = \frac{ik}{4\pi}\sum_{n=0}^{\infty}\sum_{m=0}^{n} A_{\substack{e\\o}mn}\bar{N}^{(c)}_{\substack{e\\o}mn},$$

where

$$(26)\qquad A_{\substack{e\\o}mn} = (2-\delta_0)(2n+1)\frac{(n-m)!}{(n+m)!}h_n^{(1)}(kR')P_n^m(\cos\theta')\begin{matrix}\cos\\ \sin\end{matrix} m\phi'$$

and $\bar{N}^{(c)}_{\substack{e\\o}mn}$ is defined by (2). We can now express the $\bar{N}^{(c)}_{\substack{e\\o}mn}$ functions in (25) in terms of the radial spherical vector wave functions according to the formulas given in Part C of Appendix B. The result can be written in the form

$$(27)\qquad \bar{G}_0^{(c)}(\bar{R}|\bar{R}') = \frac{ik}{4\pi}\sum_{n=1}^{\infty}\sum_{m=0}^{n}(\alpha^{(c)}_{\substack{e\\o}mn}\bar{M}_{\substack{e\\o}mn} + \beta^{(c)}_{\substack{e\\o}mn}\bar{N}_{\substack{e\\o}mn}).$$

The coefficients $\alpha^{(c)}_{\substack{e\\o}mn}$ and $\beta^{(c)}_{\substack{e\\o}mn}$ are all expressible in terms of $A_{\substack{e\\o}mn}$ of different orders. For example, we have

$$\alpha^{(x)}_{emn} = \frac{-1}{2n(n+1)}[(1+\delta_1)A_{o(m-1)n} + (n+m+1)(n-m)A_{o(m+1)n}]$$

$$\beta^{(x)}_{emn} = \frac{1}{2n(2n-1)}[(1+\delta_1)A_{e(m-1)(n-1)} - (n-m-1)(n-m)A_{e(m+1)(n-1)}],$$

where

$$\delta_1 = \begin{cases} 1, & m=1 \\ 0, & m\neq 1. \end{cases}$$

Substituting (27) into (20), we find, after another long trip of algebraic exercise, that the following identities are true:

$$\alpha^{(x)}_{\substack{e\\o}mn}\hat{x} + \alpha^{(y)}_{\substack{e\\o}mn}\hat{y} + \alpha^{(z)}_{\substack{e\\o}mn}\hat{z} = (2-\delta_0)\frac{2n+1}{n(n+1)}\frac{(n-m)!}{(n+m)!}\bar{M}'^{(1)}_{\substack{e\\o}mn}$$

and

$$\beta^{(x)}_{\substack{e\\o}mn}\hat{x} + \beta^{(y)}_{\substack{e\\o}mn}\hat{y} + \beta^{(z)}_{\substack{e\\o}mn}\hat{z} = (2-\delta_0)\frac{2n+1}{n(n+1)}\frac{(n-m)!}{(n+m)!}\bar{N}'^{(1)}_{\substack{e\\o}mn},$$

hence

$$\bar{\bar{G}}_0(\bar{R}|\bar{R}') = \frac{ik}{4\pi}\sum_{n=1}^{\infty}\sum_{m=0}^{n}(2-\delta_0)\frac{2n+1}{n(n+1)}\frac{(n-m)!}{(n+m)!}$$
$$\cdot[\bar{M}_{\substack{e\\o}mn}\bar{M}'^{(1)}_{\substack{e\\o}mn} + \bar{N}_{\substack{e\\o}mn}\bar{N}'^{(1)}_{\substack{e\\o}mn}],\qquad R<R'$$

which is the same as the one derived by the Ohm-Rayleigh method. In retrospect, the preceding exercise demonstrates very convincingly the elegance of the Ohm-Rayleigh method which bypasses all the complicated manipulations that are required by the algebraic method. The latter does, however, provide us another insight into the structure of the free-space dyadic Green's functions. It also gives us an opportunity, at least from the point of view of the author of this book, to learn much about the recurrence relations of the associated Legendre functions and the spherical Bessel functions not previously practiced.

41. DYADIC GREEN'S FUNCTIONS PERTAINING TO PERFECTLY AND IMPERFECTLY CONDUCTING SPHERES

Based on (39-18), we can construct the various kinds of dyadic Green's functions associated with a sphere by means of the method of scattering superposition. Thus, for a perfectly conducting sphere of radius equal to a with its center located at the origin of the coordinate system, we would deal with either the function of the first kind or the function of the second kind. To find the function of the first kind we let

$$\overline{\overline{G}}_1(\overline{R}|\overline{R}') = \overline{\overline{G}}_0(\overline{R}|\overline{R}') + \overline{\overline{G}}_{1s}(\overline{R}|\overline{R}') \tag{1}$$

where $\overline{\overline{G}}_0(\overline{R}|\overline{R}')$ is given by (39-18). In view of the composition of $\overline{\overline{G}}_0$ the term representing the scattered part must have the form

$$\overline{\overline{G}}_{1s}(\overline{R}|\overline{R}') = \frac{ik}{4\pi} \sum_{n=1}^{\infty} \sum_{m=0}^{n} (2-\delta_0) \frac{2n+1}{n(n+1)} \frac{(n-m)!}{(n+m)!} \left[a_{\substack{e \\ o}n} \overline{M}^{(1)}_{\substack{e \\ o}mn}(k) \overline{M}'^{(1)}_{\substack{e \\ o}mn}(k) + b_{\substack{e \\ o}n} \overline{N}^{(1)}_{\substack{e \\ o}mn}(k) \overline{N}'^{(1)}_{\substack{e \\ o}mn}(k) \right]. \tag{2}$$

The coefficients $a_{\substack{e \\ o}mn}$ and $b_{\substack{e \\ o}mn}$ are determined by applying the Dirichlet boundary condition to the entire function $\overline{\overline{G}}_1$ at the surface of the conducting sphere which yields

$$a_{\substack{e \\ o}n} = -j_n(ka) \,/\, h_n^{(1)}(ka) \tag{3}$$

$$b_{\substack{e \\ o}n} = -\frac{d}{d\rho}\left[\rho j_n(\rho)\right] \Big/ \frac{d}{d\rho}\left[\rho h_n^{(1)}(\rho)\right], \qquad \rho = ka. \tag{4}$$

Similarly, for the function of the second kind we let

$$\overline{\overline{G}}_2(\overline{R}|\overline{R}') = \overline{\overline{G}}_0(\overline{R}|\overline{R}') + \overline{\overline{G}}_{2s}(\overline{R}|\overline{R}') \tag{5}$$

with

$$\overline{\overline{G}}_{2s}(\overline{R}|\overline{R}') = \frac{ik}{4\pi} \sum_{n=1}^{\infty} \sum_{m=0}^{n} (2-\delta_0) \frac{2n+1}{n(n+1)} \frac{(n-m)!}{(n+m)!} \cdot \left[c_{\substack{e \\ o}n} \overline{M}^{(1)}_{\substack{e \\ o}mn}(k) \overline{M}'^{(1)}_{\substack{e \\ o}mn}(k) + d_{\substack{e \\ o}n} \overline{N}^{(1)}_{\substack{e \\ o}mn}(k) \overline{N}'^{(1)}_{\substack{e \\ o}mn}(k) \right]. \tag{6}$$

An application of the Neumann boundary condition to $\overline{\overline{G}}_2$ at the surface of the sphere yields

$$c_{\substack{e\\o}n} = -\frac{d}{d\rho}\left[\rho j_n(\rho)\right] \Big/ \frac{d}{d\rho}\left[\rho h_n^{(1)}(\rho)\right], \qquad \rho = ka \tag{7}$$

$$d_{\substack{e\\o}n} = -j_n(ka) \,/\, h_n^{(1)}(ka). \tag{8}$$

It is noticed that the two coefficients defined by (7) and (8) are numerically equal to the two coefficients defined by (4) and (3), respectively. This is due to the symmetrical property between the two dyadic Green's functions as discussed in Sec. 17. Once we get used to these relations, a separate derivation of $\overline{\overline{G}}_2$ is not necessary.

For the function of the third or the fourth kind associated with imperfectly conducting or dielectric spheres, it is better to adopt the following notation for the constitutive constant defined in the two regions, exterior and interior to the sphere. We let

$$k_1 = \omega\sqrt{\mu_1\varepsilon_1} \qquad \text{(Region I, } r \geqslant a)$$

$$k_2 = \omega\sqrt{\mu_2\varepsilon_2} \qquad \text{(Region II, } r \leqslant a).$$

For a dielectric sphere placed in air $\varepsilon_1 = \varepsilon_0$, $\varepsilon_2 = \varepsilon$, $\mu_1 = \mu_2 = \mu_0$, where ε denotes the permittivity of the sphere. For generality, no restriction will be placed on these constants. The constant k which appeared in $\overline{\overline{G}}_0$ as expressed by (39-18) must now be replaced by k_1. With this change of notation, we let

$$\overline{\overline{G}}_3^{(11)}(\overline{R}|\overline{R}') = \overline{\overline{G}}_0(\overline{R}|\overline{R}') + \overline{\overline{G}}_{3s}^{(11)}(\overline{R}|\overline{R}') \qquad R \geqslant a$$

$$\overline{\overline{G}}_3^{(21)}(\overline{R}|\overline{R}') = \overline{\overline{G}}_{3s}^{(21)}(\overline{R}|\overline{R}'), \qquad R \leqslant a.$$

In conformity to the superscript notations introduced in Sec. 17, the above two equations imply that we are dealing with a source located in Region I. The scattered terms now can be represented by

$$\overline{\overline{G}}_{3s}^{(11)}(\overline{R}|\overline{R}') = \frac{ik_1}{4\pi} \sum_{n=1}^{\infty} \sum_{m=0}^{n} (2-\delta_0) \frac{2n+1}{n(n+1)} \frac{(n-m)!}{(n+m)!} \cdot \left[A_{\substack{e\\o}n} \overline{M}_{\substack{e\\o}mn}^{(1)}(k_1) \overline{M}_{\substack{e\\o}mn}'^{(1)}(k_1) + B_{\substack{e\\o}n} \overline{N}_{\substack{e\\o}mn}^{(1)}(k_1) \overline{N}_{\substack{e\\o}mn}'^{(1)}(k_1)\right], \tag{9}$$

$$\overline{\overline{G}}_{3s}^{(21)}(\overline{R}|\overline{R}') = \frac{ik_1}{4\pi} \sum_{n=1}^{\infty} \sum_{m=0}^{n} (2-\delta_0) \frac{2n+1}{n(n+1)} \frac{(n-m)!}{(n+m)!} \cdot \left[C_{\substack{e\\o}n} \overline{M}_{\substack{e\\o}mn}(k_2) \overline{M}_{\substack{e\\o}mn}'^{(1)}(k_1) + D_{\substack{e\\o}n} \overline{N}_{\substack{e\\o}mn}(k_2) \overline{N}_{\substack{e\\o}mn}'^{(1)}(k_1)\right]. \tag{10}$$

Applying the boundary condition at the surface of the sphere as required by the function of the third kind, namely,

$$\left.\begin{aligned} \hat{R} \times \overline{\overline{G}}_3^{(11)} &= \hat{R} \times \overline{\overline{G}}_3^{(21)} \\ \frac{1}{\mu_1} \hat{R} \times \nabla \times \overline{\overline{G}}_3^{(11)} &= \frac{1}{\mu_2} \hat{R} \times \nabla \times \overline{\overline{G}}_3^{(21)} \end{aligned}\right\}, \qquad R = a$$

we find that the coefficients A, B, C, and D must satisfy the following system of equations.

$$j_n(\rho_1) + A_{\substack{e\\o}n} h_n^{(1)}(\rho_1) = C_{\substack{e\\o}n} j_n(\rho_2) \tag{11}$$

$$\frac{k_1}{\mu_1}\left\{\frac{[\rho_1 j_n(\rho_1)]'}{\rho_1} + A_{\substack{e\\o}n} \frac{[\rho_1 h_n^{(1)}(\rho_1)]'}{\rho_1}\right\} = \frac{k_2}{\mu_2}\left\{C_{\substack{e\\o}n} \frac{[\rho_2 j_n(\rho_2)]'}{\rho_2}\right\} \tag{12}$$

$$\frac{[\rho_1 j_n(\rho_1)]'}{\rho_1} + B_{\substack{e\\o}n} \frac{[\rho_1 h_n^{(1)}(\rho_1)]'}{\rho_1} = D_{\substack{e\\o}n} \frac{[\rho_2 j_n(\rho_2)]'}{\rho_2} \tag{13}$$

$$\frac{k_1}{\mu_1}\left[j_n(\rho_1) + B_{\substack{e\\o}n} h_n^{(1)}(\rho_1)\right] = D_{\substack{e\\o}n}\left(\frac{k_2}{\mu_2}\right) j_n(\rho_2) \tag{14}$$

where

$$\rho_1 = k_1 a, \;\; \rho_2 = k_2 a$$

and the prime denotes the derivative of the function. It is a relatively simple matter to solve these equations. By comparing the functions of the third kind for the dielectric cylinder and for the dielectric sphere we see that the spherical case is considerably simpler. The presence of the sphere does not introduce the coupling between the TE-modes ($\bar{M}$-functions) and the TM-modes ($\bar{N}$-functions) as manifested by the dielectric cylinder. We must, however, remember that the cylindrical vector wave functions are generated with the piloting vector pointed in the z-direction while the piloting vector of the spherical vector wave functions is pointed in the radial direction. Their characteristics are entirely different. The derivation of the function of the fourth kind follows, of course, the same pattern.

In regard to the applications, the examples which we used for the half-sheet are equally valid for the sphere, perhaps, with more varieties because of the availability of the functions of the third and fourth kinds. We shall, however, treat in some detail only the problem of radiation from a horizontal dipole in the presence of a sphere. Historically, the problem of a vertical dipole, electric or magnetic, and a sphere was a subject studied by many authors after the turn of the century in connection with radio-wave propagation over a spherical earth. The corresponding problem involving a horizontal dipole, however, was not resolved until many decades later by Nomura (1951). We shall now reformulate this problem using the dyadic Green's function technique and show the identity between our result and the original one obtained by Nomura. We shall also recover Mie's solution for the diffraction of a plane wave by a sphere (Mie, 1908; Stratton, 1941, p. 563), from the asymptotic solution of the horizontal dipole problem. For simplicity, we assume the sphere to be perfectly conducting. Because of the available formula for the function of the third kind, the treatment of an imperfectly conducting or dielectric sphere is no more difficult.

Figure 41-1 shows the geometry of the problem under consideration.

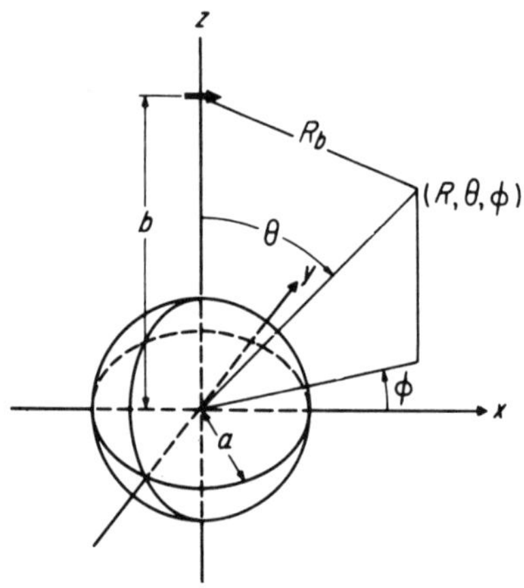

FIG. 41-1. A horizontal electric dipole in the presence of a perfectly conducting sphere.

For an infinitesimal horizontal electric dipole with current moment c pointed in the x-direction, and located at $R' = b$, $\theta' = 0$, $\phi' = 0$, we let

$$\bar{J}(\bar{R}') = c \, \frac{\delta(R' - b)\delta(\theta' - 0)\delta(\phi' - 0)}{b^2 \sin\theta'} \, \hat{x}.$$

The electric field produced by this dipole in the presence of a perfectly conducting sphere with radius equal to a is then given by

$$(15) \qquad \bar{E}(\bar{R}) = i\omega\mu_0 \iiint \bar{\bar{G}}_1(\bar{R}|\bar{R}') \cdot \bar{J}(\bar{R}') \, dV'$$

$$= i\omega\mu_0 c \bar{\bar{G}}_1(\bar{R}|\bar{R}') \cdot \hat{x}, \qquad \bar{R}' = (b, 0, 0).$$

Using the expression for $\bar{\bar{G}}_1$ given by (1) to (4), we obtain

$$(16) \qquad \bar{E}(\bar{R}) = \frac{-k\omega\mu_0 c}{4\pi} \sum_{n=1}^{\infty} \frac{2n+1}{n(n+1)}$$

$$\cdot \begin{cases} [j_n(\rho_b) + a_n h_n^{(1)}(\rho_b)]\,\bar{M}_{o1n}^{(1)}(k) \\ h_n^{(1)}(\rho_b)[\bar{M}_{o1n}(k) + a_n \bar{M}_{o1n}^{(1)}(k)] \end{cases}$$

$$\left.\begin{array}{ll} + \dfrac{1}{\rho_b}([\rho_b j_n(\rho_b)]' + b_n[\rho_b h_n(\rho_b)]')\bar{N}_{e1n}^{(1)}(k) & R > b \\ + \dfrac{[\rho_b h_n(\rho_b)]'}{\rho_b}[\bar{N}_{e1n}(k) + b_n \bar{N}_{e1n}(k)] & R < b \end{array}\right.$$

where

(17) $$a_n = -j_n(\rho_a) \,/\, h_n^{(1)}(\rho_a)$$

(18) $$b_n = -[\rho_a j_n(\rho_a)]' \,/\, [\rho_a h_n^{(1)}(\rho_a)]'$$

$$\rho_a = ka, \quad \rho_b = kb.$$

This expression is different in form from the one obtained by Nomura. He formulated the problem by the method of potential functions. As reviewed in Sec. 14, the primary or the incident field of the dipole can be expressed in the form

$$\bar{E}_i(\bar{R}) = \nabla \times \nabla \times (\Pi_i \hat{x})$$

$$\Pi_i = \frac{i\omega\mu_0 c e^{ikR_b}}{4\pi k^2 R_b}.$$

The free-space scalar Green's function contained in Π_i, according to (40-24) for $\theta' = 0$ and $\phi' = 0$, has the series expansion

(19) $$\frac{e^{ikR_b}}{4\pi R_b} = \frac{ik}{4\pi} \sum_{n=0}^{\infty} (2n+1) P_n(\cos\theta) \begin{cases} h_n^{(1)}(kR) j_n(kb), & R > b \\ j_n(kR) h_n^{(1)}(kb), & R < b. \end{cases}$$

Thus the incident field can be expressed in terms of a series of vector wave functions of the x-type defined by (40-2) with $m = 0$,

(20) $$\bar{E}_i(\bar{R}) = \frac{-k\omega\mu_0 c}{4\pi} \sum_{n=0}^{\infty} (2n+1) \begin{cases} j_n(kb)\bar{N}_{e0n}^{(x,1)}, & R > b \\ h_n^{(1)}(kb)\bar{N}_{e0n}^{(x)}, & R < b \end{cases}$$

where $\bar{N}^{(x,1)}$ denotes a spherical vector wave function of the x-type using $h_n^{(1)}(kR)$ in the generating function. Nomura found the solution for the secondary or the scattered field by letting

(21) $$\bar{E}_s(\bar{R}) = \frac{-k\omega\mu_0 c}{4\pi} \left[\sum_{n=0}^{\infty} c_n \bar{N}_{e0n}^{(x,1)}(k) + \sum_{n=1}^{\infty} d_n \bar{N}_{e1n}^{(1)}(k) \right].$$

The coefficients c_n and d_n are determined by applying the boundary condition that $\hat{R}\times(\bar{E}_i+\bar{E}_s)=0$ at the surface of the sphere which yields

$$c_n=-(2n+1)h_n^{(1)}(\rho_b)j_n(\rho_a)\,/\,h_n^{(1)}(\rho_a)$$

$$d_n=\frac{i}{[\rho_a h_n^{(1)}(\rho_a)]'}\left[\frac{h_{n-1}^{(1)}(\rho_b)}{h_{n-1}^{(1)}(\rho_a)}-\frac{h_{n+1}^{(1)}(\rho_b)}{h_{n+1}^{(1)}(\rho_a)}\right].$$

From the composition of (21), it is seen that the scattered field is derived partly from a Hertzian potential and partly from a Debye potential while the incident field is entirely derivable from a Hertzian potential. In view of our discussion of the relations between the functions of the x-type and the standard type, we can transform Nomura's formula into ours in the following manner. By taking the curl of Eq. 21 of Appendix B and letting $m=0$, we obtain

$$\bar{N}_{e0n}^{(x)}=\frac{1}{2n(n+1)}[\bar{M}_{o1n}+n(n+1)\bar{M}_{o(-1)n}]$$
$$+\frac{1}{2(n+1)(2n+1)}[\bar{N}_{e1(n+1)}-(n+1)(n+2)\bar{N}_{e(-1)(n+1)}]$$
$$-\frac{1}{2n(2n+1)}[\bar{N}_{e1(n-1)}-(n-1)n\bar{N}_{e(-1)(n-1)}]. \tag{22}$$

When m is negative, we use the formula (Sommerfeld, *Partial Diff. Eq.*, p. 129)

$$P_n^{-m}=(-1)^m\frac{(n-m)!}{(n+m)!}P_n^m$$

to obtain

$$\bar{M}_{\substack{e\\o}(-m)n}=\pm(-1)^m\frac{(n-m)!}{(n+m)!}\bar{M}_{\substack{e\\o}mn} \tag{23}$$

$$\bar{N}_{\substack{e\\o}(-m)n}=\pm(-1)^m\frac{(n-m)!}{(n+m)!}\bar{N}_{\substack{e\\o}mn} \tag{24}$$

hence (22) can be simplified to read

$$\bar{N}_{e0n}^{(x)}=\frac{1}{n(n+1)}\bar{M}_{o1n}+\frac{1}{(n+1)(2n+1)}\bar{N}_{e1(n+1)}-\frac{1}{n(2n+1)}\bar{N}_{e1(n-1)}.$$

By substituting this into (21) and changing the summation indices so that they all start with $n=1$, we can write the resultant series in the form

$$\bar{E}_s(\bar{R})=\frac{-k\omega\mu_0 c}{4\pi}\sum_{n=1}^{\infty}\left\{\frac{c_n}{n(n+1)}\bar{M}_{o1n}^{(1)}(k)\right. \tag{25}$$

$$+\left[\frac{c_{n-1}}{n(2n-1)}-\frac{c_{n+1}}{(n+1)(2n+3)}+d_n\right]\bar{N}^{(1)}_{e1n}\Bigg\} .$$

Comparing (25) with the scattered part of (16), the following relations must hold true:

$$c_n=(2n+1)h_n^{(1)}(\rho_b)a_n \tag{26}$$

$$d_n=\frac{(2n+1)}{n(n+1)}\frac{[\rho_b h_n^{(1)}(\rho_b)]'}{\rho_b}b_n+\frac{c_{n+1}}{(n+1)(2n+3)}-\frac{c_{n-1}}{n(2n-1)} \tag{27}$$

$$=\frac{(2n+1)}{n(n+1)}\frac{[\rho_b h_n^{(1)}(\rho_b)]'}{\rho_b}b_n+\frac{h_{n+1}^{(1)}(\rho_b)}{n+1}a_{n+1}+\frac{h_{n-1}^{(1)}(\rho_b)}{n-1}a_{n-1} .$$

By substituting the expressions for a_n and b_n given by (17) and (18) into (26) and (27), and applying the recurrence relations of the spherical Bessel functions of the type (40-14) and (40-15) with j_n replaced by $h_n^{(1)}$, we find that the coefficients c_n and d_n so obtained are indeed the same as the ones derived by Nomura, as they should be. The original verification of the identity between (21) and (16) was carried out by this author (Tai, 1952) without the aid of the dyadic Green's function technique. It was done by transforming the incident field as represented by (20) into two series involving both the $\bar{M}_{o1n}$ and $\bar{N}_{e1n}$ functions and then find the scattered field using the $\bar{M}^{(1)}_{o1n}$ and $\bar{N}^{(1)}_{e1n}$ functions. It seems obvious from this discussion that the use of the x-type of functions for the sphere problem makes the formulation considerably more complicated. The interpretation of the result based on Nomura's formula is also not as simple as the one offered by (16) from the eigenfunction point of view.

We shall now rederive Mie's series solution for the diffraction of a plane wave by a sphere from (16) by removing the dipole far away from the sphere. When kb is large, the spherical Hankel function has the asymptotic form,

$$h_n^{(1)}(kb)\simeq(-i)^{n+1}\frac{e^{ikb}}{kb}$$

and

$$\frac{[kb\,h_n^{(1)}(kb)]'}{kb}\simeq(-i)^n\frac{e^{ikb}}{kb} .$$

Substituting these values into (16) for $R<b$, and identifying the amplitude of the plane wave as being given by

$$E_o=\frac{i\omega\mu_o c\,e^{ikb}}{kb}\qquad(kb\gg1)$$

we obtain

$$\bar{E}(\bar{R}) = E_o \sum_{n=1}^{\infty} (-i)^n \frac{2n+1}{n(n+1)} \cdot \{[\bar{M}_{o1n}(k) + a_n \bar{M}^{(1)}_{o1n}(k)] + i[\bar{N}_{e1n}(k) + b_n \bar{N}^{(1)}_{e1n}(k)]\} . \tag{28}$$

This is Mie's series as presented by Stratton (p. 563) starting with the eigenfunction expansion of a plane wave using the spherical vector wave functions as the constituents, we note, however, that the direction of propagation of the plane wave created here is opposite to the one considered by Stratton.

Before we conclude this section, two additional formulas are given below. Their derivation is similar to the one that results in (16):

1. Vertical Electric Dipole of Current Moment $\bar{c} = c\hat{z}$ Placed at $R' = b, \theta' = 0$

$$\bar{E}(\bar{R}) = \frac{-k\omega\mu_o c}{4\pi kb} \sum_{n=1}^{\infty} \cdot \begin{cases} [j_n(kb) + b_n h_n^{(1)}(kb)]\, \bar{N}^{(1)}_{e0n}(k), & R > b \\ h_n^{(1)}(kb)\, [\bar{N}_{e0n}(k) + b_n \bar{N}^{(1)}_{e0n}(k)], & R < b \end{cases} \tag{29}$$

$$b_n = -[kaj_n(ka)]' / [ka\, h_n^{(1)}(kb)]'.$$

2. A Small Circular Aperture Located at the Top of a Conducting Sphere (Fig. 41-1) Excited by a Constant Field $\bar{E}_a = E_0\,\hat{x}, \theta' \leqslant \theta_0$

$$\bar{E}(\bar{R}) = \frac{E_0\theta_0^2}{4} \sum_{n=1}^{\infty} \frac{2n+1}{n(n+1)} \left\{ \frac{1}{h_n^{(1)}(ka)} \bar{M}^{(1)}_{o1n} + \frac{ka}{[kah_n^{(1)}(ka)]'} \bar{N}^{(1)}_{e1n} \right\} . \tag{30}$$

In deriving (30), terms of the order θ_0^4 and higher have been neglected. Equation (30), with an appropriate change of a constant of proportionality, also represents the field produced by a horizontal magnetic dipole placed at the top of the sphere and pointed in the y-direction.

42. PERFECTLY CONDUCTING CONICAL STRUCTURES

There are two distinct types of perfectly conducting conical structures of which the dyadic Green's functions of the first kind and the second kind can be found. The first type has the form of a single cone, and the second type is a bicone, formed by two single cones. These two structures are shown in Fig. 42-1. They are both infinitely long in the radial direction. For both cases, it is assumed that the axis of the cone is aligned with the z-axis. We define the conical vector wave functions in exactly the same

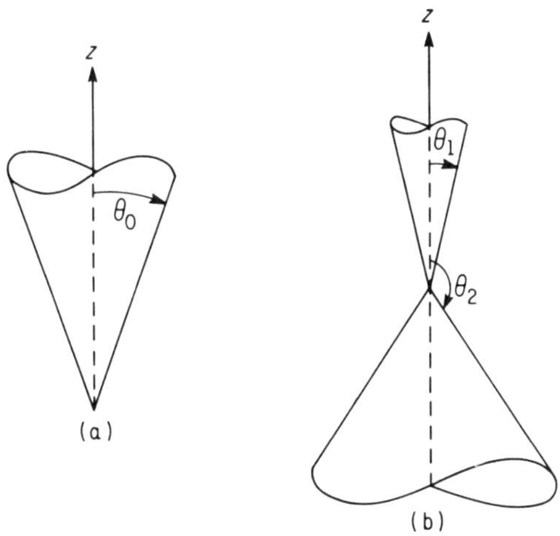

FIG. 42-1. (a) A single cone; (b) a bicone.

form as the spherical vector wave functions except that the eigenvalue n is now, in general, fractional. There are four kinds of these functions which are defined by

(1) $$\bar{M}_{\substack{e\\o}m\mu}(k)=\nabla\times(\psi_{\substack{e\\o}m\mu}\bar{R})$$

(2) $$\bar{N}_{\substack{e\\o}m\mu}(k)=\frac{1}{k}\nabla\times\nabla\times(\psi_{\substack{e\\o}m\mu}\bar{R})$$

(3) $$\bar{M}_{\substack{e\\o}m\lambda}(k)=\nabla\times(\psi_{\substack{e\\o}m\lambda}\bar{R})$$

(4) $$\bar{N}_{\substack{e\\o}m\lambda}(k)=\frac{1}{k}\nabla\times\nabla\times(\psi_{\substack{e\\o}m\lambda}\bar{R})$$

where

$$\psi_{\substack{e\\o}m\mu}=j_\mu(kR)P_\mu^m(\cos\theta)\,\substack{\cos\\ \sin}\,m\phi$$

$$\psi_{\substack{e\\o}m\lambda}=j_\lambda(kR)P_\lambda^m(\cos\theta\,\substack{\cos\\ \sin}\,m\phi.$$

The eigenvalues μ and λ are determined from the characteristic equations that for a single cone

(5) $$P_\mu^m(\cos\theta_0)=0$$

(6) $$\frac{dP_\lambda^m(\cos\theta_0)}{d\theta_0}=0$$

and for a bicone they are

(7) $$P_\mu^m(\cos\theta_1) = P_\mu^m(\cos\theta_2) = 0$$

(8) $$\frac{dP_\lambda^m(\cos\theta_1)}{d\theta_1} = \frac{dP_\lambda^m(\cos\theta_2)}{d\theta_2} = 0\,.$$

From now on we will not distinguish these two cases, as there is no difference in the formulation of the problems for the two types of cones except that the numerical values of μ and λ are different. Although we do not have a complete knowledge of these values, we are not prevented from formulating this class of problems pending further numerical computations.

The conical vector wave functions defined by (1) to (4) satisfy the boundary condition that at the surface(s) of the cone

(9) $$\hat{\theta} \times \bar{M}_{\substack{e\\o}m\lambda} = 0, \quad \hat{\theta} \times \bar{N}_{\substack{e\\o}m\mu} = 0$$

(10) $$\hat{\theta} \times \nabla \times \bar{M}_{\substack{e\\o}m\mu} = 0, \quad \hat{\theta} \times \nabla \times \bar{N}_{\substack{e\\o}m\lambda} = 0\,.$$

The functions which satisfy the Dirichlet boundary condition will be used for the construction of the dyadic Green's function of the first kind, while the ones satisfying the Neumann boundary condition are involved in the function of the second kind. To determine the normalization factor for these functions we must first give a brief review of the orthogonal relations of the associated Legendre functions.

Let us consider the case of a single cone. The associated Legendre function P_λ^m and P_μ^m satisfy, respectively, the differential equations

(11) $$\frac{1}{\sin\theta}\frac{d}{d\theta}\left(\sin\theta\,\frac{dP_\lambda^m}{d\theta}\right) + \left[\lambda(\lambda+1) - \frac{m^2}{\sin^2\theta}\right]P_\lambda^m = 0$$

(12) $$\frac{1}{\sin\theta}\frac{d}{d\theta}\left(\sin\theta\,\frac{dP_\mu^m}{d\theta}\right) + \left[\mu(\mu+1) - \frac{m^2}{\sin^2\theta}\right]P_\mu^m = 0\,.$$

By multiplying (11) by $\sin\theta\, P_\mu^m$ and (12) by $\sin\theta\, P_\lambda^m$ and integrating the difference of the two resultant equations from $\theta = \theta_0$ to π, we obtain

$$[\lambda(\lambda+1) - \mu(\mu+1)]\int_{\theta_0}^{\pi} P_\lambda^m P_\mu^m \sin\theta\, d\theta = \left[\sin\theta\, P_\lambda^m\,\frac{dP_\mu^m}{d\theta} - \sin\theta\, P_\mu^m\,\frac{dP_\lambda^m}{d\theta}\right] = 0$$

hence

(13) $$\int_{\theta_0}^{\pi} P_\lambda^m P_\mu^m \sin\theta\, d\theta = 0$$

since μ and λ are distinct. For two functions of the same species, we have

$$\int_{\theta_0}^{\pi} P_\lambda^m P_{\lambda'}^m \sin\theta \, d\theta = \begin{cases} 0, & \lambda \neq \lambda' \\ I_{m\lambda}, & \lambda = \lambda' \end{cases} \tag{14}$$

$$\int_{\theta_0}^{\pi} P_\mu^m P_{\mu'}^m \sin\theta \, d\theta = \begin{cases} 0, & \mu \neq \mu' \\ I_{m\mu}, & \mu = \mu' \end{cases} \tag{15}$$

where $I_{m\lambda}$ and $I_{m\mu}$ are two normalization constants. It is not difficult to show by integration by parts that

$$\int_{\theta_0}^{\pi} \left[\left(\frac{dP_\lambda^m}{d\theta} \right)^2 + \left(\frac{mP_\lambda^m}{\sin\theta} \right)^2 \right] \sin\theta \, d\theta = \lambda(\lambda+1) \, I_{m\lambda} \tag{16}$$

$$\int_{\theta_0}^{\pi} \left[\left(\frac{dP_\mu^m}{d\theta} \right)^2 + \left(\frac{mP_\mu^m}{\sin\theta} \right)^2 \right] \sin\theta \, d\theta = \mu(\mu+1) \, I_{m\mu} \tag{17}$$

With the aid of these orthogonal relations, we can determine the orthogonal relations of the conical vector wave functions. They are

$$\iiint \bar{M}_{\substack{e\\o}m\lambda}(k) \cdot \bar{N}_{\substack{e\\o}m'\mu'}(k') \, dV = 0 \tag{18}$$

$$\iiint \bar{M}_{\substack{e\\o}m\mu}(k) \cdot \bar{N}_{\substack{e\\o}m'\mu'}(k') \, dV = 0 \tag{19}$$

$$\iiint \left\{ \begin{matrix} \bar{M}_{\substack{e\\o}m\lambda}(k) \cdot \bar{M}_{\substack{e\\o}m'\lambda'}(k') \\ \bar{N}_{\substack{e\\o}m\lambda}(k) \cdot \bar{N}_{\substack{e\\o}m'\lambda'}(k') \end{matrix} \right\} dV \tag{20}$$

$$= \begin{cases} 0, & m \neq m', \quad \lambda \neq \lambda' \\ \dfrac{(1+\delta_0)\pi^2 \lambda(\lambda+1) I_{m\lambda}}{2k^2} \, \delta(k-k') \end{cases}$$

$$\iiint \left\{ \begin{matrix} \bar{M}_{\substack{e\\o}m\mu}(k) \cdot \bar{M}_{\substack{e\\o}m'\mu'}(k') \\ \bar{N}_{\substack{e\\o}m\mu}(k) \cdot \bar{N}_{\substack{e\\o}m'\mu'}(k') \end{matrix} \right\} dV \tag{21}$$

$$= \begin{cases} 0, & m \neq m', \quad \mu \neq \mu' \\ \dfrac{(1+\delta_0)\pi^2 \mu(\mu+1) I_{m\mu}}{2k^2} \, \delta(k-k'). \end{cases}$$

Knowing these relations for the conical vector wave functions, we can easily derive the functions of the first and the second kind. Omitting the details, we find

$$\bar{\bar{G}}_1(\bar{R}|\bar{R}') = \frac{ik}{2\pi} \sum_m (2-\delta_0) \left[\sum_\lambda \frac{1}{\lambda(\lambda+1) I_{m\lambda}} \left\{ \begin{matrix} \bar{M}^{(1)}_{\substack{e\\o}m\lambda}(k) \, \bar{M}'_{\substack{e\\o}m\lambda}(k) \\ \bar{M}_{\substack{e\\o}m\lambda}(k) \, \bar{M}'^{(1)}_{\substack{e\\o}m\lambda}(k) \end{matrix} \right\} \right. \tag{22}$$

$$+ \sum_{\mu} \frac{1}{\mu(\mu+1)I_{m\mu}} \left\{ \begin{matrix} \bar{N}^{(1)}_{\substack{e\\o}m\mu}(k)\, \bar{N}'_{\substack{e\\o}m\mu}(k) \\ \\ \bar{N}_{\substack{e\\o}m\mu}(k)\, \bar{N}'^{(1)}_{\substack{e\\o}m\mu}(k) \end{matrix} \right\} \Bigg] \qquad \begin{matrix} R > R' \\ \\ R < R'. \end{matrix}$$

In the function of the second kind, we replace $\bar{M}_{\substack{e\\o}m\lambda}$ and $\bar{N}_{\substack{e\\o}m\mu}$ in (22), respectively, by $\bar{N}_{\substack{e\\o}m\lambda}$ and $\bar{M}_{\substack{e\\o}m\mu}$.

43. CONE WITH A SPHERICAL SECTOR

When a spherical sector is attached to a cone (Fig. 43-1), the dyadic Green's functions pertaining to such a composite body can also be derived.

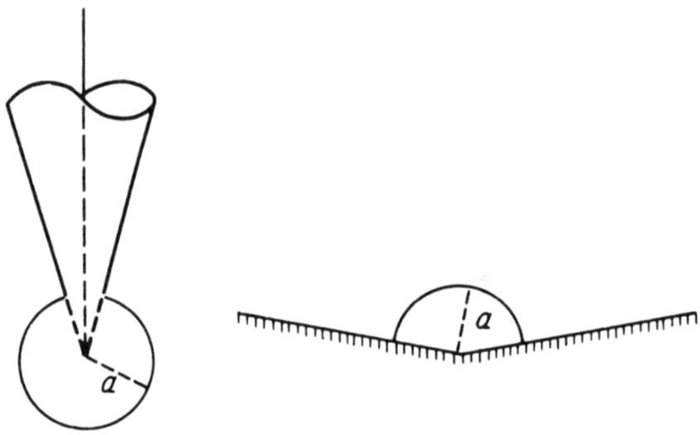

FIG. 43-1. Composite bodies made of a cone and a spherical sector.

Thus for a perfectly conducting sphere, we obtain, by the method of scattering superposition, for $R > R'$,

(1) $$\bar{\bar{G}}_1(\bar{R}|\bar{R}') = \frac{ik}{2\pi} \sum_m (2-\delta_0)$$

$$\cdot \left\{ \sum_{\lambda} \frac{1}{\lambda(\lambda+1)I_{m\lambda}} \bar{M}^{(1)}_{\substack{e\\o}m\lambda}(k)\left[\bar{M}'_{\substack{e\\o}m\lambda}(k) + \alpha_{\substack{e\\o}m\lambda} \bar{M}'^{(1)}_{\substack{e\\o}m\lambda}(k)\right]\right.$$

$$\left. + \sum_{\mu} \frac{1}{\mu(\mu+1)I_{m\mu}} \bar{N}^{(1)}_{\substack{e\\o}m\mu}(k)\left[N'_{\substack{e\\o}m\mu}(k) + \beta_{\substack{e\\o}m\mu} \bar{N}'^{(1)}_{\substack{e\\o}m\mu}(k)\right] \right\}$$

where

$$\alpha_{\substack{e\\o}m\lambda} = -j_\lambda(ka) \,/\, h^{(1)}_\lambda(ka)$$

$$\beta_{\substack{e\\o}m\mu} = -\left[ka\, j_\mu(ka)\right]' \,/\, \left[ka\, h^{(1)}_\mu(ka)\right]'$$

chapter **12**

Inhomogeneous Media and Moving Media

This chapter contains some generalization of the dyadic Green's function technique to more complex media, including particularly the inhomogeneous media and the moving isotropic media. Vector wave functions for plane stratified and spherically stratified media are introduced. Several spherical lens functions are treated in detail, as they have not previously been covered in books either on electromagnetic theory or on differential equations. The last part of the chapter deals with the topics on moving isotropic media. Both the free-space dyadic Green's function and the function of the first kind for a rectangular waveguide filled with a moving isotropic medium are derived. The two examples should illustrate quite well the application of the dyadic Green's function technique to this new class of boundary-value problems involving a moving medium.

44. VECTOR WAVE FUNCTIONS FOR PLANE STRATIFIED MEDIA

When the permittivity and the permeability of a medium are functions of position, we refer to such a medium as inhomogeneous. For an inhomogeneous isotropic medium Maxwell's equations for a harmonically oscillating field read

$$\nabla \times \bar{E} = i\omega\mu_0\mu_r(\bar{R})\bar{H}$$

$$\nabla \times \bar{H} = \bar{J} - i\omega\varepsilon_0\varepsilon_r(\bar{R})\bar{E}$$

where $\mu_r(\bar{R})$ and $\varepsilon_r(\bar{R})$ denote, respectively, the relative permeability and permittivity functions of the medium. In many practical cases of interest the inhomogeneity is normally due to the variation of the permittivity only so that $\mu_r(\bar{R})$ is constant. In this section and that which follows, we shall restrict ourselves to this case only and let $\mu_r(\bar{R})$ be equal to unity. Under this condition, the differential equations for $\bar{E}$ and $\bar{H}$ become

$$\nabla\times\nabla\times\bar{E}-k^2\varepsilon_r(\bar{R})\bar{E}=i\omega\mu_0\bar{J} \tag{1}$$

$$\nabla\times\left[\frac{1}{\varepsilon_r(\bar{R})}\nabla\times\bar{H}\right]-k^2\bar{H}=\nabla\times\left[\frac{\bar{J}}{\varepsilon_r(\bar{R})}\right]. \tag{2}$$

We are seeking the dyadic Green's functions pertaining to these two equations under various boundary conditions, particularly, the eigenfunction expansions of these functions for bodies made of such an inhomogeneous medium but of different shapes. To that end, we must first discuss the vector wave functions which are solutions for the inhomogeneous vector wave equations of the form

$$\nabla\times\nabla\times\bar{E}-k^2\varepsilon_r(\bar{R})\bar{E}=0 \tag{3}$$

$$\nabla\times\left[\frac{1}{\varepsilon_r(\bar{R})}\nabla\times\bar{H}\right]-k^2\bar{H}=0. \tag{4}$$

If $\varepsilon_r(\bar{R})$ is a function of all three variables in an orthogonal system, we have so far no general method of finding these vector wave functions. In fact, we have only two classes of inhomogeneous media for which we know how to find the appropriate vector wave functions. The first class consists of plane stratified media in which $\varepsilon_r(\bar{R})$ is a function of one of the Cartesian coordinate variables only, such as z. In the second class, the medium is stratified in the radial direction in the spherical coordinate system so that the permittivity is a function of R only. Strangely enough, we have not been able to find the general vector wave functions for a cylindrically stratified medium, except under the condition that the problems under consideration are either rotationally symmetric or when the field is a two-dimensional one being independent of the longitudinal axis. For this reason we shall discuss only the vector wave functions for plane and spherically stratified media.

For plane stratified media, we assume that the stratification occurs along the z-direction so that $\varepsilon_r(\bar{R})=\varepsilon_r(z)$. Under that condition, we have

$$\nabla\times\nabla\times\bar{E}-k^2\varepsilon_r(z)\bar{E}=0 \tag{5}$$

$$\nabla\times\left[\frac{1}{\varepsilon_r(z)}\nabla\times\bar{H}\right]-k^2\bar{H}=0. \tag{6}$$

It can easily be verified that the following two sets of vector wave functions are solutions of (5):

(7) $$\bar{M}^{(m)} = \nabla \times (\Psi \hat{z})$$

and

(8) $$\bar{N}^{(e)} = \frac{1}{k\varepsilon_r(z)} \nabla \times \nabla \times (\Phi \hat{z})$$

where the two generating functions Ψ and Φ satisfy, respectively, the differential equations

(9) $$\nabla^2 \Psi + k^2 \varepsilon_r(z) \Psi = 0$$

and

(10) $$\nabla^2 \Phi - \frac{1}{\varepsilon_r} \frac{d\varepsilon_r}{dz} \frac{\partial \Phi}{\partial z} + k^2 \varepsilon_r(z) \Phi = 0 .$$

The superscript m attached to the M-function signifies that it is of the magnetic or transverse-electric type with respect to the z-axis, while the superscript e represents the electric or transverse-magnetic type. Similarly, the vector wave functions which are solutions of (6) can be represented by

(11) $$\bar{M}^{(e)} = \nabla \times (\Phi \hat{z})$$

and

(12) $$\bar{N}^{(m)} = \frac{1}{k} \nabla \times \nabla \times (\Psi \hat{z}) .$$

These four kinds of vector wave functions satisfy the symmetrical relations that

$$\bar{N}^{(e)} = \frac{1}{k\varepsilon_r} \nabla \times \bar{M}^{(e)}$$

$$\bar{M}^{(e)} = \frac{1}{k} \nabla \times \bar{N}^{(e)}$$

$$\bar{N}^{(m)} = \frac{1}{k} \nabla \times \bar{M}^{(m)}$$

$$\bar{M}^{(m)} = \frac{1}{k\varepsilon_r} \nabla \times \bar{N}^{(m)}$$

We repeat that $\bar{M}^{(m)}$ and $\bar{N}^{(e)}$ are solutions for the vector wave equation satisfied by the electric E-field, that is

$$\nabla \times \nabla \times \begin{Bmatrix} \bar{M}^{(m)} \\ \bar{N}^{(e)} \end{Bmatrix} - k^2 \varepsilon_r(z) \begin{Bmatrix} \bar{M}^{(m)} \\ \bar{N}^{(e)} \end{Bmatrix} = 0$$

and $\bar{M}^{(e)}$ and $\bar{N}^{(m)}$ are solutions for the vector wave equation satisfied by the magnetic H-field, that is

$$\nabla \times \frac{1}{\varepsilon_r(z)} \nabla \times \begin{Bmatrix} \bar{M}^{(e)} \\ \\ \bar{N}^{(m)} \end{Bmatrix} - k^2 \begin{Bmatrix} \bar{M}^{(e)} \\ \\ \bar{N}^{(m)} \end{Bmatrix} = 0 .$$

We cannot find a better notation for these functions without causing some kind of confusion. To find the eigenfunctions to be used in the dyadic Green's function expansion, we must specify not only the functional representation for $\varepsilon_r(z)$ but also the geometrical shape of the structure to be considered. For example, if we have a rectangular waveguide filled with a medium stratified in the z-direction, corresponding to the longitudinal axis of the guide, then (9) and (10) must be solved in the rectangular coordinate system. The appropriate solutions for Ψ and Φ are

$$\Psi = \Psi_{emn} = \cos\frac{m\pi x}{a} \cos\frac{n\pi y}{b} F_1(z)$$

and

$$\Phi = \Phi_{omn} = \sin\frac{m\pi x}{a} \sin\frac{n\pi y}{b} F_2(z)$$

where $F_1(z)$ and $F_2(z)$ satisfy, respectively, the equations

$$\frac{d^2F_1(z)}{dz^2} + \left[k^2\varepsilon_r(z) - \left(\frac{m\pi}{a}\right)^2 - \left(\frac{n\pi}{b}\right)^2\right] F_1(z) = 0 \tag{13}$$

$$\frac{d^2F_2(z)}{dz^2} - \frac{1}{\varepsilon_r(z)} \frac{d\varepsilon_r(z)}{dz} \frac{dF_2(z)}{dz} + \left[k^2\varepsilon_r(z) - \left(\frac{m\pi}{a}\right)^2 - \left(\frac{n\pi}{b}\right)^2\right] F_2(z) = 0 . \tag{14}$$

The two complementary solutions of F_1 or F_2 play the same role as the experimental functions $e^{\pm ihz}$ for a homogeneous medium. Only a limited number of stratified profiles have so far been investigated whereby the solutions can be expressed in terms of well-known functions. A few of them are found in Wait (1962, Ch. 3). No attempt is made here to cover this large class of problems in this book. It suffices to point out that once we have at our disposal the available solutions for $F_1(z)$ and $F_2(z)$, we can follow the same procedure as in the homogeneous case for the construction of the dyadic Green's functions. The above remark also applies to problems involving a cylindrical waveguide with a longitudinal stratification or for an inhomogeneous flat earth. For a plane stratified flat earth the appropriate solutions for Ψ and Φ would be

$$\Psi_{\substack{e\\o}n\lambda} = J_n(\lambda r) \begin{matrix}\cos\\ \sin\end{matrix} n\phi F_1(z)$$

$$\Phi_{\substack{e\\o}n\lambda} = J_n(\lambda_r) \begin{matrix}\cos\\ \sin\end{matrix} n\phi F_2(z)$$

where F_1 and F_2 are again solutions for (13) and (14). We, of course, use the ones which would satisfy the radiation condition at $z = -\infty$ in the construction of $\bar{\bar{G}}_3^{(11)}$ and $\bar{\bar{G}}_3^{(21)}$. This concludes our brief discussion for the general method of finding the dyadic Green's functions for plane stratified media.

45. VECTOR WAVE FUNCTIONS FOR SPHERICALLY STRATIFIED MEDIA

When the permittivity is a function of the spherical variable R only, the wave equations for $\bar{E}$ and $\bar{H}$ according to (44-3) and (44-4) become

$$\nabla \times \nabla \times \bar{E} - k^2 \varepsilon_r(R) \bar{E} = 0 \tag{1}$$

$$\nabla \times \left[\frac{1}{\varepsilon_r(R)} \nabla \times \bar{H} \right] - k^2 \bar{H} = 0. \tag{2}$$

We now define four inhomogeneous, spherical vector wave functions (Tai, 1958a) which are generalizations of the functions defined for homogeneous media. They are

$$\bar{M}^{(m)} = \nabla \times (\Psi \bar{R}) \tag{3}$$

$$\bar{N}^{(m)} = \frac{1}{k} \nabla \times \nabla \times (\Psi \bar{R}) \tag{4}$$

$$\bar{M}^{(e)} = \nabla \times (\Phi \bar{R}) \tag{5}$$

$$\bar{N}^{(e)} = \frac{1}{k \varepsilon_r(R)} \nabla \times \nabla \times (\Phi \bar{R}). \tag{6}$$

It can be verified that (3) and (6) are solutions to (1), and (4) and (5) are solutions to (2) provided that Ψ and Φ satisfy the following scalar equations:

$$\nabla^2 \Psi + k^2 \varepsilon_r(R) \Psi = 0 \tag{7}$$

$$\nabla^2 \Phi - \frac{1}{\varepsilon_r(R)} \frac{d\varepsilon_r(R)}{dR} \frac{\partial \Phi}{\partial R} + k^2 \varepsilon_r(R) \Phi = 0. \tag{8}$$

The four sets of vector wave functions have certain symmetrical relations. They are

(9) $$\bar{N}^{(e)} = \frac{1}{k\varepsilon_r(R)} \nabla \times \bar{M}^{(e)}$$

(10) $$\bar{M}^{(e)} = \frac{1}{k} \nabla \times \bar{N}^{(e)}$$

(11) $$\bar{N}^{(m)} = \frac{1}{k} \nabla \times \bar{M}^{(m)}$$

and

(12) $$\bar{M}^{(m)} = \frac{1}{k\varepsilon_r(R)} \nabla \times \bar{N}^{(m)}$$

By applying the method of separation of variables to (7) and (8) in spherical coordinate systems, we find that the eigenfunctions can be written in the form

(13) $$\Psi_{\substack{e\\o}mn} = \frac{1}{R} S_n(kR) P_n^m(\cos\theta) \begin{matrix}\cos\\ \sin\end{matrix} m\phi$$

(14) $$\Phi_{\substack{e\\o}mn} = \frac{1}{R} T_n(kR) P_n^m(\cos\theta) \begin{matrix}\cos\\ \sin\end{matrix} m\phi$$

where $S_n(kR)$ and $T_n(kR)$ satisfy, respectively, the differential equations

(15) $$\frac{d^2 S_n}{dR^2} + \left[k^2\varepsilon_r(R) - \frac{n(n+1)}{R^2}\right] S_n = 0$$

(16) $$\varepsilon_r(R) \frac{d}{dR}\left[\frac{1}{\varepsilon_r(R)} \frac{dT_n}{dR}\right] + \left[k^2\varepsilon_r(R) - \frac{n(n+1)}{R^2}\right] T_n = 0 .$$

The functions S_n and T_n play the same role as $Rj_n(kR)$ in the homogeneous case. It is obvious that when $\varepsilon_r(R)$ is equal to unity both (15) and (16) reduce to the equation satisfied by $Rj_n(kR)$ as discussed in Sec. 39. By using (13) and (14) as the generating functions the complete expressions for the four vector wave functions can be found. They are given by

(17) $$\bar{M}^{(m)}_{\substack{e\\o}mn} = \frac{S_n(kR)}{R} \bar{m}_{\substack{e\\o}mn}$$

(18) $$\bar{N}^{(m)}_{\substack{e\\o}mn} = \frac{1}{k}\left[\frac{S_n(kR)}{R^2} \bar{l}_{\substack{e\\o}mn} + \frac{1}{R} \frac{\partial S_n(kR)}{\partial R} \hat{R} \times \bar{m}_{\substack{e\\o}mn}\right]$$

(19) $$\bar{M}^{(e)}_{\substack{e\\o}mn} = \frac{T_n(kR)}{R} \bar{m}_{\substack{e\\o}mn}$$

$$\bar{N}^{(e)}_{\substack{e\\o}mn} = \frac{1}{k\varepsilon_r(R)}\left[\frac{T_n(kR)}{R^2}\,\bar{l}_{\substack{e\\o}mn} + \frac{1}{R}\frac{\partial T_n(kR)}{\partial R}\,\hat{R}\times\bar{m}_{\substack{e\\o}mn}\right] \tag{20}$$

where

$$\bar{l}_{\substack{e\\o}mn} = n(n+1)P_n^m(\cos\theta)\,{}^{\cos}_{\sin}\, m\phi\,\hat{R}$$

$$\bar{m}_{\substack{e\\o}mn} = \mp\frac{mP_n^m}{\sin\theta}\,{}^{\sin}_{\cos}\, m\phi\,\hat{\theta} - \frac{\partial P_n^m}{\partial\theta}\,{}^{\cos}_{\sin}\, m\phi\,\hat{\theta}.$$

These vector zonal harmonic functions are analogous to the $\bar{P},\bar{C},\bar{B}$ functions used by Morse and Feshbach (pp. 1898-1899) in their presentation of the spherical vector wave functions, except that our normalization constants are different from theirs. These vector wave functions will now be used to construct the eigenfunction expansion of the dyadic Green's function of the third or the fourth kind pertaining to the inhomogeneous spherical lenses.

46. INHOMOGENEOUS SPHERICAL LENSES

The various spherical lenses to be discussed are characterized by certain profiles of the permittivity function. Most of these profiles were originally discovered by the investigators whose names are associated with them and are based on Fermat's principle in the geometrical theory of optics. The Nomura Takaku distribution, however, was used in a wave propagation study. Table 46-1 lists some of the profiles. The wave theory of these distributions will be discussed in this section.

TABLE 46-1

Profiles of Some Permittivity Functions

Maxwell fish-eyes (1860)........................	$4\Big/\left[1+\left(\frac{R}{a}\right)^2\right]^2, \quad a\geqslant R\geqslant 0$
Luneberg lens (1944)...........................	$2-(R/a)^2, \quad a\geqslant R\geqslant 0$
Conical lens of Luneberg (1944)..................	$K[1+(R_1/R)]$
Eaton lens (1952)..............................	$(R/a)^2, \quad a\geqslant R\geqslant 0$
Nomura-Takaku distribution (1955)................	$(R/a)^{2q}$

We shall treat one profile in great detail showing how to find the solutions for S_n and T_n associated with a specific profile. The work (Tai, 1958a) contains some analyses which are not found in most books on the theory of differential equations or on electromagnetic theory. The case to be considered is the spherical Luneberg lens with the permittivity profile given by

(1) $$\varepsilon_r(R) = 2 - (R/a)^2, \qquad a \geqslant R \geqslant a$$

where a denotes the radius of the inhomogeneous lens. According to Luneberg's original theory (1944), this lens will focus the rays, originated from a point source placed at the rim, to form a collimated beam when passing through the lens. Since the original theory is based on a scalar formulation, the polarization status of the source does not enter into the formulation. From the electromagnetic theory point of view, it is an important factor to be considered. A complete theory would also remove some of the uncertain characteristics obtained from the geometrical theory of diffraction.

By substituting (1) into (45-15), and making the following changes of variables

$$kR = \rho, \; ka = \rho_a$$

$$S_n(kR) = \rho^{n+1} \, e^{-\rho^2/2\rho_a} \, U_n(\rho)$$

we find that the function $U_n(\rho)$ must satisfy the following differential equation

(2) $$\frac{d^2 U_n}{d\rho^2} + 2\left(\frac{n+1}{\rho} - \frac{\rho}{\rho_0}\right)\frac{dU_n}{d\rho} + \left(2 - \frac{2n+3}{\rho_a}\right) U_n = 0 .$$

By letting

$$\rho^2/\rho_a = z$$

$$\nu = n + \frac{3}{2}$$

$$\alpha = \frac{1}{2}\left(n + \frac{3}{2} - \rho_a\right)$$

the above equation can be transformed into the standard form of the confluent hypergeometric equation (Copson, 1948, Chap. 10), that is

(3) $$z \frac{d^2 U_n(z)}{dz^2} + (\nu - z)\frac{dU_n(z)}{dz} - \alpha U_n(z) = 0 .$$

For the analysis of the Luneberg lens with the source placed at the rim we need the function which is regular at $R = 0$, that is represented by Kummer's functions,

(4) $$U_n(z) = {}_1F_1(\alpha, \nu, z) = 1 + \frac{\alpha}{\nu} z + \frac{\alpha(\alpha+1)}{\nu(\nu+1)} \frac{z^2}{2!} + \cdots$$

The other independent solution represented by

$$z^{1-\nu}{}_1F_1(\alpha-\nu+1, 2-\nu, z)$$

is not needed. Thus, we identify the $S_n(\rho)$ function to be

$$S_n(\rho)=\rho^{n+1}e^{-\rho^2/2\rho_a}{}_1F_1\left(\alpha, \nu, \frac{\rho^2}{\rho_a}\right). \tag{5}$$

For the $T_n(kR)$ function the analysis is much more complicated. In the first place, (45-16) with $\varepsilon_r(R)$ given by (1) cannot be reduced to a differential equation of the standard type. If we let

$$kR=\rho, \ ka=\rho_a$$

$$T_n(\rho)=(2\rho_a^2-\rho^2)\rho^{n+1}e^{-\rho^2/2\rho_a}V_n(\rho) \tag{6}$$

then the function $V_n(\rho)$ satisfies the following equation

$$\frac{d^2V_n}{d\rho^2}+2\left(\frac{n+1}{\rho}-\frac{\rho}{\rho_a}\right)\frac{dV_n}{d\rho}+\left[2-\frac{2n+3}{\rho_a}-\frac{1}{2\rho_a^2-\rho^2}-\frac{3\rho^2}{(2\rho_a^2-\rho^2)^2}\right]V_n=0.$$

A further transformation of the independent variable $z=\rho^2/\rho_a$ converts it into

$$\frac{d^2V_n}{dz^2}+\left(\frac{\gamma}{z}-1\right)\frac{dV_n}{dz}-\left[\frac{\alpha_1}{z}+\frac{2\alpha_2}{z-a_2}-\frac{2a_2\alpha_3}{(z-a_2)^2}\right]V_n=0, \tag{7}$$

where the various constants are defined by

$$\gamma=n+\frac{3}{2},$$

$$\alpha_1=\frac{1}{2}\left(n+\frac{3}{2}-\rho_a+\frac{1}{4\rho_a}\right),$$

$$\alpha_2=-\frac{1}{16\rho_a}, \ \alpha_3=-\frac{3}{16\rho_a}, \ a_2=2\rho_a.$$

Equation (7) is similar to the confluent hypergeometric equation defined by (3), except that it has another regular singularity at $z=a_2$. The constants in the equation have purposely been so arranged in order to simplify the series solution discussed later. In order to appreciate more fully the nature of the solution to (7) it is profitable to digress to a discussion of the general characteristics of this equation. Like the hypergeometric equation it has a regular singularity at $z=0$ with exponents 0 and $1-\gamma$ (Copson, 1948, Chap. 10) and an irregular singularity at ∞. The exponents of the regular singularity at $z=a_2$ are equal to 3/2 and $-1/2$. From the theory of differential equations it is known that the confluent hypergeometric equation can be obtained by the confluence of two regular singularities of the hypergeometric equation. The confluence is more easily seen by starting with the Papperitz equation or the Reimann P-equation (Morse and Feshbach, p. 539). It is therefore, anticipated that the new equation defined by (7) perhaps can be obtained by the suitable confluence of a number of regular singularities of a second-order differential equation with at least four or more regular singularities. It can be shown that unless the constants α_1, α_2, and α_3 are related in a certain special way, in general, we need a second-order differential equation with five regular singularities to start with. The equation which fulfills this requirement is given by Whittaker and Watson (1943, p. 203) and can be written in the form

(8)*

$$\frac{d^2u}{dz^2} + \sum_{r=1}^{4} \frac{1-\lambda_r-\lambda'_r}{z-a_r}\frac{du}{dz} + \frac{1}{\prod_{r=1}^{4}(z-a_r)}\left[Az^2 + Bz + C + \sum_{r=1}^{4}\frac{C_r}{z-a_r}\right]u = 0.$$

where a_1, a_2, a_3 and a_4 denote the four regular singularities. Their exponents are λ_r and λ'_r. The point $z=\infty$ is the remaining regular singularity with exponents designated by μ and μ'. The constants A and the C_r's are given by

$$\alpha^2 - \left[3 - \sum_{r=1}^{4}(\mu_r + \lambda'_r)\right]\alpha + A = 0, \qquad \alpha = \mu \text{ or } \mu',$$

$$C_r = \frac{\lambda_r\lambda'_r}{\prod_s (a_r - a_s)}, \; s = 1, 2, \ldots, s \neq r.$$

The constants B and C are arbitrary so far. It is observed that because we have put (8) in a different form, the indicial equation for μ and μ' is also changed in appearance. Equation (8) is most conveniently presented by using the Reimann scheme

$$P\begin{Bmatrix} a_1 & a_2 & a_3 & a_4 & \infty & \\ \lambda_1 & \lambda_2 & \lambda_3 & \lambda_4 & \mu & z \\ \lambda'_1 & \lambda'_2 & \lambda'_3 & \lambda'_4 & \mu' & \end{Bmatrix}$$

* The constants A, B, C in the present equation are not the same as those in Whittaker and Watson's equation.

In order to reduce (8) into (7) we first let

$$a_1 = 0, \ \lambda_1 = 0, \ \lambda_1' = 1 - \nu$$

$$a_2 = a_2, \ \lambda_2 = \frac{3}{2}, \ \lambda_2' = -\frac{1}{2},$$

which correspond to the two regular points and their exponents of (7). Without restricting ourselves to the exact procedure of the limiting process, if we assume then when a_3 and a_4 approach infinity

$$\frac{1 - \lambda_3 - \lambda_3'}{a_3} + \frac{1 - \lambda_4 - \lambda_4'}{a_4} \to -1,$$

$$\frac{A}{a_3 a_4} \to 0,$$

$$\frac{B}{a_3 a_4}, \ \frac{C}{a_3 a_4}, \ \frac{\lambda_3 \lambda_3'(a_3 - a_4)}{a_4}, \ \frac{\lambda_4 \lambda_4'(a_3 - a_4)}{a_3} \to \text{finite},$$

then (8) immediately reduces to the form

$$\frac{d^2u}{dz^2} + \left(\frac{\gamma}{z} - 1\right)\frac{du}{dz} + \left[\frac{\beta}{z - a_2} + \frac{\delta}{z(z - a_2)} + \frac{\lambda_2 \lambda_2'}{z(z - a_2)^2}\right] u = 0 \tag{9}$$

where β and δ are two arbitrary constants. By rearranging the terms in the last brackets and introducing two new arbitrary constants α_1 and α_2, we can write (9) in the form

$$\frac{d^2u}{dz^2} + \left(\frac{\gamma}{z} - 1\right)\frac{du}{dz} - \left[\frac{\alpha_1}{z} + \frac{2\alpha_2}{z - a_2} - \frac{\lambda_2 \lambda_2'}{(z - a_2)^2}\right] u = 0, \tag{10}$$

which is identical in form to (7), observing that

$$\lambda_2 \lambda_2' = 2a_2\alpha_3 = -\frac{3}{4}.$$

The foregoing discussion on the analytical nature of the function V_n is simply to show that while the confluent hypergeometric equation is obtained by the confluence of two regular singularities of the Papperitz equation, the "generalized" confluent hypergeometric equation as defined by (7) can be obtained by the confluence of three regular singularities of (8). It may be remarked that the above procedure is not the only way to deduce (7) from a more general second-order linear differential equation. An alternative procedure is to start with an equation of five regular singularities and with the point at infinity considered to be an ordinary point. The same result can thus be achieved. If one uses the notation of Ince (1944, pp. 497–

504), (7) can be classified as of the type $(0,2,1_2)$ which can be derived, by confluence of singularities, from the equations of the type (8,0,0). It is, therefore, concluded that at least five regular singularities are needed to execute the proper confluence in obtaining (7). The foregoing discussion demonstrates the fact that analytically the differential equation for T_n is basically different from the differential equation for S_n and no simple connection exists between them. It also shows that the "generalized" confluent hypergeometric function is an entirely new function which is not related to other well-known functions.

For the Luneberg lens problem it is necessary to obtain a series expansion for V_n, hence T_n, which is finite at $z=0$. Recall that the two exponents at $z=0$ are 0 and $1-\gamma$, where γ is greater than one for this problem. The series solution which remains finite at the origin is therefore associated with the exponent equal to zero. On the other hand, the exponent $1-\gamma$ leads to a solution which is singular at the origin. The desirable solution plays the same role as the Kummer function in (45-15). The series solution of interest is of the form

$$V_n = \sum_{m=0}^{\infty} A_m z^m . \tag{11}$$

By substituting (11) into (7), one finds that

$$\frac{A_1}{A_0} = \frac{\alpha_1}{\gamma},$$

$$\frac{A_2}{A_0} = \frac{1}{2}\,\frac{\alpha_1(\alpha_1+1)}{\gamma(\gamma+1)} - \frac{\alpha_2+\alpha_3}{(\gamma+1)a_2},$$

$$\frac{A_3}{A_0} = \frac{1}{3!}\,\frac{\alpha_1(\alpha_1+1)(\alpha_1+2)}{\gamma(\gamma+1)(\gamma+2)} - \frac{1}{3(\gamma+2)}\left[\frac{(\alpha_1+2)(\alpha_2+\alpha_3)}{(\gamma+1)a_2} + \frac{2\alpha_1(\alpha_2+\alpha_3)}{\gamma a_2} + \frac{2(\alpha_2+2\alpha_3)}{a_2^2}\right].$$

For $m \geqslant 3$, the coefficient can be obtained from the following four-term recurrence relationship

$$a_2^2(m+1)(m+\gamma)A_{m+1} - a_2[a_2(m+\alpha_1)+2n(\gamma-m+1)]A_m + [2a_2(\alpha_1+\alpha_2+\alpha_3+m-1)+(n-1)(\gamma+m-2)]A_{m-1} - (\alpha_1+2\alpha_2+m-2)A_{m-2} = 0 .$$

The series solution for V_n converges uniformly and absolutely for $z < a_2$ or $\rho < 2\rho_a$. Actually, in the Luneberg lens problem the values of ρ never exceed ρ_a. It is observed that for large values of ρ_a the leading term of

A_m/A_0 is practically the same as the corresponding coefficient of the series expansion of the Kummer function. For example, when ρ_a is large,

$$\alpha_1 \simeq \alpha = \frac{1}{2}\left(n + \frac{3}{2} - \rho_a\right)$$

and

$$-\frac{\alpha_2 + \alpha_3}{a_2} = \frac{5}{32\rho_a^2}.$$

The second term of A_2/A_0 is, therefore, much smaller than the first term. The T_n function, in view of (6), is then practically equal to the S_n function at $\rho = \rho_a$. The same is true for the derivative of T_n with respect to r when ρ_a is sufficiently large.

Knowing the two radial functions S_n and T_n, we can construct the vector wave functions according to (45-17) to (45-20), which can then be used to find the dyadic Green's functions pertaining to the spherical Luneberg lens. According to the method of scattering superposition, we let

$$\overline{\overline{G}}_3^{(11)}(\bar{R}|\bar{R}') = \overline{\overline{G}}_0(\bar{R}|\bar{R}') + \overline{\overline{G}}_{3s}^{(11)} \tag{12}$$

$$\overline{\overline{G}}_3^{(21)}(\bar{R}|\bar{R}') = \overline{\overline{G}}_{3s}^{(21)} \tag{13}$$

where $\overline{\overline{G}}_0$ is represented by (39-18). The composition, of course, implies that the source is located outside of the lens. Following the same procedure as described in Sec. 41 for the dielectric sphere, we find that

$$\overline{\overline{G}}_{3s}^{(11)}(\bar{R}|\bar{R}') = \frac{ik}{4\pi} \sum_{n=1}^{\infty} \sum_{m=0}^{n} (2-\delta_0) \frac{2n+1}{n(n+1)} \frac{(n-m)!}{(n+m)!} \cdot \left[A_{\substack{e\\o}n} \bar{M}_{\substack{e\\o}mn}^{(1)}(k) \bar{M}_{\substack{e\\o}mn}^{\prime(1)}(k) + B_{\substack{e\\o}n} \bar{N}_{\substack{e\\o}mn}^{(1)}(k) \bar{N}_{\substack{e\\o}mn}^{\prime(1)}(k)\right] \tag{14}$$

$$\overline{\overline{G}}_{3s}^{(21)}(\bar{R}|\bar{R}') = \frac{ik}{4\pi} \sum_{n=1}^{\infty} \sum_{m=0}^{n} (2-\delta_0) \frac{2n+1}{n(n+1)} \frac{(n-m)!}{(n+m)!} \cdot \left[C_{\substack{e\\o}n} \bar{M}_{\substack{e\\o}mn}^{(m)}(k) \bar{M}_{\substack{e\\o}mn}^{\prime(1)}(k) + D_{\substack{e\\o}n} \bar{N}_{\substack{e\\o}mn}^{(e)}(k) \bar{N}_{\substack{e\\o}mn}^{\prime(1)}(k)\right] \tag{15}$$

where the coefficients A_n, B_n, C_n, D_n are determined from the following system of equations resulting from the matching of boundary conditions for the function of the third kind:

$$\begin{aligned} P_a + A_n Q_a &= C_n S_a \\ P_a' + A_n Q_a' &= C_n S_a' \\ P_a + B_n Q_a &= D_n T_a \\ P_a' + B_n Q_a' &= D_n T_a' \end{aligned}$$

where

$$P_a = \rho_a j_n(\rho_a), \qquad P'_a = \frac{d}{d\rho_a}[\rho_a j_n(\rho_a)]$$

$$Q_a = \rho_a h_n^{(1)}(\rho_a), \qquad Q'_a = \frac{d}{d\rho_a}[\rho_a h_n^{(1)}(\rho_a)]$$

$$S_a = S_n(\rho_a), \qquad S'_a = \frac{d}{d\rho_a} S_n(\rho_a)$$

$$T_a = T_n(\rho_a), \qquad T'_a = \frac{d}{d\rho_a} T_n(\rho_a)$$

According to the excitation requirement of a Luneberg lens based on the geometrical theory of optics, the source should be located at the rim of the spherical lens. If we let it be represented by an electric dipole pointed in the x-direction with current moment equal to c so that

$$\bar{J}(\bar{R}') = c(x'-0)\delta(y'-0)\delta(z'+a)\,\hat{x}$$

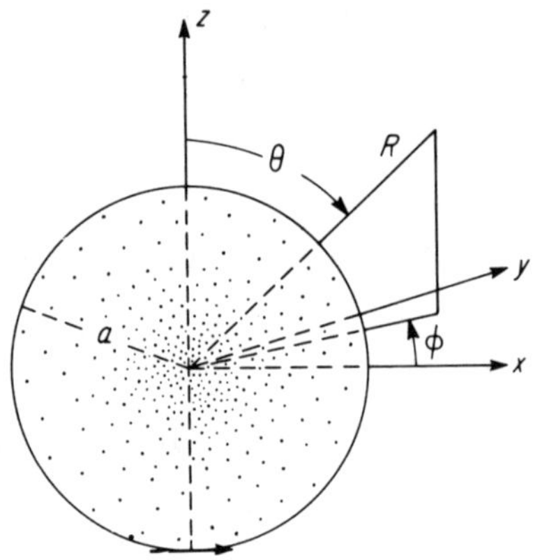

FIG. 46-1. Horizontal dipole placed at the surface of a spherical Luneberg lens.

as shown in Fig. 46-1 then the total electric field outside of the lens is given by

$$\bar{E}(\bar{R}) = i\omega\mu_0 \iiint \bar{\bar{G}}_3^{(11)}(\bar{R}|\bar{R}') \cdot \bar{J}(\bar{R}')\, dV' \tag{16}$$

$$= \frac{-i\omega\mu_0 c}{4\pi a} \sum_{n=1}^{\infty} (-1)^{n+1} \frac{2n+1}{n(n+1)}$$

$$\cdot \frac{1}{Q_a}\left[\frac{\bar{M}^{(1)}_{o1n}}{\left(\frac{Q'_a}{Q_a}-\frac{S'_a}{S_a}\right)}-\frac{\bar{N}^{(1)}_{e1n}}{\left(\frac{T_a}{T'_a}\frac{Q'_a}{Q_a}-1\right)}\right].$$

The far-zone expression for $\bar{E}(\bar{R})$, valid for large values of kR, is obtained by using the asymptotic expressions for $\bar{M}^{(1)}_{o1n}$ and $\bar{N}^{(1)}_{e1n}$:

$$\bar{M}^{(1)}_{o1n} \simeq (-i)^{n+1}\frac{e^{ikR}}{kR}\,\bar{m}_{o1n}$$

$$\bar{N}^{(1)}_{e1n} \simeq (-i)^{n}\frac{e^{ikR}}{kR}\,\bar{n}_{e1n}.$$

These are the basic formulas which are needed for the evaluation of the radiation pattern of a spherical Luneberg lens when excited by an infinitesimal electric dipole. It is obvious that a similar expression can be derived if the exciting source is a magnetic dipole and the result will be slightly different.

As far as the wave theory of the spherical Luneberg lens is concerned, it is not necessary to arrive at this result by way of a dyadic Green's formulation. By expanding the dipole field in terms of the spherical vector wave functions and then applying the scattering superposition theorem we can derive the same result. In fact, this was the approach taken in the author's original paper quoted previously. The electromagnetic theory of a spherical Luneberg lens was examined independently by Wilcox (1956). There is, however, considerable difference in the two approaches. Wilcox found a series solution for T_n directly without examining in detail the analytical behavior of this function from the point of view of the theory of differential equations.

Although it is not yet feasible to construct the general dyadic Green's functions for cylindrically stratified lenses, the electromagnetic theory for these lenses can be formulated if the field is a two-dimensional one, independent of the longitudinal variable or if it is a rotationally symmetrical field. The first electromagnetic theory of a Luneberg lens, in fact, was founded by Jasik (1954) using a cylindrically stratified lens with an electric linc current as the source of excitation. The radial functions encountered in his work are expressed in terms of the confluent hypergeometric functions. A few calculations are found in Jasik's original work showing the radiation pattern of a moderately sized cylindrical Luneberg lens. The ideal characteristics of such a lens as deduced from the geometrical theory of optics are, of course, not preserved. Most important of all, the lens would emit a cylindrical wave in the far zone with a pattern function depending on the size of the lens instead of a collimated beam as predicted from the geometrical theory of optics. A cylindrical Luneberg lens excited by a magnetic line source can be analyzed in a similar manner (Tai, 1956). In that case we encountered again a radial function which is of the generalized confluent hypergeometric type.

For the conical lens of Luneberg, it can be shown that the S_n-function can be expressed in terms of the confluent hypergeometric function while the T_n-function is of the generalized confluent hypergeometric type. It is probably a coincidence that the same type of function is involved in Luneberg's two distinct lenses.

For the Maxwell fish-eyes, the permittivity function is described by

$$\varepsilon_r(R) = 4 \Big/ \left[1 - \left(\frac{R}{a}\right)^2\right]^2, \qquad a \geqslant R \geqslant 0.$$

According to Maxwell's original theory, the lens will focus the rays emitted from a point source located at the surface of the lens to another focus at the opposite side of the lens—hence—the name "fish-eyes." If we let (Tai, 1958b)

$$\rho = kR, \ \rho_a = ka, \quad \xi = -(\rho/\rho_a)^2$$

and

$$S_n(\xi) = \xi^{\frac{n+1}{2}} (\xi - 1)^{\mu} U_n(\xi) \tag{17}$$

$$T_n(\xi) = \xi^{\frac{n+1}{2}} (\xi - 1)^{\mu - 1} V_n(\xi) \tag{18}$$

where

$$\mu = \frac{1}{2}\left(1 + \sqrt{1 + 4\rho_a^2}\right)$$

then the functions $U_n(\xi)$ and $V_n(\xi)$ satisfy, respectively, the following two equations:

$$\xi(\xi - 1) \frac{d^2 U_n}{d\xi^2} + [(2\mu + \beta)\xi - \beta] \frac{dU_n}{d\xi} + \alpha U_n = 0 \tag{19}$$

$$\xi(\xi - 1) \frac{d^2 V_n}{d\xi^2} + [(2\mu + \beta)\xi - \beta] \frac{dV_n}{d\xi} + \left(\alpha - \frac{1}{2}\right) V_n = 0 \tag{20}$$

where

$$\beta = n + \frac{3}{2}$$

$$\alpha = \beta\mu + \rho_a^2$$

Equations (19) and (20) are identical, in form, to the hypergeometric equation normally written in the form (Copson, 1948, p. 247):

$$\xi(\xi - 1)\,\frac{d^2w}{d\xi^2} + [(a + b + 1)\xi - c]\,\frac{dw}{d\xi} + abw = 0. \tag{21}$$

It is relatively easy to determine the constants a, b, and c in terms of α, β, and μ for the two cases. For the analysis of the Maxwell fish-eyes, we only need the function $F(a,b,c,\xi)$ which is regular at $\xi = 0$. It is interesting to observe that the generalized confluent hypergeometric function is not encountered in the electromagnetic theory of Maxwell's fish-eyes.

In regard to the Nomura-Takaku distribution, as shown by these two authors, the differential equations for S_n and T_n are special cases of Malmsten's equation (Watson, 1922, p. 99). These functions can be expressed in terms of the fractional order Bessel functions provided that $q \neq -1$ (see Table 46-1 for the permittivity function). The explicit relations are

$$S_n(\rho) = \rho^{1/2} J_\nu \left[\frac{\rho^{q+1}}{(q+1)\rho_a^q}\right] \tag{22}$$

$$T_n(\rho) = \rho^{q+1/2} J_{\nu'} \left[\frac{\rho^{q+1}}{(q+1)\rho_a^q}\right] \tag{23}$$

where

$$\rho = kR, \quad \rho_a = ka$$

$$\nu = \left(n + \frac{1}{2}\right) \Big/ (q+1)$$

$$\nu' = \left[\left(n + \frac{1}{2}\right)^2 + q^2 + q\right]^{1/2} \Big/ \, (q+1).$$

For $q = 1$, the profile reduces to that of Eaton's lens. For $q = -1$, the S_n and T_n functions are expressible in terms of elementary functions. Two typical solutions are

$$S_n = \rho^{1/2} \sin\left[\left(\rho_a^2 - n^2 - n - \frac{1}{4}\right)^{1/2} \ln \rho\right] \tag{24}$$

$$T_n = \rho^{-1/2} \sin\left[\left(\rho_a^2 - n^2 - n - \frac{1}{4}\right)^{1/2} \ln \rho\right]. \tag{25}$$

Because of that particular profile, which is physically unrealizable if we allow R to be extended to zero, the function T_n has a singularity at $\rho=0$.

Although we have not done any numerical work based on these formulas, the theoretical foundation provided here for various types of lenses should be useful to investigate not only the radiation pattern of these lenses but also the exact nature of a focal point, since the geometrical theory of optics fails to offer a valid description of the energy distribution at or in the neighborhood of such a point.

47. FREE-SPACE DYADIC GREEN'S FUNCTION AND HUYGENS' PRINCIPLE IN MOVING ISOTROPIC MEDIA

Based on Minkowski's relativistic theory of electrodynamics (Minkowski, 1908; Sommerfeld, *Electrodynamics*, 1952) the constitutive relations between the field vectors in a moving isotropic medium can be described in a rather compact form (Tai, 1965a, 1967). They are

$$\bar{B}=\mu\bar{\bar{\alpha}}\cdot\bar{H}-\bar{\Omega}\times\bar{E} \tag{1}$$

$$\bar{D}=\varepsilon\bar{\bar{\alpha}}\cdot\bar{E}+\bar{\Omega}\times\bar{H} \tag{2}$$

where $\mu, \varepsilon =$ permeability and permittivity of the medium when it is stationary

$$\bar{\Omega}=\frac{(n^2-1)\bar{v}}{(1-n^2\beta^2)c^2}$$

$\bar{v}=v\hat{z}=$ velocity of the moving medium, assuming to be directed in the z-direction

$c=(\mu_0\varepsilon_0)^{-1/2}=$ velocity of light

$$n=\left(\frac{\mu\varepsilon}{\mu_0\varepsilon_0}\right)^{1/2}=\text{index of refraction}$$

$$\beta=v/c$$

$$\bar{\bar{\alpha}}=a(\hat{x}\hat{x}+\hat{y}\hat{y})+\hat{z}\hat{z}$$

$$a=\frac{1-\beta^2}{1-n^2\beta^2}\cdot$$

The dyad $\bar{\bar{\alpha}}$ is a characteristic function of the moving medium. When $v=0$ or $n=1$, (1) and (2) reduce to the constitutive relations for a stationary isotropic medium or that of the vacuum as it should be. For a harmonically oscillating field, the definite form of Maxwell's equation, therefore, reads

(3) $$\nabla \times \bar{E} = i\omega(\mu\bar{\bar{\alpha}} \cdot \bar{H} - \bar{\Omega} \times \bar{E})$$

(4) $$\nabla \times \bar{H} = \bar{J} - i\omega(\varepsilon\bar{\bar{\alpha}} \cdot \bar{E} + \bar{\Omega} \times \bar{H}).$$

We now introduce two auxiliary field vectors $\bar{E}^{(b)}$ and $\bar{H}^{(b)}$ defined by

(5) $$\bar{E} = e^{-i\omega\Omega z}\bar{\bar{b}} \cdot \bar{E}^{(b)}$$

(6) $$\bar{H} = e^{-i\omega\Omega z}\bar{\bar{b}} \cdot \bar{H}^{(b)}$$

where

$$\bar{\bar{b}} = \frac{1}{a}(\hat{x}\hat{x} + \hat{y}\hat{y}) + \hat{z}\hat{z}\,.$$

The dyad $\bar{\bar{b}}$ is reciprocal to $\bar{\bar{\alpha}}$ in the sense that

$$\bar{\bar{b}} \cdot \bar{\bar{\alpha}} = \bar{\bar{I}}$$

where $\bar{\bar{I}}$ denotes the unit dyad or the idem factor. By substituting (5) and (6) into (3) and (4) we find that $\bar{E}^{(b)}$ and $\bar{H}^{(b)}$ satisfy the following pairs of equations:

(7) $$\nabla \times [\bar{\bar{b}} \cdot \bar{E}^{(b)}] = i\omega\mu\bar{H}^{(b)}$$

(8) $$\nabla \times [\bar{\bar{b}} \cdot \bar{H}^{(b)}] = \bar{J}e^{i\omega\Omega z} - i\omega\varepsilon\bar{E}^{(b)}\,.$$

By eliminating $\bar{H}^{(b)}$ between these two equations, we obtain the wave equation for the auxiliary electric field $\bar{E}^{(b)}$,

(9) $$\nabla \times [\bar{\bar{b}} \cdot \nabla \times (\bar{\bar{b}} \cdot \bar{E}^{(b)})] - k^2\bar{E}^{(b)} = i\omega\mu\bar{J}e^{i\omega\Omega z}.$$

In a previous work (Tai, 1965b), we introduced an auxiliary field $\bar{E}^{(a)}$ defined by

$$\bar{E} = e^{-i\omega\Omega z}\bar{E}^{(a)}$$

There is some convenience in using $\bar{E}^{(b)}$ in the analysis, instead of $\bar{E}^{(a)}$. The difference, however, is not essential. We point this out merely to avoid possible confusion. To integrate (9) we introduce a free-space dyadic Green's function $\bar{\bar{G}}_0^{(b)}(\bar{R}|\bar{R}')$ which satisfies the differential equation

(10) $$\nabla \times [\bar{\bar{b}} \cdot \nabla \times (\bar{\bar{b}} \cdot \bar{\bar{G}}_0^{(b)})] - k^2\bar{\bar{G}}_0^{(b)} = \bar{\bar{I}}\delta(\bar{R} - \bar{R}')\,.$$

The function obeys a yet unspecified radiation condition at infinity. We now consider a vector function defined by

$$\bar{A} = \bar{\bar{b}} \cdot \bar{E}^{(b)} \times [\bar{\bar{b}} \cdot \nabla \times (\bar{\bar{b}} \cdot \bar{P})] + \bar{\bar{b}} \cdot \nabla \times (\bar{\bar{b}} \cdot \bar{E}^{(b)}) \times (\bar{\bar{b}} \cdot \bar{P})$$

where

$$\bar{P} = \bar{\bar{G}}_0^{(b)}(\bar{R}|\bar{R}') \cdot \bar{c}$$

and $\bar{c}$ denotes an arbitrary constant vector. The divergence of the vector function $\bar{A}$ is given by

$$\nabla \cdot \bar{A} = \nabla \times (\bar{\bar{b}} \cdot \bar{E}^{(b)}) \cdot [\bar{\bar{b}} \cdot \nabla \times (\bar{\bar{b}} \cdot \bar{P})] - \bar{\bar{b}} \cdot \bar{E}^{(b)} \cdot \nabla \times [\bar{\bar{b}} \cdot \nabla \times (\bar{\bar{b}} \cdot \bar{P})]$$
$$+ \nabla \times [\bar{\bar{b}} \cdot \nabla \times (\bar{\bar{b}} \cdot \bar{E}^{(b)})] \cdot (\bar{\bar{b}} \cdot \bar{P}) - \bar{\bar{b}} \cdot \nabla \times (\bar{\bar{b}} \cdot \bar{E}^{(b)}) \cdot \nabla \times (\bar{\bar{b}} \cdot \bar{P}).$$

Since $\tilde{\bar{\bar{b}}} = \bar{\bar{b}}$, the first and the last terms cancel each other; hence

$$\nabla \cdot \bar{A} = \nabla \times [\bar{\bar{b}} \cdot \nabla \times (\bar{\bar{b}} \cdot \bar{E}^{(b)})] \cdot (\bar{\bar{b}} \cdot \bar{P}) - \bar{\bar{b}} \cdot \bar{E}^{(b)} \cdot \nabla \times [\bar{\bar{b}} \cdot \nabla \times (\bar{\bar{b}} \cdot \bar{P})].$$

In view of (9) and (10), this can be written in the form

$$\nabla \cdot \bar{A} = i\omega\mu e^{i\omega\Omega z} \bar{J} \cdot \bar{\bar{b}} \cdot \bar{P} - \bar{\bar{b}} \cdot \bar{E}^{(b)} \cdot \bar{c}\delta(\bar{R} - \bar{R}').$$

By taking a volume integration of $\nabla \cdot \bar{A}$ in a closed region and making use of the divergence theorem we obtain

$$\bar{\bar{b}} \cdot \bar{E}^{(b)}(\bar{R}') \cdot \bar{c} = i\omega\mu \iiint e^{i\omega\Omega z} \bar{J}(\bar{R}) \cdot \bar{\bar{b}} \cdot \bar{\bar{G}}_0(\bar{R}|\bar{R}') \cdot \bar{c}\, dV$$
$$+ \oiint \hat{n}_i \cdot \{ \bar{\bar{b}} \cdot \bar{E}^{(b)}(\bar{R}) \times [\bar{\bar{b}} \cdot \nabla \times [\bar{\bar{b}} \cdot \bar{\bar{G}}_0^{(b)}(\bar{R}|\bar{R}')] \cdot \bar{c}]$$
$$+ \bar{\bar{b}} \cdot \nabla \times [\bar{\bar{b}} \cdot \bar{E}^{(b)}(\bar{R})] \times \bar{\bar{b}} \cdot \bar{\bar{G}}_0^{(b)}(\bar{R}|\bar{R}') \cdot \bar{c} \}\, ds$$

where $\hat{n}_i$ denotes the inward normal to a closed region. By interchanging the roles of $\bar{R}$ and $\bar{R}'$ and deleting the posterior product with respect to $\bar{c}$, due to the fact that it is an arbitrary constant vector, we obtain

$$\bar{\bar{b}} \cdot \bar{E}^{(b)}(\bar{R}) = i\omega\mu \iiint e^{i\omega\Omega z'} \bar{J}(\bar{R}') \cdot \bar{\bar{b}} \cdot \bar{\bar{G}}_0^{(b)}(\bar{R}'|\bar{R})\, dV'$$
$$+ \oiint \hat{n}_i \cdot \{ \bar{\bar{b}} \cdot \bar{E}^{(b)}(\bar{R}') \times [\bar{\bar{b}} \cdot \nabla' \times \bar{\bar{b}} \cdot \bar{\bar{G}}_0^{(b)}(\bar{R}'|\bar{R})]$$
$$+ \bar{\bar{b}} \cdot \nabla' \times [\bar{\bar{b}} \cdot \bar{E}^{(b)}(\bar{R}')] \times \bar{\bar{b}} \cdot \bar{\bar{G}}_0^{(b)}(\bar{R}'|\bar{R}) \}\, dS'. \tag{11}$$

It can be shown that the dyadic Green's function $\bar{\bar{G}}_0^{(b)}(\bar{R}|\bar{R}')$ possesses the following symmetrical properties:

$$\widetilde{\bar{\bar{b}} \cdot \bar{\bar{G}}_0^{(b)}(\bar{R}|\bar{R}')} = \bar{\bar{b}} \cdot \bar{\bar{G}}_0^{(b)}(\bar{R}'|\bar{R}) \tag{12}$$

$$\widetilde{\bar{\bar{b}} \cdot \nabla \times \bar{\bar{G}}_0^{(b)}(\bar{R}|\bar{R}')} = \bar{\bar{b}} \cdot \nabla' \times \bar{\bar{G}}_0^{(b)}(\bar{R}'|\bar{R}). \tag{13}$$

Consequently, (11) can be written in the form

$$\bar{b} \cdot \bar{E}^{(b)}(\bar{R}) = i\omega\mu \iiint \bar{b} \cdot \bar{\bar{G}}_0^{(b)}(\bar{R}|\bar{R}') \cdot \bar{J}(\bar{R}') e^{i\omega\Omega z'} \, dV'$$
$$+ \oiint \{[\bar{b} \cdot \nabla \times \bar{b} \cdot \bar{\bar{G}}_0^{(b)}(\bar{R}|\bar{R}')] \cdot [\hat{n}_i \times \bar{b} \cdot \bar{E}^{(b)}(\bar{R}')]$$
$$+ i\omega\mu[\bar{b} \cdot \bar{\bar{G}}_0^{(b)}(\bar{R}|\bar{R}')] \cdot [\hat{n}_i \times \bar{b} \cdot \bar{H}^{(b)}(\bar{R}')]\} \, dS'.$$

In terms of the actual field, as stated by (5) and (6), we obtain

$$\bar{E}(\bar{R}) = i\omega\mu \iiint \bar{b} \cdot \bar{\bar{G}}_0^{(b)}(\bar{R}|\bar{R}') \cdot \bar{J}(\bar{R}') e^{-i\omega\Omega(z-z')} \, dV' \tag{14}$$
$$+ \iint_S \{[\bar{b} \cdot \nabla \times \bar{b} \cdot \bar{\bar{G}}_0^{(b)}(\bar{R}|\bar{R}')] \cdot [\hat{n}_i \times \bar{E}(\bar{R}')]$$
$$+ i\omega\mu[\bar{b} \cdot \bar{\bar{G}}_0^{(b)}(\bar{R}|\bar{R}')] \cdot [\hat{n}_i \times \bar{H}(\bar{R}')]\} \, e^{-i\omega\Omega(z-z')} \, dS'$$

whereby we have already omitted the surface integral at infinity as a result of the radiation condition. In a region where $\bar{J}(\bar{R}') = 0$, only the surface integral remains. The representation then provides us a mathematical statement of Huygens' Principle in the presence of a moving isotropic medium. Having discussed that much about the integration of the wave equation, we shall now give a derivation of the expression of the free-space dyadic Green's function using the operational method of Levine and Schwinger. The success of that method as applied to the present case owes to one key identity, namely,

$$\nabla \times [\bar{b} \cdot \nabla \times (\bar{b} \cdot \bar{A}] = \frac{1}{a} [\nabla_a \nabla \cdot \bar{A} - (\nabla_a \cdot \nabla) \bar{A}] \tag{15}$$

where $\bar{A}$ denotes any vector function with continuous derivatives and

$$\nabla_a = \hat{x} \frac{\partial}{\partial x} + \hat{y} \frac{\partial}{\partial y} + \hat{z} \frac{\partial}{a \partial z} = \frac{\bar{\bar{\alpha}} \cdot \nabla}{a}.$$

This identity can be verified by expanding these terms in the Cartesian coordinate system. By taking the divergence of (10), we have

$$\nabla \cdot \bar{\bar{G}}_0^{(b)}(\bar{R}|\bar{R}') = -\frac{1}{k^2} \nabla \delta(\bar{R} - \bar{R}'). \tag{16}$$

As a result of (15) and (16), (10) can be written in the form

$$(\nabla_a \cdot \nabla) \bar{\bar{G}}_0^{(b)}(\bar{R}|\bar{R}') + k^2 a \bar{\bar{G}}_0^{(b)}(\bar{R}|\bar{R}') \tag{17}$$
$$= -a \left[\bar{\bar{I}} + \frac{1}{k^2 a^2} \bar{\bar{\alpha}} \cdot \nabla\nabla \right] \delta(\bar{R} - \bar{R}').$$

If we relate $\bar{\bar{G}}_0^{(b)}$ with a scalar function g_0 such that

$$\bar{\bar{G}}_o^{(b)}(\bar{R}|\bar{R}') = a\left[\bar{\bar{I}} + \frac{1}{k^2a^2}\,\bar{\bar{\alpha}}\cdot\nabla\nabla\right] g_o(\bar{R}|\bar{R}'). \tag{18}$$

Then (18) is a solution to (17) if g_0 satisfies the differential equation

$$(\nabla_a\cdot\nabla)g_0(\bar{R}|\bar{R}') + k^2 a g_0(\bar{R}|\bar{R}') = -\delta(\bar{R}-\bar{R}')$$

or

$$\left(\frac{\partial^2}{\partial x^2} + \frac{\partial^2}{\partial y^2} + \frac{1}{a}\frac{\partial^2}{\partial z^2} + k^2 a\right) g_0(\bar{R}|\bar{R}') = -\delta(\bar{R}-\bar{R}'). \tag{19}$$

The solution for g_0 depends on the value of a, particularly its sign.

Case 1. $a>0$, or $v<\dfrac{c}{n}$

$$g_0(\bar{R}|\bar{R}') = \frac{a^{1/2}e^{ika^{1/2}R_a}}{4\pi R_a} \tag{20}$$

where

$$R_a = [(x-x')^2 + (y-y')^2 + a(z-z')^2]^{1/2}.$$

This solution can be obtained by simply introducing a new variable $\xi = a^{1/2}z$ and treating (19) as an ordinary scalar wave equation in a pseudo-coordinate system (x,y,ξ).

Case 2. $a<0$, or $v>\dfrac{c}{n}$

In this case $a=-|a|$. Equation (19) can be treated as a two-dimensional Klein-Gordon equation. Its solution is given by (Cohen, 1961):

$$g_0(\bar{R}|\bar{R}') = \begin{cases} 0, & |a|^{1/2}(z-z')<r \\ \dfrac{|a|^{1/2}\cos k|a|^{1/2}R_a'}{2\pi R_a'}, & |a|^{1/2}(z-z')>r \end{cases} \tag{21}$$

where

$$R_a' = [|a|(z-z')^2 - r^2]^{1/2}$$
$$r^2 = (x-x')^2 + (y-y')^2.$$

The discontinuous behavior of the function is a manifestation of the Čerenkov phenomenon. The solutions as given by (20) and (21) can be obtained alternatively by applying a Fourier transform and a Hankel transform to (15) so that g_0 may be represented by a double integral which can be evaluated in a closed form by the method of contour integration. We assign this problem as an exercise to the reader. This completes our discussion of the derivation of the free-space dyadic Green's function for a moving medium. In the remaining part of this section we shall give the complete expressions of the electromagnetic field of a Hertzian dipole placed in a moving medium based on the above formulation. For simplicity, we consider just the case corresponding to $v < \frac{c}{n}$. The pertinent formulas to be used are

$$\bar{b} \cdot \bar{E}^{(b)}(\bar{R}) = i\omega\mu \iiint \bar{b} \cdot \bar{\bar{G}}_0^{(b)}(\bar{R}|\bar{R}') \cdot \bar{J}(\bar{R}') e^{i\omega\Omega z'} \, dV' \tag{22}$$

$$\bar{H}^{(b)}(\bar{R}) = \frac{1}{i\omega\mu} \nabla \times [\bar{b} \cdot \bar{E}^{(b)}] \tag{23}$$

where

$$\bar{\bar{G}}_0^{(b)} = a\left[\bar{\bar{I}} + \frac{1}{k^2 a^2} \bar{\bar{\alpha}} \cdot \nabla\nabla\right] g_0(\bar{R}|\bar{R}') \tag{24}$$

$$g_0(\bar{R}|\bar{R}') = \frac{a^{1/2} e^{i k_a R_a}}{4\pi R_a} \tag{25}$$

$$k_a = k a^{1/2}$$

$$R_a = \sqrt{(x - x')^2 + (y - y')^2 + a(z - z')^2}$$

The above formulation may also be cast in terms of a potential function if one so desires. Thus we can define a vector potential function

$$\bar{A}(\bar{R}) = \mu \iiint \bar{J}(\bar{R}') e^{i\omega\Omega z'} g_0(\bar{R}|\bar{R}') \, dV' \tag{26}$$

then

$$\bar{E}(\bar{R}) = e^{-i\omega\Omega z} \bar{b} \cdot \bar{E}^{(b)}(\bar{R}) = i\omega a e^{-i\omega\Omega z}\left[\bar{b} \cdot \bar{A} + \frac{1}{k^2 a^2} \nabla\nabla \cdot \bar{A}\right] \tag{27}$$

and

$$\bar{H}(\bar{R}) = e^{-i\omega\Omega z} \bar{b} \cdot \bar{H}^{(b)}(\bar{R}) = \frac{a}{\mu} e^{-i\omega\Omega z} \bar{b} \cdot \nabla \times (\bar{b} \cdot \bar{A}). \tag{28}$$

As a result of the motion of the medium, the electromagnetic field of an oscillating dipole depends on the orientation of the dipole with respect to the direction of motion. It is sufficient for us to consider two distinct cases.

Case A. Dipole Parallel to the Direction of Motion

$$\bar{J}(\bar{R}') = c\delta(\bar{R}' - 0)\hat{z}.$$

In this case, we find that the expressions for the field are given by

$$E_R = \frac{\eta c k^2 a^2 e^{i(D - \omega\Omega z)}}{2\pi D^2}\left(1 + \frac{i}{D}\right)\cos\theta$$

$$E_\theta = \frac{-i\eta c k^2 a^2 e^{i(D - \omega\Omega z)}}{4\pi f^2 D}\cdot\left\{1 + \frac{i}{D}\left(1 + \frac{i}{D}\right)[1 - 2(a-1)\cos^2\theta]\right\}\sin\theta$$

$$H_\phi = \frac{-ick^2 a^{3/2} e^{i(D - \omega\Omega z)}}{4\pi f D}\left(1 + \frac{i}{D}\right)\sin\theta$$

where $D = ka^{1/2} fR$,

$$f = [1 + (a-1)\cos^2\theta]^{1/2}$$

$$R = (x^2 + y^2 + z^2)^{1/2}$$

$$\eta = (\mu/\varepsilon)^{1/2}$$

$c =$ current moment

Case B. Dipole Perpendicular to the Direction of Motion

$$\bar{J}(\bar{R}') = c\delta(\bar{R}' - 0)\hat{x}.$$

$$E_R = \frac{\eta c k^2 a e^{i(D - \omega\Omega z)}}{2\pi D^2}\left(1 + \frac{i}{D}\right)\sin\theta\cos\phi$$

$$E_\theta = \frac{i\eta c k^2 a^2 e^{i(D - \omega\Omega z)}}{4\pi f^2 D}\cdot\left\{1 + \frac{i}{D}\left(1 + \frac{i}{D}\right)\left[1 + 2\left(1 - \frac{1}{a}\right)\sin^2\theta\right]\right\}\cos\theta\cos\phi$$

$$E_\phi = \frac{-i\eta c k^2 a e^{i(D - \omega\Omega z)}}{4\pi D}\left(1 + \frac{i}{D} - \frac{i}{D^2}\right)\sin\phi$$

$$\bar{H} = \frac{ick^2 a^{3/2} e^{i(D - \omega\Omega z)}}{4\pi f D}\left(1 + \frac{i}{D}\right)(\sin\phi\hat{\theta} + \cos\theta\cos\phi\hat{\phi}).$$

In a terrestrial environment, a is very near to unity and Ω is a very small quantity. The effect of the motion of the medium is, therefore, not significant. The theory presented here, however, does provide us with a complete understanding of the radiation process in such a medium.

The expressions derived here can also be obtained in a more formal manner by the method of potentials and by the way of Fourier transform (Lee and Papas, 1964) at the expense of a more elaborate analysis.

48. RECTANGULAR WAVEGUIDE FILLED WITH A MOVING ISOTROPIC MEDIUM

In the final section of this book we shall present the dyadic Green's function for a rectangular waveguide filled with a moving isotropic medium. The treatment is to show that the method of eigenfunction expansion is still applicable to this new class of boundary-value problems. To derive the function of the first kind, we need the vector wave functions which are solutions to the equation

(1) $$\nabla \times [\bar{\bar{b}} \cdot \nabla \times (\bar{\bar{b}} \cdot \bar{F})] - \kappa^2 \bar{F} = 0.$$

The functions which satisfy the Dirichlet boundary condition at the inner surface of the waveguide, corresponding to $x = 0$ and x_0, $y = 0$ and y_0, are

(2) $$\bar{M}_{emn}(h) = \nabla \times [\psi_{emn}(h)\hat{z}]$$

(3) $$\bar{N}_{omn}(h) = \frac{1}{\kappa} \nabla \times [\bar{\bar{b}} \cdot \bar{M}_{omn}(h)] = \frac{1}{\kappa a} \nabla \times \nabla \times [\psi_{omn}(h)\hat{z}]$$

where

$$\psi_{e\,mn}(h) = \cos\left(\frac{m\pi x}{x_0}\right) \cos\left(\frac{n\pi y}{y_0}\right) e^{ihz}$$

$$\psi_{omn}(h) = \sin\left(\frac{m\pi x}{x_0}\right) \sin\left(\frac{n\pi y}{y_0}\right) e^{ihz}$$

$$\kappa^2 a^2 = h^2 + a k_c^2$$

$$k_c^2 = \left(\frac{m\pi}{x_0}\right)^2 + \left(\frac{n\pi}{y_0}\right)^2.$$

The scalar functions $\psi_{\substack{e \\ o}mn}$ are solutions of the equation

(4) $$\left(\frac{\partial^2}{\partial x^2} + \frac{\partial^2}{\partial y^2} + \frac{1}{a}\frac{\partial^2}{\partial z^2} + \kappa^2 a\right)\psi = 0.$$

These vector wave functions have the orthogonal property that

(5) $$\iiint \bar{M}_{emn}(h) \cdot \bar{N}_{om'n'}(-h') dV = 0$$

(6)

$$\iiint \bar{M}_{emn}(h) \cdot \bar{M}_{em'n'}(-h')\, dV = \begin{cases} 0, & m \neq m', \quad n \neq n' \\ \dfrac{(1+\delta_0)}{2} \pi k_c^2 x_0 y_0 \delta(h-h'), & m = m', \quad n = n' \end{cases}$$

where $\delta_0 = 1$ when m or n is zero.

(7)

$$\iiint \bar{N}_{omn}(h) \cdot \bar{N}_{om'n'}(-h')\, dV = \begin{cases} 0, & m \neq m', n \neq n' \\ \dfrac{\pi}{2} k_c^2 x_0 y_0 \dfrac{h^2 + k_c^2}{h^2 + a k_c^2} \delta(h-h'), & m = m', n = n'. \end{cases}$$

The normalization constants for the two sets of functions are not the same. It can be shown that the following symmetrical relations hold true:

(8) $$\nabla \times (\bar{\bar{b}} \cdot \bar{M}_{\substack{e\\o}mn}) = \kappa \bar{N}_{\substack{e\\o}mn}$$

and

(9) $$\nabla \times (\bar{\bar{b}} \cdot \bar{N}_{\substack{e\\o}mn}) = \kappa \bar{M}_{\substack{e\\o}mn} .$$

We now define the dyadic Green's function of the first kind which satisfies the equation

(10) $$\nabla \times \{\bar{\bar{b}} \cdot \nabla \times [\bar{\bar{b}} \cdot \bar{\bar{G}}_1^{(b)}(\bar{R}|\bar{R}')]\} - k^2 \bar{\bar{G}}_1^{(b)}(\bar{R}|\bar{R}') = \bar{\bar{I}} \delta(\bar{R} - \bar{R}').$$

By its classification, it satisfies the Dirichlet boundary condition at the surface of the waveguide. This function and its companion function have the following symmetrical properties.

(11) $$\widetilde{\bar{\bar{b}} \cdot \bar{\bar{G}}_1^{(b)}(\bar{R}|\bar{R}')} = \bar{\bar{b}} \cdot \bar{\bar{G}}_1^{(b)}(\bar{R}'|\bar{R})$$

(12) $$\widetilde{\bar{\bar{b}} \cdot \nabla \times \bar{\bar{G}}_1^{(b)}(\bar{R}|\bar{R}')} = \bar{\bar{b}} \cdot \nabla' \times \bar{\bar{G}}_2^{(b)}(\bar{R}'|\bar{R})$$

where $\bar{\bar{G}}_2^{(b)}(\bar{R}'|\bar{R})$ denotes the function of the second kind which satisfies the same differential equation as $\bar{\bar{G}}_1^{(b)}$ but obeys the Neumann boundary condition at the surface of the waveguide.

Following the method developed previously, we expand the dyadic delta function in terms of the $\bar{M}$ and $\bar{N}$ functions. As will be seen shortly, it is necessary to generalize the path of integration to a closed contour C_+. or C_- as illustrated in Fig. 48-1. The resulting expansion is

(13) $$\bar{\bar{I}} \delta(\bar{R} - \bar{R}') = \int_{C_+ \text{ or } C_-} dh \sum_m \sum_n [\bar{M}_{emn}(h) \bar{A}_{emn}(h) + \bar{N}_{omn}(h) \bar{B}_{omn}(h)] .$$

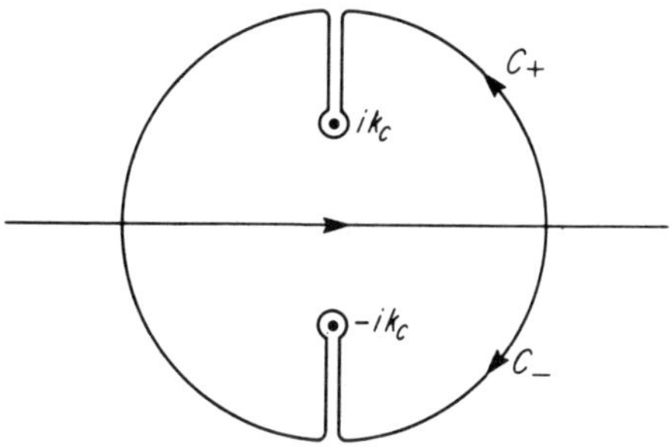

FIG. 48-1. Contour of integration for Eq. (48-13).

Again performing the same routine as done in Chapter 5 we obtain

$$\bar{A}_{emn}(h) = \frac{2-\delta_0}{\pi x_0 y_0 k_c^2} \bar{M}'_{emn}(-h)$$

$$\bar{B}_{omn}(h) = \frac{2-\delta_0}{\pi x_0 y_0 k_c^2} \frac{h^2 + ak_c^2}{h^2 + k_c^2} \bar{N}'_{omn}(-h)$$

where the primed functions are defined with respect to the primed coordinates.

We note that the coefficient $\bar{B}_{omn}(h)$ contain a pair of poles at $h = \pm ik_c$. If these poles are not excluded from the path of integration they would give rise to a term of the form

$$\bar{\bar{f}}(x, y) e^{-k_c|z-z'|}$$

which clearly is not a proper function to describe the dyadic delta function since the function does not vanish for $z \neq z'$. Thus we see that the poles at $\pm ik_c$ must be excluded.

The expansion for $\bar{\bar{G}}_1^{(b)}(\bar{R}|\bar{R}')$ then becomes

$$\bar{\bar{G}}_1^{(b)}(\bar{R}|\bar{R}') = \int_{C_+ \text{ or } C_-} dh \sum_m \sum_n \frac{2-\delta_0}{\pi x_0 y_0} \frac{a^2}{(h^2 - a^2k^2 + ak_c^2)} \cdot \left[\bar{M}_{emn}(h)\bar{M}'_{emn}(-h) + \frac{h^2 + ak_c^2}{h^2 + k_c^2} \bar{N}_{omn}(h)\bar{N}'_{omn}(-h) \right].$$

Performing the indicated contour integration, we find that

$$(14) \quad \bar{\bar{G}}_1^{(b)}(\bar{R}|\bar{R}') = \frac{i}{x_0 y_0} \sum_m \sum_n \frac{(2-\delta_0)a^2}{k_c^2 k_g} \cdot \begin{cases} \bar{M}_{emn}(k_g)\bar{M}'_{emn}(-k_g) + \Gamma \bar{N}_{omn}(k_g)\bar{N}'_{omn}(-k_g), & z > z' \\ \bar{M}_{emn}(-k_g)\bar{M}'_{emn}(k_g) + \Gamma \bar{N}_{omn}(-k_g)\bar{N}'_{omn}(k_g), & z < z' \end{cases}$$

where

$$\Gamma = \frac{k^2a^2}{k^2a^2 + (1-a)k_c^2}$$

$$k_g = (k^2a^2 - k_c^2 a)^{1/2}.$$

The wave number k_g behaves normally when a is positive. For negative values of a, k_g is always positive, hence, there is no cutoff phenomenon in this case. Once $\overline{\overline{G}}_1^{(b)}$ is known the electric field for any distribution of current inside the waveguide can be determined by the formula

$$\bar{E}(\bar{R}) = i\omega\mu \iiint \bar{\bar{b}} \cdot \overline{\overline{G}}_1^{(b)}(\bar{R}|\bar{R}') \cdot \bar{J}(\bar{R}') e^{-i\omega\Omega(z-z')}\, dV'. \tag{15}$$

The dyadic Green's function formulation discussed here encompasses the previous work done on this problem based on the method of potentials (Collier and Tai, 1965; Du and Compton, 1966). It is obvious that one can extend this treatment to a cylindrical waveguide or a semi-infinite moving medium in the lower-half region using the same technique. The proper vector wave functions can be constructed for these cases without any difficulty.

In conclusion, we have assembled in this book the dyadic Green's functions pertaining to some simple-shaped bodies and the ones related to some special media. It is hoped that once this technique becomes familiar to the reader, it can be used quite effectively in formulating various boundary-value problems pertaining to these bodies or media.

appendix A

Vector Identities and Formulas

1. $\bar{a}\cdot(\bar{b}\times\bar{c})=\bar{b}\cdot(\bar{c}\times\bar{a})=\bar{c}\cdot(\bar{a}\times\bar{b})$
2. $\bar{a}\times(\bar{b}\times\bar{c})=(\bar{a}\cdot\bar{c})\bar{b}-(\bar{a}\cdot\bar{b})\bar{c}$
3. $\nabla\times\nabla\psi=0$
4. $\nabla\cdot\nabla\times\bar{A}=0$
5. $\nabla\cdot(\psi\bar{A})=\bar{A}\cdot\nabla\psi+\psi\nabla\cdot\bar{A}$
6. $\nabla\times(\psi\bar{A})=\nabla\psi\times\bar{A}+\psi\nabla\times\bar{A}$
7. $\nabla\cdot(\bar{A}\times\bar{B})=\bar{B}\cdot\nabla\times\bar{A}-\bar{A}\cdot\nabla\times\bar{B}$
8. $\nabla(\bar{A}\cdot\bar{B})=(\bar{A}\cdot\nabla)\bar{B}+(\bar{B}\cdot\nabla)\bar{A}+\bar{A}\times\nabla\times\bar{B}+\bar{B}\times\nabla\times\bar{A}$
9. $\nabla\times(\bar{A}\times\bar{B})=(\bar{B}\cdot\nabla)\bar{A}-(\bar{A}\cdot\nabla)\bar{B}+\bar{A}\nabla\cdot\bar{B}-\bar{B}\nabla\cdot\bar{A}$
10. $\nabla^2\bar{A}=\nabla\nabla\cdot\bar{A}-\nabla\times\nabla\times\bar{A}$

The last expression does not have the usual meaning of an identity. It should be treated as the definition of the quantity $\nabla^2\bar{A}$.

GRADIENT, DIVERGENCE, CURL AND LAPLACIAN IN RECTANGULAR, CYLINDRICAL, SPHERICAL AND GENERAL ORTHOGONAL SYSTEMS

(a) Rectangular system; x, y, z.

11. $$\nabla\psi=\frac{\partial\psi}{\partial x}\hat{x}+\frac{\partial\psi}{\partial y}\hat{y}+\frac{\partial\psi}{\partial z}\hat{z}$$

12. $$\nabla\cdot\bar{A}=\frac{\partial A_x}{\partial x}+\frac{\partial A_y}{\partial y}+\frac{\partial A_z}{\partial z}$$

13. $$\nabla\times\bar{A}=\left(\frac{\partial A_z}{\partial y}-\frac{\partial A_y}{\partial z}\right)\hat{x}+\left(\frac{\partial A_x}{\partial z}-\frac{\partial A_z}{\partial x}\right)\hat{y}+\left(\frac{\partial A_y}{\partial x}-\frac{\partial A_x}{\partial y}\right)\hat{z}$$

14. $\nabla^2\psi = \frac{\partial^2\psi}{\partial x^2} + \frac{\partial^2\psi}{\partial y^2} + \frac{\partial^2\psi}{\partial z^2}$

(b) Cylindrical system; r, ϕ, z

15. $\nabla\psi = \frac{\partial\psi}{\partial r}\hat{r} + \frac{1}{r}\frac{\partial\psi}{\partial\phi}\hat{\phi} + \frac{\partial\psi}{\partial z}\hat{z}$

16. $\nabla\cdot\bar{A} = \frac{1}{r}\frac{\partial}{\partial r}(rA_r) + \frac{1}{r}\frac{\partial A_\phi}{\partial\phi} + \frac{\partial A_z}{\partial z}$

17. $\nabla\times\bar{A} = \left(\frac{1}{r}\frac{\partial A_z}{\partial\phi} - \frac{\partial A_\phi}{\partial z}\right)\hat{r} + \left(\frac{\partial A_r}{\partial z} - \frac{\partial A_z}{\partial r}\right)\hat{\phi} + \frac{1}{r}\left(\frac{\partial}{\partial r}(rA_\phi) - \frac{\partial A_r}{\partial\phi}\right)\hat{z}$

18. $\nabla^2\psi = \frac{1}{r}\frac{\partial}{\partial r}\left(r\frac{\partial\psi}{\partial r}\right) + \frac{1}{r^2}\frac{\partial^2\psi}{\partial\phi^2} + \frac{\partial^2\psi}{\partial z^2}$

(c) Spherical system; R, θ, ϕ

19. $\nabla\psi = \frac{\partial\psi}{\partial R}\hat{R} + \frac{1}{R}\frac{\partial\psi}{\partial\theta}\hat{\theta} + \frac{1}{R\sin\theta}\frac{\partial\psi}{\partial\phi}\hat{\phi}$

20. $\nabla\cdot\bar{A} = \frac{1}{R^2}\frac{\partial}{\partial R}(R^2A_R) + \frac{1}{R\sin\theta}\frac{\partial}{\partial\theta}(\sin\theta A_\theta) + \frac{1}{R\sin\theta}\frac{\partial A_\phi}{\partial\phi}$

21. $\nabla\times\bar{A} = \frac{1}{R\sin\theta}\left[\frac{\partial}{\partial\theta}(\sin\theta A_\phi) - \frac{\partial A_\theta}{\partial\phi}\right]\hat{R} + \frac{1}{R}\left[\frac{1}{\sin\theta}\frac{\partial A_R}{\partial\phi} - \frac{\partial}{\partial R}(RA_\phi)\right]\hat{\theta}$

$$+\frac{1}{R}\left[\frac{\partial}{\partial R}(RA_\phi) - \frac{\partial A_R}{\partial\theta}\right]\hat{\phi}$$

22. $\nabla^2\psi = \frac{1}{R^2}\frac{\partial}{\partial R}\left(R^2\frac{\partial\psi}{\partial R}\right) + \frac{1}{R^2\sin\theta}\frac{\partial}{\partial\theta}\left(\sin\theta\frac{\partial\psi}{\partial\theta}\right) + \frac{1}{R^2\sin^2\theta}\frac{\partial^2\psi}{\partial\phi^2}$.

(d) General orthogonal system; x_1, x_2, x_3

The metric coefficients (h_1, h_2, h_3) in any orthogonal system are defined by

23. $ds_i = h_i\,dx_i$; $i = 1$ or 2, or 3

where ds_i denotes a differential length measured in the direction of dx_i. The variable, x_i, of course, may not have the dimension of length. A convenient method of finding the metric coefficients is to express the rectangular variables in terms of the variables of that system

$$x = x(x_1, x_2, x_3)$$
$$y = y(x_1, x_2, x_3)$$
$$z = z(x_1, x_2, x_3)$$

then

24. $ds_i = \left[\left(\frac{\partial x}{\partial x_i}\right)^2 + \left(\frac{\partial y}{\partial x_i}\right)^2 + \left(\frac{\partial z}{\partial x_i}\right)^2\right]^{1/2} dx_i, \quad i = 1,2,3$

hence

25. $h_i = \left[\left(\frac{\partial x}{\partial x_i}\right)^2 + \left(\frac{\partial y}{\partial x_i}\right)^2 + \left(\frac{\partial z}{\partial x_i}\right)^2\right]^{1/2}.$

For example, in an elliptical cylinder system, we have

26. $x = c \cosh u \cos v$

27. $y = c \sinh u \sin v.$

Let (x_1, x_2, x_3) represent (u, v, z), then, by applying (25) we obtain

28. $h_1 = h_2 = c(\sinh^2 u \cos^2 v + \cosh^2 u \sin^2 v)^{1/2} = c(\cosh^2 u - \cos^2 v)^{1/2}$

29. $h_3 = 1.$

In general, for any orthogonal system, we have

30. $\nabla\psi = \sum_{i=1}^{3} \frac{1}{h_i} \frac{\partial \psi}{\partial x_i} \hat{x}_i$

31. $\nabla \cdot \bar{A} = \frac{1}{\Delta} \sum_{i=1}^{3} \frac{\partial}{\partial x_i}\left(\frac{\Delta A_i}{h_i}\right), \quad \Delta = h_1 h_2 h_3$

32. $\nabla \times \bar{A} = \frac{1}{\Delta} \begin{vmatrix} h_1\hat{x}_1 & h_2\hat{x}_2 & h_3\hat{x}_3 \\ \frac{\partial}{\partial x_1} & \frac{\partial}{\partial x_2} & \frac{\partial}{\partial x_3} \\ h_1 A_1 & h_2 A_2 & h_3 A_3 \end{vmatrix}$

33. $\nabla^2\psi = \frac{1}{\Delta} \sum_{i=1}^{3} \frac{\partial}{\partial x_i}\left(\frac{\Delta}{h_i^2} \frac{\partial \psi}{\partial x_i}\right)$

appendix **B**

Vector Wave Functions and Their Mutual Relations

A. RECTANGULAR VECTOR WAVE FUNCTIONS

$$\psi_{\substack{e\\o}mn}(h) = \begin{Bmatrix} \cos\dfrac{m\pi x}{a}\cos\dfrac{n\pi y}{b} \\ \\ \sin\dfrac{m\pi x}{a}\sin\dfrac{n\pi y}{b} \end{Bmatrix} e^{ihz}$$

$$m, n = 0, 1, 2\cdots$$

$$\left(\frac{m\pi}{a}\right)^2 + \left(\frac{n\pi}{b}\right)^2 + h^2 = k^2$$

$$\nabla^2\psi_{\substack{e\\o}mn}(h) + k^2\psi_{\substack{e\\o}mn}(h) = 0$$

(1) $\bar{M}_{\substack{e\\o}mn}(h) = \nabla \times [\psi_{\substack{e\\o}mn}(h)\hat{z}]$

$$= \left\{ \mp\frac{n\pi}{b}\begin{bmatrix} \cos\dfrac{m\pi x}{a}\sin\dfrac{n\pi y}{b} \\ \sin\dfrac{m\pi x}{a}\cos\dfrac{n\pi y}{b} \end{bmatrix}\hat{x} \pm \frac{m\pi}{a}\begin{bmatrix} \sin\dfrac{m\pi x}{a}\cos\dfrac{n\pi y}{b} \\ \cos\dfrac{m\pi x}{a}\sin\dfrac{n\pi y}{b} \end{bmatrix}\hat{y} \right\} e^{ihz}$$

(2) $\bar{N}_{\substack{e\\o}mn}(h) = \dfrac{1}{k}\nabla \times \nabla \times [\psi_{\substack{e\\o}mn}(h)\hat{z}]$

$$= \frac{1}{k}\left\{ \mp ih\frac{m\pi}{a}\begin{bmatrix} \sin\dfrac{m\pi x}{a}\cos\dfrac{n\pi y}{b} \\ \cos\dfrac{m\pi x}{a}\sin\dfrac{n\pi y}{b} \end{bmatrix}\hat{x} \mp ih\frac{n\pi}{b}\begin{bmatrix} \cos\dfrac{m\pi x}{a}\sin\dfrac{n\pi y}{b} \\ \sin\dfrac{m\pi x}{a}\cos\dfrac{n\pi y}{b} \end{bmatrix}\hat{y} \right.$$

$$+\left[\left(\frac{m\pi}{a}\right)^2+\left(\frac{n\pi}{b}\right)^2\right]\begin{bmatrix}\cos\dfrac{m\pi x}{a}\cos\dfrac{n\pi y}{b}\\ \sin\dfrac{m\pi x}{a}\sin\dfrac{n\pi y}{b}\end{bmatrix}\hat{z}\Bigg\}e^{ihz}$$

(3) $$\nabla\times\bar{M}_{{}^{e}_{o}mn}(h)=k\bar{N}_{{}^{e}_{o}mn}(h)$$

(4) $$\nabla\times\bar{N}_{{}^{e}_{o}mn}(h)=k\bar{M}_{{}^{e}_{o}mn}(h)$$

(5) $$\bar{M}^{(x)}_{mn}(h)=\nabla\times\left[\sin\frac{m\pi x}{a}\cos\frac{n\pi y}{b}\,e^{ihz}\,\hat{x}\right]$$

$$=\left(ih\sin\frac{m\pi x}{a}\cos\frac{n\pi y}{b}\,\hat{y}+\frac{n\pi}{b}\sin\frac{m\pi x}{a}\sin\frac{n\pi y}{b}\,\hat{z}\right)e^{ihz}$$

(6) $$\bar{M}^{(y)}_{mn}(h)=\nabla\times\left[\cos\frac{m\pi x}{a}\sin\frac{n\pi y}{b}\,e^{ihz}\,\hat{y}\right]$$

$$=\left(-ih\cos\frac{m\pi x}{a}\sin\frac{n\pi y}{b}\,\hat{x}-\frac{m\pi}{a}\sin\frac{m\pi x}{a}\sin\frac{n\pi y}{b}\,\hat{z}\right)e^{ihz}$$

(7) $$\bar{N}^{(x)}_{mn}(h)=\frac{1}{k}\nabla\times\nabla\times\left(\cos\frac{m\pi x}{a}\sin\frac{n\pi y}{b}\,e^{ihz}\,\hat{x}\right)$$

$$\frac{1}{k}\left[\left(k^2-\frac{m^2\pi^2}{a^2}\right)\cos\frac{m\pi x}{a}\sin\frac{n\pi y}{b}\,\hat{x}-\left(\frac{m\pi}{a}\right)\left(\frac{n\pi}{b}\right)\sin\frac{m\pi x}{a}\cos\frac{n\pi y}{b}\,\hat{y}\right.$$

$$\left.-ih\left(\frac{m\pi}{a}\right)\sin\frac{m\pi x}{a}\sin\frac{n\pi y}{b}\,\hat{z}\right]e^{ihz}$$

(8) $$\bar{N}^{(y)}_{mn}(h)=\frac{1}{k}\nabla\times\nabla\times\left(\sin\frac{m\pi x}{a}\cos\frac{n\pi y}{b}\,e^{ihz}\,\hat{y}\right)$$

$$=\frac{1}{k}\left[-\left(\frac{m\pi}{a}\right)\left(\frac{n\pi}{b}\right)\cos\frac{m\pi x}{a}\sin\frac{n\pi y}{b}\,\hat{x}+\left(k^2-\frac{n^2\pi^2}{b^2}\right)\sin\frac{m\pi x}{a}\cos\frac{n\pi y}{b}\,\hat{y}\right.$$

$$\left.-ih\left(\frac{n\pi}{b}\right)\sin\frac{m\pi x}{a}\sin\frac{n\pi y}{b}\,\hat{z}\right]e^{ihz}$$

(9) $$\bar{M}^{(x)}_{mn}(h)=\frac{1}{k^2-h^2}\left[k\left(\frac{n\pi}{b}\right)\bar{N}_{omn}(h)+ih\left(\frac{m\pi}{a}\right)\bar{M}_{emn}(h)\right]$$

(10) $$\bar{M}^{(y)}_{mn}(h)=\frac{1}{k^2-h^2}\left[-k\left(\frac{m\pi}{a}\right)\bar{N}_{omn}(h)+ih\left(\frac{n\pi}{b}\right)\bar{M}_{emn}(h)\right]$$

(11) $$\bar{N}^{(x)}_{mn}(h)=\frac{1}{k^2-h^2}\left[-ih\left(\frac{m\pi}{a}\right)\bar{N}_{omn}(h)-k\left(\frac{n\pi}{b}\right)\bar{M}_{emn}(h)\right]$$

$$\bar{N}^{(y)}_{mn}(h)=\frac{1}{k^2-h^2}\left[-ih\left(\frac{m\pi}{b}\right)\bar{N}_{omn}(h)+k\left(\frac{m\pi}{a}\right)\bar{M}_{emn}(h)\right]. \tag{12}$$

It is to be noticed that $\bar{N}^{(x)}_{mn}$ is not equal to $\frac{1}{k}\nabla\times\bar{M}^{(x)}_{mn}$ because the generating functions used to define them are different. The functions defined by (5) through (8) all satisfy the vector Dirichlet boundary condition at the side walls of the rectangular guide, corresponding to $x=0$ and a, $y=0$ and b. Another four functions that satisfy the vector Neumann boundary condition can be found in a similar manner.

B. CYLINDRICAL VECTOR WAVE FUNCTIONS WITH DISCRETE (OR CONTINUOUS) EIGENVALUES

$$\psi_{\substack{e\\o}n\lambda}(h)=J_n(\lambda r)\begin{matrix}\cos\\ \sin\end{matrix}n\phi\, e^{ihz}$$

$$J_n(\lambda_{nm}a)=0 \quad\text{or}\quad \lambda_{nm}a=p_{nm}$$

$$\psi_{\substack{e\\o}n\mu}(h)=J_n(\mu r)\begin{matrix}\cos\\ \sin\end{matrix}n\phi\, e^{ihz}$$

$$\frac{dJ_n(\mu_{mn}a)}{d(\mu_{nm}a)}=0 \quad\text{or}\quad \mu_{nm}a=q_{nm}$$

$$\nabla^2\psi_{\substack{e\\o}n\mu}+(\mu^2+h^2)\psi_{\substack{e\\o}n\mu}=0$$

$$\nabla^2\psi_{\substack{e\\o}n\lambda}+(\lambda^2+h^2)\psi_{\substack{e\\o}n\lambda}=0$$

$$\begin{aligned}\bar{M}_{\substack{e\\o}n\mu}(h)&=\nabla\times[\psi_{\substack{e\\o}n\mu}(h)\hat{z}]\\&=\left[\mp\frac{nJ_n(\mu r)}{r}\begin{matrix}\sin\\ \cos\end{matrix}n\phi\,\hat{r}-\frac{\partial J_n(\mu r)}{\partial r}\begin{matrix}\cos\\ \sin\end{matrix}n\phi\,\hat{\phi}\right]e^{ihz}\end{aligned} \tag{13}$$

$$\begin{aligned}\bar{N}_{\substack{e\\o}n\lambda}(h)&=\frac{1}{\sqrt{\lambda^2+h^2}}\nabla\times\nabla\times[\psi_{\substack{e\\o}n\lambda}(h)\hat{z}]=\frac{1}{\sqrt{\lambda^2+h^2}}\left[ih\frac{\partial J_n(\lambda r)}{\partial r}\begin{matrix}\cos\\ \sin\end{matrix}n\phi\,\hat{r}\right.\\&\left.\mp\frac{ihn}{r}J_n(\lambda r)\begin{matrix}\sin\\ \cos\end{matrix}n\phi\,\hat{\phi}+\lambda^2J_n(\lambda r)\begin{matrix}\cos\\ \sin\end{matrix}n\phi\,\hat{z}\right]e^{ihz}\end{aligned} \tag{14}$$

The functions $\bar{M}_{\substack{e\\o}n\lambda}$, and $\bar{N}_{\substack{e\\o}n\mu}$ are obtained respectively by replacing μ by λ in (13) and λ by μ in (14).

When the eigenfunctions are continuous, we discard the constraints on λ and μ so that they may assume any value. In that case, we have only two sets of functions, namely, $\bar{M}_{\substack{e\\o}n\lambda}$ and $\bar{N}_{\substack{e\\o}n\lambda}$, instead of four sets. Other types of cylindrical vector wave functions can also be used to represent the electromagnetic fields. Thus we have

(15) $$\bar{M}^{(x)}_{\substack{e\\o}n\lambda}(h) = \nabla \times \left[J_n(\lambda r) \substack{\cos\\ \sin} n\phi \, e^{ihz} \hat{x}\right]$$

(16) $$\bar{N}^{(x)}_{\substack{e\\o}n\lambda}(h) = \frac{1}{k} \nabla \times \nabla \times \left[J_n(\lambda r) \substack{\cos\\ \sin} n\phi \, e^{ihz} \hat{x}\right], \quad k^2 = h^2 + \lambda^2.$$

These functions are related to the functions of the z-type. The relations between them are

(17) $$\bar{M}^{(x)}_{\substack{e\\o}n\lambda}(h) = \frac{1}{2} \left\{ \frac{-ih}{\lambda} \left[\bar{M}_{\substack{e\\o}(n+1)\lambda}(h) - \bar{M}_{\substack{e\\o}(n-1)\lambda}(h)\right] \pm \frac{k}{\lambda} \left[\bar{N}_{\substack{o\\e}(n+1)\lambda}(h) + \bar{N}_{\substack{o\\e}(n-1)\lambda}(h)\right] \right\}$$

(18) $$\bar{N}^{(x)}_{\substack{e\\o}n\lambda}(h) = \frac{1}{2} \left\{ \frac{-ih}{\lambda} \left[\bar{N}_{\substack{e\\o}(n+1)\lambda}(h) - \bar{N}_{\substack{e\\o}(n-1)\lambda}(h)\right] \pm \frac{k}{\lambda} \left[\bar{M}_{\substack{o\\e}(n+1)\lambda}(h) + \bar{M}_{\substack{o\\e}(n-1)\lambda}(h)\right] \right\}.$$

C. SPHERICAL VECTOR WAVE FUNCTIONS

$$\psi_{\substack{e\\o}mn}(k) = j_n(kR) P_n^m(\cos\theta) \substack{\cos\\ \sin} m\phi$$

$$\nabla^2 \psi_{\substack{e\\o}mn}(k) + k^2 \psi_{\substack{e\\o}mn}(k) = 0$$

(19) $$\bar{M}_{\substack{e\\o}mn}(k) = \nabla \times \left[\psi_{\substack{e\\o}mn}(k) \bar{R}\right] = j_n(kR) \left[\mp \frac{m}{\sin\theta} P_n^m(\cos\theta) \substack{\sin\\ \cos} m\phi \, \hat{\theta} - \frac{\partial P_n^m(\cos\theta)}{\partial\theta} \substack{\cos\\ \sin} m\phi \, \hat{\phi} \right]$$

(20) $$\bar{N}_{\substack{e\\o}mn}(k) = \frac{1}{k} \nabla \times \nabla \times \left[\psi_{\substack{e\\o}mn}(k) \bar{R}\right] = \frac{n(n+1)}{kR} j_n(kR) P_n^m(\cos\theta) \substack{\cos\\ \sin} m\phi \, \hat{R} + \frac{1}{kR} \frac{\partial}{\partial R} \left[R j_n(kR)\right] \left[\frac{\partial P_n^m}{\partial\theta} \substack{\cos\\ \sin} m\phi \, \hat{\theta} \mp \frac{m}{\sin\theta} P_n^m(\cos\theta) \substack{\sin\\ \cos} m\phi \, \hat{\phi} \right]$$

(21) $$\begin{aligned} \bar{M}^{(x)}_{\substack{e\\o}mn}(k) &= \frac{1}{k} \nabla \times \left[\psi_{\substack{e\\o}mn}(k) \hat{x}\right] \\ &= \frac{1}{k} \nabla \times \left[\psi_{\substack{e\\o}mn}(k) (\sin\theta \cos\phi \hat{R} + \cos\theta \cos\phi \hat{\theta} - \sin\phi \hat{\phi})\right] \\ &= \pm \frac{1}{2n(n+1)} \left[\bar{N}_{\substack{o\\e}(m+1)n} + (n+m)(n-m+1) \bar{N}_{\substack{o\\e}(m-1)n}\right] \\ &+ \frac{1}{2(n+1)(2n+1)} \left[\bar{M}_{\substack{e\\o}(m+1)(n+1)} - (n-m+1)(n-m+2) \bar{M}_{\substack{e\\o}(m-1)(n+1)}\right] \end{aligned}$$

$$-\frac{1}{2n(2n+1)}\left[\bar{M}_{\substack{e\\o}(m+1)(n-1)}-(n+m-1)(n+m)\,\bar{M}_{\substack{e\\o}(m-1)(n-1)}\right]$$

where the functional dependence (k) for $\bar{N}$ and $\bar{M}$ is implied.

$$\text{(22)}\quad \bar{M}^{(y)}_{\substack{e\\o}mn}(k)=\frac{1}{k}\nabla\times[\psi_{\substack{e\\o}mn}(k)\,\hat{y}]$$

$$=\frac{1}{k}\nabla\times[\psi_{\substack{e\\o}mn}(k)\,(\sin\theta\sin\phi\hat{R}+\cos\theta\sin\phi\hat{\theta}+\cos\phi\hat{\phi})]$$

$$=-\frac{1}{2n(n+1)}\left[\bar{N}_{\substack{e\\o}(m+1)n}-(n+m)(n-m+1)\bar{N}_{\substack{e\\o}(m-1)n}\right]$$

$$\pm\frac{1}{2(n+1)(2n+1)}\left[\bar{M}_{\substack{o\\e}(m+1)n+1)}+(n-m+1)(n-m+2)\bar{M}_{\substack{o\\e}(m-1)(n+1)}\right]$$

$$\mp\frac{1}{2n(2n+1)}\left[\bar{M}_{\substack{o\\e}(m+1)(n-1)}+(n+m-1)(n+m)\,\bar{M}_{\substack{o\\e}(m-1)(n-1)}\right]$$

$$\text{(23)}\quad \bar{M}^{(z)}_{\substack{e\\o}mn}(k)=\frac{1}{k}\nabla\times[\psi_{\substack{e\\o}mn}(k)\hat{z}]=\frac{1}{k}\nabla\times[\psi_{\substack{e\\o}mn}(k)\,(\cos\theta\hat{R}-\sin\theta\hat{\theta})]$$

$$=\mp\frac{m}{n(n+1)}\,\bar{N}_{\substack{o\\e}mn}+\frac{1}{2n+1}\left[\frac{n-m+1}{n+1}\,\bar{M}_{\substack{e\\o}m(n+1)}+\frac{n+m}{n}\,\bar{M}_{\substack{e\\o}m(n-1)}\right].$$

The relations between $\bar{N}^{(x)}_{\substack{e\\o}mn}$, $\bar{N}^{(y)}_{\substack{e\\o}mn}$, $\bar{N}^{(z)}_{\substack{e\\o}mn}$ and $\bar{M}_{\substack{e\\o}mn}$, $\bar{N}_{\substack{e\\o}mn}$ can be obtained by taking the curl of (21), (22) and (23). The result is the same by interchanging the roles of $\bar{M}$ and $\bar{N}$ in (21) to (23). For example, we have

$$\text{(24)}\quad \bar{N}^{(z)}_{\substack{e\\o}mn}(k)=\mp\frac{m}{n(n+1)}\,\bar{M}_{\substack{o\\e}mn}+\frac{1}{2n+1}\left[\frac{n-m+1}{n+1}\,\bar{N}_{\substack{e\\o}m(n+1)}+\frac{n+m}{n}\,\bar{N}_{\substack{e\\o}m(n-1)}\right].$$

D. CONICAL VECTOR WAVE FUNCTIONS

$$\psi_{\substack{e\\o}m\mu}(k)=j_\mu(kR)P^m_\mu\,(\cos\theta)\substack{\cos\\ \sin}\,m\phi$$

Characteristic equation for μ:

$$\left.\frac{\partial P^m_\mu(\cos\theta)}{\partial\theta}\right|_{\theta=\theta_0}=0$$

$$\Psi_{\substack{e\\o}m\nu}(k)=j_\nu(kR)P^m_\nu\,(\cos\theta)\substack{\cos\\ \sin}\,m\phi$$

Characteristic equation for ν:

$$P_\nu^m(\cos\theta_0) = 0$$

$$(\nabla^2 + k^2)\begin{pmatrix}\psi_{\substack{e\\o}m\mu}\\ \Psi_{\substack{e\\o}m\nu}\end{pmatrix} = 0\,.$$

The expressions for $\bar{M}_{\substack{e\\o}m\mu}$, $\bar{N}_{\substack{e\\o}m\mu}$, $\bar{M}_{\substack{e\\o}n\nu}$ and $\bar{N}_{\substack{e\\o}n\nu}$ have the same form as the spherical vector wave functions defined by (19) and (20) with n replaced by μ or ν.

appendix **C**

The Functions $K_{1/2}$, $K_{3/2}$, $I_{1/2}$, and $I_{3/2}$

$$K_{1/2}(x)=\frac{1}{x^{3/2}(1+x)^{1/2}}\left[-C(\rho_1)+C(\rho_2)\right]$$

$$K_{3/2}(x)=\frac{1}{x^{1/2}}\left\{\frac{-4\sin\left(\frac{\pi}{2}x\right)}{\pi x^2}+\frac{1}{(1+x)^{1/2}}\left[1-3\left(\frac{1+x}{x}\right)+2\left(\frac{1+x}{x}\right)^2\right]S(\rho_1)\right.$$
$$\left.+\frac{1}{(1-x)^{1/2}}\left[-1-3\left(\frac{1-x}{x}\right)-2\left(\frac{1-x}{x}\right)^2\right]S(\rho_2)\right\}$$

$$I_{1/2}(x)=\frac{2}{x^{1/2}}\left[(1+x)^{1/2}C(\rho_1)-(1-x)^{1/2}C(\rho_2)\right]$$

$$I_{3/2}(x)=\frac{2}{x^{1/2}}\left\{\frac{4\sin\left(\frac{\pi}{2}x\right)}{3\pi x}+(1+x)^{1/2}\left[1-\frac{2}{3}\left(\frac{1+x}{x}\right)S(\rho_1)\right]\right.$$
$$\left.+(1-x)^{1/2}\left[1+\frac{2}{3}\left(\frac{1-x}{x}\right)S(\rho_2)\right]\right\}$$

where

$$S(\rho)=\int_0^{\rho}\frac{\sin t}{(2\pi t)^{1/2}}\,dt,\quad C(\rho)=\int_0^{\rho}\frac{\cos t}{(2\pi t)^{1/2}}\,dt$$

$$\rho_1=\frac{\pi}{2}(1+x)$$

$$\rho_2=\frac{\pi}{2}(1-x)$$

$$x=\sin\theta\,.$$

appendix D

Exercises

CHAPTER 1

1. Starting from Gauss' theorem derive the gradient identity (2-6) by letting $\bar{A} = \psi\bar{c}$ where $\bar{c}$ denotes a constant vector.

2. The relationship between the unit vectors of two coordinate systems can be found quite conveniently by the method of gradient. For example, we have

$$x = R \sin\theta \cos\phi .$$

By taking the gradient of the above equation in the rectangular system and in the spherical system, respectively, we obtain

$$\hat{x} = \sin\theta \cos\phi \hat{R} + \cos\theta \cos\phi \hat{\theta} - \sin\phi \hat{\phi}$$

Applying this method derive the transformation relations between the unit vectors in the rectangular system and those in

(a) the prolate spheroidal system,

(b) the parabolic system.

The scalar relations between the coordinate variables are:

(a) $$x = \frac{1}{2} d \sinh u \sin v \cos\phi$$

$$y = \frac{1}{2} d \sinh u \sin v \sin\phi$$

$$z = \frac{1}{2} d \cosh u \cos v$$

(b) $$x = 2\sqrt{\xi\eta} \cos\phi$$

$$y = 2\sqrt{\xi\eta} \sin\phi$$

$$z = \xi - \eta .$$

The metric coefficients pertaining to these systems can be found according to the method described in Section (d) of Appendix A.

3. Derive the vector Green's first identity (17-49), using Gauss' theorem by letting $\bar{A}=\bar{Q}\times\nabla\times\bar{P}$.

CHAPTER 2

1. Verify (6-20) and (6-21) by applying Gauss' theorem to (6-4) and (6-5).

2. Prove that if

$$\begin{aligned}\bar{E}_c = E_x \cos(\omega t-\Phi_x)\,\hat{x}\\ +E_y \cos(\omega t-\Phi_y)\,\hat{y}\\ +E_z \cos(\omega t-\Phi_z)\,\hat{z}\end{aligned}$$

is a particular solution to the vector wave equation

$$\nabla\times\nabla\times\bar{E}+\frac{1}{c^2}\frac{\partial^2\bar{E}}{\partial t^2}=-\mu_0\frac{\partial\bar{J}}{\partial t}$$

corresponding to

$$\begin{aligned}\bar{J}_c = J_x \cos(\omega t-\phi_x)\hat{x}\\ +J_y \cos(\omega t-\phi_y)\hat{y}\\ +J_z \cos(\omega t-\phi_z)\hat{z}\end{aligned}$$

then

$$\begin{aligned}\bar{E}_s = E_x \sin(\omega t-\Phi_x)\hat{x}\\ +E_y \sin(\omega t-\Phi_y)\hat{y}\\ +E_z \sin(\omega t-\Phi_z)\hat{z}\end{aligned}$$

would be a particular solution if $\bar{J}$ varies as

$$\begin{aligned}\bar{J} = J_x \sin(\omega t-\phi_x)\hat{x}\\ +J_y \sin(\omega t-\phi_y)\hat{y}\\ +J_z \sin(\omega t-\phi_z)\hat{z}\end{aligned}$$

3. The Hertzian potential functions obey the gauge condition $\nabla\cdot\bar{A}=i\omega\mu_0\varepsilon_0\psi$ as described by (8-5). The Debye potential functions were introduced under the assumption that

$$\bar{A}=A_R\hat{R}$$

and the corresponding gauge condition that

$$\frac{\partial A_R}{\partial R}=i\omega\mu_0\varepsilon_0\psi\,.$$

Find the differential equations for A_R and ψ for the case that $\bar{J}=J_R\hat{R}$. Show that the electromagnetic field for a Hertzian dipole can be found by using A_R instead of A_z. This exercise demonstrates very clearly the nonuniqueness of the potential functions as far as the solutions for $\bar{E}$ and $\bar{H}$ are concerned. It should be pointed out that the gauge condition for the Debye potentials was incorrectly presented by Aharoni (1946) in his book on antennas.

4. Show that the gauge condition for the Hertzian potential functions as described by (8-5) is satisfied by the free-space solutions for $\bar{A}$ and ψ given by (8-8) and (8-9).

CHAPTER 3

1. An infinite line is excited by a current source described by

$$K(x) = \begin{cases} K_0 & |x| \leqslant x_0 \\ 0 & |x| > x_0 . \end{cases}$$

Apply the Green's function technique to determine the voltage $V(x)$ along the entire line and verify your result by the classical method based on the theory of differential equations. This exercise shows that (9-9) is valid for the point of observation, x, located either outside or inside the source region.

2. Derive (10-30) directly from the differential equations for $g_3^{(1)}$ and $g_3^{(2)}$ without resorting to the explicit expressions for $V_2(x)$.

3. Derive the expression for the three-dimensional free-space Green's function given by (12-6) by the method of Hankel-transform.

4. Derive the expression for the two-dimensional free-space Green's function given by (12-7) by the method of Fourier-transform.

CHAPTER 4

1. The free-space vector Green's function introduced by Stratton and Chu (Stratton, p. 464) is constructed by letting

$$\bar{G} = G_0 \bar{a}$$

where G_0 denotes the three-dimensional scalar Green's function which satisfies the equation

$$\nabla^2 G_0 + k^2 G_0 = -\delta(\bar{R} - \bar{R}').$$

As a result of such a construction, the expression describing the Huygens' principle involves not only $\hat{n} \times \bar{E}$, $\hat{n} \times \bar{H}$, but also $\hat{n} \cdot \bar{E}$. The term involving $\hat{n} \cdot \bar{E}$ can be eliminated if we choose a vector Green's function of the form (Franz (1948); Mentzer (1955), p. 14)

$$\bar{G} = \nabla G_0 \times \bar{a}.$$

Derive (14-21) by means of the vector Green's theorem with this modified vector Green's function.

2. The dyadic Green's function pertaining to half-space (18-9), can be constructed by using the vector Green's function for half-space as derived by Toraldo di Francia (1955, p. 218). Rederive (18-9) by this approach.

3. Show that the dyadic Green's function of the first kind defined for a half-space satisfies the radiation condition at infinity.

4. Find the expression for the dyadic Green's function of the second kind pertaining to a half-space.

CHAPTER 5

1. An infinite rectangular waveguide is filled with a dielectric from $z=-\infty$ to $z=0$. Determine the expressions for $\bar{\bar{G}}_3^{(11)}$ and $\bar{\bar{G}}_3^{(21)}$ where region I is defined from $z=0$ to $z=\infty$ and region II is defined from $z=-\infty$ to $z=0$.

2. A rectangular waveguide is partially filled with a dielectric slab from $x=0$ to $x=d$. Find the proper vector wave functions which can be used to describe the electromagnetic field in such a guide. What is the characteristic equation that determines the eigenvalues of these functions? (Collin (1960), Chap. 6).

3. Find the dyadic Green's function of the first kind for a semi-infinite rectangular waveguide. The guide is short-circuited at $z=0$ and extends to $z=\infty$.

4. Find the dyadic Green's function of the first kind pertaining to a space bounded by two parallel conducting plates located at $z=0$ and $z=d$.

CHAPTER 6

1. Derive the expression for the dyadic Green's function of the first kind pertaining to a coaxial line of inner radius a and outer radius b. In addition to the TEM mode, the radial eigenfunctions involved in this problem are of the type

$$S_r(\lambda r)=AJ_n(\lambda r)+BN_n(\lambda r)$$

$$T_r(\mu r)=CJ_n(\mu r)+DN_n(\mu r).$$

The eigenvalues λ and μ are solutions of the following characteristic equations:

$$J_n(\lambda a)N_n(\lambda b)-J_n(\lambda b)N_n(\lambda a)=0$$

$$J_n'(\mu a)N_n'(\mu b)-J_n'(\mu b)N_n'(\mu a)=0.$$

Solutions to these equations are found in Dwight (1948) and Abramowitz and Stegun (1964).

2. A circular cylindrical waveguide is bifurcated by a conducting strip extended from the center of the guide to its wall. It is located at $\phi=0$. Find the dyadic Green's function of the first kind for such a guide. (*Hint*: use half-order Bessel functions as well as integral-order Bessel functions for the radial eigenfunctions. See the construction of the Green's functions for a half-sheet discussed in Chapter 10.)

3. Find the dyadic Green's of the first kind for a semicircular cylindrical waveguide.

4. A coaxial line is terminated in a conducting plane at $z=0$. Find the integral equation for the aperture electric field when the line is excited by a TEM mode initiated

at $z = -\infty$. The problem can be formulated by using the dyadic Green's function for the half-space and the one for the coaxial line. A direct formulation using scalar Green's functions is given by Levine and Papas (1951).

5. Reformulate the waveguide problems treated by Chien et al. (1949) using the dyadic Green's function.

CHAPTER 7

1. A plane wave is incident obliquely on a conducting cylinder. The incident electric field is assumed to be polarized in the x-z plane. Find the scattered field.

This problem can be solved conveniently by finding the scattered field due to an electric dipole with a current distribution given by

$$\bar{J}(\bar{R}') = c\delta(R' - R_0)\delta(\theta' - \theta_0)\delta(\phi' - 0)\hat{\theta}'$$

where c denotes the current moment. The scattered field due to a plane wave can be obtained by letting $R_0 \to \infty$. The relationship between the incident dipole field and the amplitude of the corresponding plane wave is given by (37-3) and (37-4). Sec. 37 treats a similar problem from this point of view for the case when the diffracting body is a half-sheet.

2. With the aid of the dyadic Green's function derive the far-zone field of an electric dipole placed near a conducting cylinder at $r = b$, $\phi = 0$ and $z = 0$ for the following orientations.

(a) dipole axis perpendicular to the axis of the cylinder
(b) dipole axis parallel to the axis of the cylinder
(c) dipole axis parallel to the surface of the cylinder and pointed in the ϕ-direction.

Compare your result with the ones derived by Carter (1943) and Wait (1959, p. 23).

3. Repeat the same problem for a magnetic dipole. In this case, it is convenient to use the formula

$$\bar{H} = k^2 \iiint \bar{\bar{G}}_2 \cdot \bar{M}\, dV'$$

where $\bar{M}$ denotes the magnetization vector as defined in Sec. 7. For a magnetic dipole we let $\bar{M} = \bar{m}\delta(\bar{R} - \bar{R}')$ where $\bar{m}$ denotes the magnetic dipole moment.

4. Find the functions $\bar{\bar{G}}_3^{(22)}$ and $\bar{\bar{G}}_3^{(12)}$ pertaining to a dielectric cylinder. The superscripts imply that the source is located in region II corresponding to the interior region of the dielectric cylinder. The solution can be found either by following the procedures described in Sec. 24 or by making use of the symmetrical properties of the dyadic Green's functions of the third kind.

5. Construct the dyadic Green's function of the third kind for a conducting cylinder covered by a sheath of dielectric material. The desired function can be found by using $\bar{\bar{G}}_3$ for a dielectric cylinder as the basic function then apply the method of scattering superposition to take care of the scattered field due to the presence of the conducting cylinder.

6. Apply (25-5) to find the far-zone field of a half-wave dipole placed in the outside of a conducting cylinder and parallel to its axis. We assume

$$\bar{J}(\bar{R}') = I_0 \cos\left(\frac{2\pi}{\lambda} z'\right)\delta(r' - b)\delta(\phi' - 0)\hat{z}\,, \qquad -\frac{\lambda}{4} \leqslant z' \leqslant \frac{\lambda}{4}\,.$$

7. Apply (25-6) to find the far-zone field of a longitudinal slot antenna cut on a conducting cylinder assuming that the aperture field is represented adequately by

$$E_\phi = E_0 \cos\left(\frac{2\pi}{\lambda} z'\right)\delta(r' - a)\delta(\phi' - 0), \qquad -\frac{\lambda}{4} \leqslant z' \leqslant \frac{\lambda}{4}\,.$$

8. Construct the dyadic Green's function of the first kind for a semicircular conducting cylinder placed on a conducting ground plane.

9. Explain more explicitly why the two sets of functions $\bar{M}$ and $\bar{N}$ are coupled in the construction of $\bar{\bar{G}}_3$ for a dielectric cylinder while such a coupling does not occur for the functions pertaining to a flat earth.

10. A plane wave is incident obliquely upon a dielectric circular cylinder. Find the scattered field. Two distinct polarizations associated with the incident wave should be considered. As suggested in Prob. 1, the solution of this problem can be obtained by considering the field of a dipole in the presence of the cylinder. By removing the dipole to infinity in a fixed direction one can obtain the plane wave solution.

CHAPTER 8

1. Find the dyadic Green's function of the third kind for a dielectric slab placed on a conducting plane, assuming that the source is located in the space outside of the slab.

2. Find the expression for the magnetic field of a small loop (magnetic dipole) placed above a flat earth. Let the axis of the dipole lie, (a) in the x-direction (horizontal magnetic dipole), and (b) in the z-direction (vertical magnetic dipole).

3. When σ approaches infinity the flat earth becomes a perfectly conducting plane, hence, $\bar{\bar{G}}_3^{(11)}$ reduces to $\bar{\bar{G}}_1$. Determine the eigenfunction expansion of $\bar{\bar{G}}_1$ pertaining to a half-space from this approach.

4. Derive the integral equation that governs the current distribution on a circular ground plane of radius a placed above a flat earth at a height h from the earth. The ground plane is excited by a vertical Hertzian dipole placed at the center of the top surface.

5. Show that the field of a horizontal magnetic dipole placed above a flat earth can be derived by using one electric Hertzian potential function and one magnetic Hertzian potential function.

CHAPTER 9

1. Prove the orthogonal relation stated by (29-24).

2. Find the asymptotic expression for $\overline{\overline{G}}_1$ as given by (30-4).

CHAPTER 10

1. Derive (33-11) according to the procedure described in the text.

2. Derive (34-4), using (34-2) and (34-3) with the aid of the saddle point method of integration.

3. Discuss the singular behavior of both $\bar{E}$ and $\bar{H}$ near the edge of a half-sheet with the aid of series expansion of the Fresnel integrals as given by (37-12). Consider first the case where the incident E field is parallel to the edge and then perpendicular to the edge. Show that the result is compatible with Meixner's integrability condition (Meixner, 1949).

4. A plane wave is incident on a composite body made of a circular cylinder and a half-sheet (Fig. 38-1). Determine the scattered field, assuming that the incident E field is perpendicular to the axis of the cylinder and impinges on the body in an oblique direction.

CHAPTER 11

1. Prove (39-9).

2. Find the dyadic Green's function $\overline{\overline{G}}_3^{(11)}$ for a perfectly conducting sphere covered by a concentric layer of dielectric material. Region I denotes the space outside the composite body.

3. Derive (41-20).

4. State precisely the condition under which (41-21) is derived.

5. Find the expression for the electromagnetic field due to a longitudinal slot or a transverse slot on a single cone (Bailin and Silver, 1956). The error made in their original paper would have been avoided if the dyadic Green's function technique were used.

6. Find the expression for the back scattering cross section of a hemisphere resting on an infinitely conducting ground plane. Assume that the incident wave is polarized with the E-vector parallel to the surface of the ground plane, but is impinging on the sphere at an oblique angle. By definition, the back scattering cross section is defined as

$$\sigma_b = \lim_{R \to 0} \frac{4\pi R^2 |E_s|^2}{|E_i|^2}$$

where E_i denotes the amplitude of the incident wave and E_s is the amplitude of the scattered wave in the backward direction.

7. Find the dyadic Green's function of the third kind for a spherical dielectric shell of inner radius a and outer radius b.

CHAPTER 12

1. Find the complete expression for $\overline{\overline{G}}_3^{(11)}$ and $\overline{\overline{G}}_3^{(21)}$ for a plane stratified medium using each of the following profiles for the dielectric constant:

$$\text{(a)}\quad \varepsilon(z)=\begin{cases}\varepsilon_0, & z>0\\ (\varepsilon_1-\varepsilon_2)e^{az}+\varepsilon_2, & z<0\end{cases}$$

$$\text{(b)}\quad \varepsilon(z)=\begin{cases}\varepsilon_0, & z>0\\ \varepsilon_0(1+bz), & z<0.\end{cases}$$

2. Verify the recurrence relations for the coefficients A_m given by the equations following (46-11).

3. Find the functions S_n and T_n for the conical lens of Luneberg with a permittivity profile given by

$$\varepsilon_r(R)=K\left(1+\frac{R_1}{R}\right),\qquad R<R_2.$$

4. Prove the symmetrical property of the Green's function for a moving isotropic medium as stated by (47-12).

5. Verify the identity given by (47-15).

6. Verify (48-12).

7. Derive the expressions for $g_0\,(\bar{R}|\bar{R}')$ as given by (47-20) and (47-21) by applying the method of Fourier-Hankel transform to (47-19).

appendix **E**

References

(a) BOOKS

Abramowitz, M., and I. A. Stegun (1964). *Handbook of Mathematical Functions.* National Bureau of Standards, Government Printing Office.

Aharoni, J. (1946). *Antennae.* Oxford University Press, London.

Baker, B. B., and E. T. Copson (1950). *The Mathematical Theory of Huygens' Principle.* Oxford University Press, London.

Baños, A. (1966). *Dipole Radiation in the Presence of a Conducting Half-Plane.* Pergamon Press, New York.

Bateman, H. (1955). *Electrical and Optical Wave-Motion.* Dover Publications, Inc., New York.

Bowman, J. J., T. B. A. Senior, and P. L. E. Uslenghi (1969). *Electromagnetic and Acoustic Scattering by Simple Shapes.* North-Holland Publishing Company, Amsterdam.

Collin, R. E. (1960). *Field Theory of Guided Waves.* McGraw-Hill Book Company, New York.

Copson, E. T. (1948). *Theory of Functions of a Complex Variable.* Oxford University Press, London.

Courant, R. (1936). *Differential and Integral Calculus.* Vol. II. Interscience Publishers, New York.

Feynberg, Y. L. (1961). *The Propagation of Radio Waves Along the Surface of the Earth.* English trans. of 1961 Russian text published by the Foreign Technology Division, Wright-Patterson Air Force Base, Dayton, Ohio.

Ince, E. L. (1944). *Ordinary Differential Equations.* Dover Publications, Inc., New York.

Luneberg, R. K. (1944). *The Mathematical Theory of Optics.* Brown University Press, Providence, R.I.

Maxwell, J. C. (1860), *Scientific Papers, I.* Dover Publications, Inc., New York, pp. 76-79.

Mentzer, J. R. (1955). *Scattering and Diffraction of Radio Waves.* Pergamon Press, New York.

Morse, P. M., and H. Feshbach (1953). *Methods of Theoretical Physics.* Part II. McGraw-Hill Book Company, New York.

Ramo, S., J. R. Whinnery, and T. Van Duzer (1965). *Fields and Waves in Communication Electronics.* John Wiley & Sons, Inc., New York.

Reitz, J. R., and F. J. Milford (1960). *Foundations of Electromagnetic Theory.* Addison-Wesley Publishing Company, Inc., Reading, Mass.

Sommerfeld, A. (1949). *Partial Differential Equations.* Academic Press, Inc., New York.

Sommerfeld, A. (1952). *Electrodynamics.* Academic Press, Inc., New York.

Sommerfeld, A. (1954). *Optics.* Academic Press, Inc., New York.

Stratton, J. A. (1941). *Electromagnetic Theory.* McGraw-Hill Book Company, New York.

Toraldo di Francia, G. (1955). *Electromagnetic Waves.* Interscience Publishers, Inc., New York.

Wait, J. R. (1959). *Electromagnetic Radiation from Cylindrical Structures.* Pergamon Press, New York.

Wait, J. R. (1962). *Electromagnetic Waves in Stratified Media.* Pergamon Press, New York.

Watson, G. N. (1922). *Theory of Bessel Functions.* Cambridge University Press, Cambridge.

Whittaker, E. T., and G. N. Watson (1943). *Modern Analysis.* Cambridge University Press, Cambridge.

(b) ARTICLES AND REPORTS

Bailin, L. L., and S. Silver (1956), "Exterior Electromagnetic Boundary Value Problems for Spheres and Cones," *IRE Trans., AP-4*, pp. 5-15; *AP-5*, p. 313 (correction).

Carter, P. S. (1943), "Antenna Arrays Around Cylinders," *Proc. IRE, 31*, pp. 671-693.

Chien, W. Z., L. Infeld, J. R. Pounder, A. F. Stevenson, and J. L. Synge (1949). "Contributions to the Theory of Wave Guides," *Canadian J. Research, 27*, pp. 69–129.

Cohen, M. (1961), "Radiation in Plasma," Part I, *Phys. Rev., 123*, pp. 711–721.

Collier, J. R., and C-T Tai (1965), "Guided Waves in Moving Media," *IEEE Trans., MTT-13*, pp. 441–445.

Costen, R. C., and D. Adamson (1965), "Three Dimensional Derivation of the Electromagnetic Jump Conditions and Momentum-Energy Laws at a Moving Boundary," *Proc. IEEE, 53*, No. 9, pp. 1181–1196.

Du, L. J., and R. T. Compton, Jr. (1966), "Cut-Off Phenomena for Guided Waves in Moving Media," *IEEE Trans., MTT-14*, pp. 358–363.

Dwight, H. B. (1948), "Tables of Roots for Natural Frequencies in Coaxial Type Cavities," *J. Math. Phys., 27*, pp. 84–89.

Eaton, J. E. (1952), "On Spherically Symmetrical Lens," *IRE Trans., AP-4*, pp. 66-71.

Franz, W. (1948), "Zur Formulierung des Huygenschen Prinzips," Z. Naturforchung, Vol. 3a, pp. 500-506.

Hansen, W. W. (1935), "A New Type of Expansion in Radiation Problems," *Phys. Rev.*, *47*, pp. 139–143.

Hansen, W. W. (1936), "Directional Characteristics of Any Antenna over a Plane Earth," *J. Appl. Phys.*, *7*, pp. 460–465.

Hansen, W. W. (1937), "Transformations Useful in Certain Antenna Calculations," *J. Appl. Phys.*, *8*, pp. 282–286.

Hargreaves, R. (1918), "A Diffraction Problem and an Asymptotic Theorem in Bessel Series," *Phil. Mag.*, *36*, pp. 191–199; Supplementary Note, *36*, pp. 317–319.

Harrington, R. F. (1953), "Current Element Near the Edge of a Conducting Half-Plane," *J. Appl. Phys.*, *24*, pp. 547-550.

Itoh, M. (1959), "The Unified Electromagnetic Equation and Its Properties in Curvilinear Coordinate Systems," *Rev. Matemat. Fis. Teor.*, Tucuman, Argentina, pp. 85–100.

Jasik, H. (1954), "Electromagnetic Theory of Luneberg Lens," Dissertation, Department of Electrical Engineering, Polytechnic Institute of Brooklyn, New York.

Lee, K. S. H., and C. H. Papas (1964), "Electromagnetic Radiation in the Presence of Moving Simple Media," *J. Math. Phys.*, *5*, pp, 1668–1672.

Levine, H., and C. H. Papas (1951), "Theory of Circular Diffraction Antenna," *J. Appl. Phys.*, *22*, pp. 29–43.

Levine, H., and J. Schwinger (1950), "On the Theory of Electromagnetic Wave Diffraction by an Aperture in an Infinite Plane Conducting Screen," *Comm. Pure and Appl. Math.*, *III*, pp. 355-391.

Meixner, J. (1949), "Die Kantenbedingung in der Theorie der Beugung elektromagnetischer Wellen an volkommen leitenden ebenen Schirmen," *Ann. Phys.*, *441*, pp. 2–9.

Mie, G. (1908), "Beiträge zur Optik Trüber Medien, Speziell Kolloidaler Metallösungen," *Ann. der Phys.*, *25*, pp. 377-445.

Minkowski, H. (1908), "Die Grundgleichungen für die Elektromagnetischen Vorgange in Bewegten Korpern," *Gottingen Nachrichten*, pp. 53–116.

Nomura, Y. (1951), "On the Propagation of Electric Waves from a Horizontal Dipole over the Surface of the Earth Sphere," *Science Report of Tohoku University (Japan)*, *1-2*, pp. 25–49.

Nomura, Y., and K. Takaku (1955), "On the Propagation of the Electromagnetic Waves in an Inhomogeneous Atmosphere," *Science Report of Tohoku University (Japan)*, *7*, pp. 107–148.

Senior, T. B. A. (1953), "The Diffraction of a Dipole Field by a Perfectly Conducting Half-Plane," *Q.J.Mech. and Appl. Math.*, *6*, pp. 101–114.

Tai, C-T (1952), "On the Theory of Diffraction of Electromagnetic Waves by a Sphere," Technical Report No. 29, Stanford Research Institute, Stanford, Calif.

Tai, C-T (1954a), "Radiation from Current Elements and Apertures in the Presence of a Perfectly Conducting Half-Plane Sheet," Technical Report No. 45, Stanford Research Institute, Stanford, Calif.

Tai, C-T (1954b), "A Glossary of Dyadic Green's Functions," Technical Report No. 46, Stanford Research Institute, Stanford, Calif.

Tai, C-T (1956), "Theory of the Cylindrical Luneberg Lens Excited by a Line Magnetic Current," Technical Report 678-3, Antenna Laboratory, The Ohio State University, Columbus, Ohio.

Tai, C-T (1958a), "Electromagnetic Theory of Spherical Luneberg Lens," *Appl. Sci. Res.*, Sec. B, *7*, pp. 113–130.

Tai, C-T (1958b), "Maxwell Fish-eye Treated by Maxwell Equations," *Nature*, *182*, pp. 1600–1601.

Tai, C-T (1965a), "Huygens' Principle in a Moving Isotropic, Homogeneous and Linear Medium," *Applied Optics*, *4*, pp. 1347-1349.

Tai, C-T (1965b), "The Dyadic Green's Function for a Moving Isotropic Medium," *IEEE Trans.*, *AP-13*, pp. 322–323.

Tai, C-T (1967), "Present Views on Electrodynamics of Moving Media," *Radio Science*, *2*, pp. 245–248.

Wilcox, C. H. (1956), "Electromagnetic Theory of Luneberg Lens," Technical Report No. MSD-1802, Lockheed Aircraft Corporation, Burbank, Calif.

Indexes

AUTHOR INDEX

SUBJECT INDEX